AF618524

Semiconducting Silicon Nanowires for Biomedical Applications

Related titles:

Porous silicon for biomedical applications
(ISBN 978-0-85709-711-8)

Biomedical imaging: Applications and advances
(ISBN 978-0-85709-127-7)

Implantable sensor systems for medical applications
(ISBN 978-1-84569-987-1)

Woodhead Publishing Series in Biomaterials: Number 73

Semiconducting Silicon Nanowires for Biomedical Applications

Edited by

Jeffery L. Coffer

AMSTERDAM • BOSTON • CAMBRIDGE • HEIDELBERG • LONDON
NEW YORK • OXFORD • PARIS • SAN DIEGO
SAN FRANCISCO • SINGAPORE • SYDNEY • TOKYO
Woodhead Publishing is an imprint of Elsevier

Woodhead Publishing is an imprint of Elsevier
80 High Street, Sawston, Cambridge, CB22 3HJ, UK
225 Wyman Street, Waltham, MA 02451, USA
Langford Lane, Kidlington, OX5 1GB, UK

British Library Cataloguing-in-Publication Data
A catalogue record for this book is available from the British Library

Library of Congress Control Number: 2013955415

ISBN 978-0-85709-766-8 (print)
ISBN 978-0-85709-771-2 (online)

For information on all Woodhead Publishing publications
visit our website at http://store.elsevier.com/

Typeset by RefineCatch Limited, Bungay, Suffolk

Contents

Contributor contact details

(* = main contact)

Editor and chapters 1 and 6

J. L. Coffer
Department of Chemistry
Texas Christian University
Fort Worth, TX 76129, USA

E-mail: j.coffer@tcu.edu

Chapter 2

Gengfeng Zheng* and Ming Xu
Laboratory of Advanced Materials
Department of Chemistry
Fudan University
Shanghai, People's Republic of China

E-mail: gfzheng@fudan.edu.cn

Chapter 3

Y. Coffinier and R. Boukherroub*
Interdisciplinary Research Institute, USR CNRS 3078
University of Lille1
Parc de la Haute Borne
50 avenue de Halley – BP 70478
59658 Villeneuve d'Ascq, France

E-mail: yannick.coffinier@iri.univ-lille1.fr; rabah.boukherroub@iri.univ-lille1.fr

Chapter 4

L. Marcon and R. Boukherroub*
Interdisciplinary Research Institute, USR CNRS 3078
University of Lille1
Parc de la Haute Borne
50 avenue de Halley – BP 70478
59658 Villeneuve d'Ascq, France

E-mail: lionel.marcon@iri.univ-lille1.fr; rabah.boukherroub@iri.univ-lille1.fr

Chapter 5

W. Zhang
Department of Chemistry
Purdue University
560 Oval Drive
West Lafayette, IN 47907, USA

C. Yang*
Department of Chemistry and Department of Physics
Purdue University
560 Oval Drive
West Lafayette, IN 47907, USA

E-mail: yang@purdue.edu

Chapter 7

T.-J. Yen and H.-I Lin*
Department of Materials Science and Engineering
National Tsing Hua University
101 Sec. 2, Kuang-Fu Road
Hsinchu 30013, Taiwan R. O. C.

E-mail: tjyen@mx.nthu.edu.tw; samual52422001@gmail.com

Chapter 8

C. Chiappini* and C. Almeida
Department of Materials
Imperial College London
Prince Consort Road
London SW7 2AZ, UK

E-mail: c.chiappini@imperial.ac.uk

Chapter 9

J. Wu
Department of Chemistry
Georgia Southern University
250 Forest Drive
Statesboro, GA 30460, USA

E-mail: jwu@georgiasouthern.edu

Chapter 10

C. Villard
Institut Néel, CNRS and Université Joseph Fourier
25 rue des Martyrs
BP 166
38042 Grenoble Cedex 9, France

E-mail: catherine.villard@grenoble.cnrs.fr

Chapter 11

E. Segal* and Y. Bussi
Department of Biotechnology and Food Engineering
Russell Berrie Nanotechnology Institute
Technion – Israel Institute of Technology
Haifa 32000, Israel

E-mail: esegal@tx.technion.ac.il

Chapter 12

A. De and S. Chen
University of Twente
MESA+ Institute for Nanotechnology
Enschede 7500 AE, The Netherlands

E-mail: a.de@utwente.nl; s.chen@utwente.nl

E. T. Carlen*
Institute of Materials Science
University of Tsukuba
1-1-1 Tennodai
Ibaraki, Tsukuba 305-8573, Japan

and

University of Twente
MESA+ Institute for Nanotechnology
Enschede 7500 AE, The Netherlands

E-mail: ecarlen@ims.tsukuba.ac.jp; e.t.carlen@utwente.nl

Woodhead Publishing Series in Biomaterials

1. **Sterilisation of tissues using ionising radiations**
   Edited by J. F. Kennedy, G. O. Phillips and P. A. Williams
2. **Surfaces and interfaces for biomaterials**
   Edited by P. Vadgama
3. **Molecular interfacial phenomena of polymers and biopolymers**
   Edited by C. Chen
4. **Biomaterials, artificial organs and tissue engineering**
   Edited by L. Hench and J. Jones
5. **Medical modelling**
   R. Bibb
6. **Artificial cells, cell engineering and therapy**
   Edited by S. Prakash
7. **Biomedical polymers**
   Edited by M. Jenkins
8. **Tissue engineering using ceramics and polymers**
   Edited by A. R. Boccaccini and J. Gough
9. **Bioceramics and their clinical applications**
   Edited by T. Kokubo
10. **Dental biomaterials**
    Edited by R. V. Curtis and T. F. Watson
11. **Joint replacement technology**
    Edited by P. A. Revell
12. **Natural-based polymers for biomedical applications**
    Edited by R. L. Reiss et al.
13. **Degradation rate of bioresorbable materials**
    Edited by F. J. Buchanan
14. **Orthopaedic bone cements**
    Edited by S. Deb
15. **Shape memory alloys for biomedical applications**
    Edited by T. Yoneyama and S.Miyazaki
16. **Cellular response to biomaterials**
    Edited by L. Di Silvio
17. **Biomaterials for treating skin loss**
    Edited by D. P. Orgill and C. Blanco

18 **Biomaterials and tissue engineering in urology**
Edited by J. Denstedt and A. Atala
19 **Materials science for dentistry**
B. W. Darvell
20 **Bone repair biomaterials**
Edited by J. A. Planell, S. M. Best, D. Lacroix and A. Merolli
21 **Biomedical composites**
Edited by L. Ambrosio
22 **Drug–device combination products**
Edited by A. Lewis
23 **Biomaterials and regenerative medicine in ophthalmology**
Edited by T. V. Chirila
24 **Regenerative medicine and biomaterials for the repair of connective tissues**
Edited by C. Archer and J. Ralphs
25 **Metals for biomedical devices**
Edited by M. Ninomi
26 **Biointegration of medical implant materials: Science and design**
Edited by C. P. Sharma
27 **Biomaterials and devices for the circulatory system**
Edited by T. Gourlay and R. Black
28 **Surface modification of biomaterials: Methods, analysis and applications**
Edited by R. Williams
29 **Biomaterials for artificial organs**
Edited by M. Lysaght and T. Webster
30 **Injectable biomaterials: Science and applications**
Edited by B. Vernon
31 **Biomedical hydrogels: Biochemistry, manufacture and medical applications**
Edited by S. Rimmer
32 **Preprosthetic and maxillofacial surgery: Biomaterials, bone grafting and tissue engineering**
Edited by J. Ferri and E. Hunziker
33 **Bioactive materials in medicine: Design and applications**
Edited by X. Zhao, J. M. Courtney and H. Qian
34 **Advanced wound repair therapies**
Edited by D. Farrar
35 **Electrospinning for tissue regeneration**
Edited by L. Bosworth and S. Downes
36 **Bioactive glasses: Materials, properties and applications**
Edited by H. O. Ylänen
37 **Coatings for biomedical applications**
Edited by M. Driver
38 **Progenitor and stem cell technologies and therapies**
Edited by A. Atala
39 **Biomaterials for spinal surgery**
Edited by L. Ambrosio and E. Tanner
40 **Minimized cardiopulmonary bypass techniques and technologies**
Edited by T. Gourlay and S. Gunaydin
41 **Wear of orthopaedic implants and artificial joints**
Edited by S. Affatato

42 **Biomaterials in plastic surgery: Breast implants**
Edited by W. Peters, H. Brandon, K. L. Jerina, C. Wolf and V. L. Young
43 **MEMS for biomedical applications**
Edited by S. Bhansali and A. Vasudev
44 **Durability and reliability of medical polymers**
Edited by M. Jenkins and A. Stamboulis
45 **Biosensors for medical applications**
Edited by S. Higson
46 **Sterilisation of biomaterials and medical devices**
Edited by S. Lerouge and A. Simmons
47 **The hip resurfacing handbook: A practical guide to the use and management of modern hip resurfacings**
Edited by K. De Smet, P. Campbell and C. Van Der Straeten
48 **Developments in tissue engineered and regenerative medicine products**
J. Basu and J. W. Ludlow
49 **Nanomedicine: Technologies and applications**
Edited by T. J. Webster
50 **Biocompatibility and performance of medical devices**
Edited by J. P. Boutrand
51 **Medical robotics: Minimally invasive surgery**
Edited by P. Gomes
52 **Implantable sensor systems for medical applications**
Edited by A. Inmann and D. Hodgins
53 **Non-metallic biomaterials for tooth repair and replacement**
Edited by P. Vallittu
54 **Joining and assembly of medical materials and devices**
Edited by Y. (Norman) Zhou and M. D. Breyen
55 **Diamond-based materials for biomedical applications**
Edited by R. Narayan
56 **Nanomaterials in tissue engineering: Fabrication and applications**
Edited by A. K. Gaharwar, S. Sant, M. J. Hancock and S. A. Hacking
57 **Biomimetic biomaterials: Structure and applications**
Edited by A. J. Ruys
58 **Standardisation in cell and tissue engineering: Methods and protocols**
Edited by V. Salih
59 **Inhaler devices: Fundamentals, design and drug delivery**
Edited by P. Prokopovich
60 **Bio-tribocorrosion in biomaterials and medical implants**
Edited by Y. Yan
61 **Microfluidic devices for biomedical applications**
Edited by X. J. James Li and Y. Zhou
62 **Decontamination in hospitals and healthcare**
Edited by J. T. Walker
63 **Biomedical imaging: Applications and advances**
Edited by P. Morris
64 **Characterization of biomaterials**
Edited by M. Jaffe, W. Hammond, P. Tolias and T. Arinzeh
65 **Biomaterials and medical tribology**
Edited by J. Paolo Davim

66 **Biomaterials for cancer therapeutics: Diagnosis, prevention and therapy**
Edited by K. Park
67 **New functional biomaterials for medicine and healthcare**
E. P. Ivanova, K. Bazaka and R. J. Crawford
68 **Porous silicon for biomedical applications**
Edited by H. A. Santos
69 **A practical approach to spinal trauma**
Edited by H. N. Bajaj and S. Katoch
70 **Rapid prototyping of biomaterials: Principles and applications**
Edited by R. Narayan
71 **Cardiac regeneration and repair: Volume 1: Pathology and therapies**
Edited by R.-K. Li and R. D. Weisel
72 **Cardiac regeneration and repair: Volume 2: Biomaterials and tissue engineering**
Edited by R.-K. Li and R. D. Weisel
73 **Semiconducting silicon nanowires for biomedical applications**
Edited by J. L. Coffer
74 **Silk biomaterials for tissue engineering and regenerative medicine**
Edited by S. Kundu
75 **Biomaterials for bone regeneration: Novel techniques and applications**
Edited by P. Dubruel and S. Van Vlierberghe
76 **Biomedical foams for tissue engineering applications**
Edited by P. Netti
77 **Precious metals for biomedical applications**
Edited by N. Baltzer and T. Copponnex
78 **Bone substitute biomaterials**
Edited by K. Mallick
79 **Regulatory affairs for biomaterials and medical devices**
Edited by S. Amato and R. Ezzell
80 **Joint replacement technology. Second edition**
Edited by P. A. Revell
81 **Computational modelling of biomechanics and biotribology in the musculoskeletal system: Biomaterials and tissues**
Edited by Z. Jin
82 **Biophotonics for medical applications**
Edited by I. Meglinski
83 **Modelling degradation of bioresorbable polymeric medical devices**
Edited by J. Pan
84 **Perspectives in total hip arthroplasty: Advances in biomaterials and their tribological interactions**
S. Affatato
85 **Tissue engineering using ceramics and polymers. Second edition**
Edited by A. R. Boccaccini and P. X. Ma

Foreword

Silicon has been central to many technological innovations for decades, and remains an unsubstitutable key material for the electronics industry. Micro- and nanofabrication techniques have enabled a highly reliable production of well-defined and sophisticated structures down to sub-50 nm scale. The conventional top-down fabrication, however, is running into fundamental limitations in the fabrication of molecular scale nanostructures. Alternatively, bottom-up synthetic chemistry is able to produce complex nanostructures via self-assembly of nanocrystals, nanowires or nanotubes. Silicon nanostructures have garnered the greatest attention in the past decades for a variety of applications including nanoelectronics, energy conversion and bio–nano interfaces.

Silicon nanowires represents one of the most important research subjects in the nanowire research community. In the past two decades, many interesting physical properties have been discovered including, for example, the giant piezoresistance effect, significantly reduced thermoconductivity and enhanced thermoelectric performance. Horizontal and vertical silicon nanowire field effect transistors (FETs), complementary logic gates, nanoelectromechanical systems (NEMS), and various energy conversion devices have been reported in the past decade. These fundamentally new properties could eventually lead to a significant breakthrough in their commercial applications. There are, however, still many important issues remaining to be addressed in the following years. These include, for example, low-cost large-scale production and assembly of high quality nanowires, precise doping and heterostructure formation in nanowires, reproducible surface and defect engineering of the nanowires, and nano–macro interface and addressability issues. When our synthetic control on these nanostructures improves, novel and unexpected chemical and physical properties will arise. These novel nanostructures could have significant impact in electronics, photonics, energy conversion, biomedical as well as other unexplored territories.

Meanwhile, it is easy to notice that nanotechnology has received increased attention in the biological research field. The important examples are:

1. the usage of nanoparticles in optical and magnetic resonance imaging;
2. the demonstration of potential application of metal nanoshells and carbon nanotubes for the treatment of tumor and cancer cells;
3. the application of nanowire-based transistors to electrically detect specific biomolecules.

Also, direct interconnection of the living cells to the external world by interfacing nano-materials may afford great opportunities to probe and manipulate biological processes occurring inside the cells, across the membranes, and between neighboring cells. For instance, silicon nanowires ($d = 1 \sim 100$ nm) are a few orders of magnitude smaller in diameter than mammalian cells ($d_{cell} \sim$ on the order of 10 μm), yet comparable to the size of various intracellular biomolecules. The nanowires have high-aspect ratio ($< 10^3$) and yet are sufficiently rigid to be mechanically manipulated. The nanometer scale diameter and the high-aspect ratio of silicon nanowires make them readily accessible to the interiors of living cells, which may facilitate the study of the complex regulatory and signaling patterns at the molecular level. This research direction of interfacing nanowires and living cells is certainly one of the most exciting topics at the moment, and is quickly unfolding. While it is true that the nanowire in this case is serving as a versatile technological tool or platform, many new discoveries are expected when such platforms are used to tackle real biological problems.

The review articles in this book represent a snapshot of a very active research field, namely the biomedical applications of silicon nanowires. Nanowire research has a great future, but there is still a great deal of research that remains to be done on both the fundamental and applied levels, namely, more precise structural control and assembly and novel property exploration. Expect great science in this direction in the years to come!

Peidong Yang
S. K. and Angela Chan Distinguished Chair in Energy
Department of Chemistry
Department of Materials Science and Engineering
University of California, Berkeley, CA 94720, USA

Part I

Introduction to silicon nanowires for biomedical applications

1

Overview of semiconducting silicon nanowires for biomedical applications

J. L. COFFER, Texas Christian University, USA

DOI: 10.1533/9780857097712.1.3

Abstract: This chapter highlights the broad fundamental and technological significance of semiconductor nanowires in general, and silicon nanowires in particular. After a brief discussion of the historical development of silicon in nanowire form, the scope of topics covered in this book and its organization are presented, along with brief comments regarding the significance of each and its relevance to biomedical applications.

Key words: silicon, nanowire, nanotechnology, biomedicine, biosensors, drug delivery, tissue engineering.

1.1 Introduction

Silicon remains the unquestionable mainstay of the electronic device industry, with a constant scrutiny of its use under the lens of Moore's Law and an ongoing reduction in feature size and corresponding device dimensions (Mack, 2011). At the same time, knowledge of its fundamental properties have also benefited from the expansive growth in nanoscience and nanotechnology, with a broad spectrum of investigations being reported for this elemental semiconductor in one-dimensional nanowire form. Early experiments probed its ability to act as a chemical sensor (Cui *et al.*, 2001) and high-density p-n junction and transistor array (Cui and Lieber, 2001; Huang *et al.*, 2001). However, while bulk crystalline Si is traditionally viewed as bio-inert, the unique geometry of Si nanowires (SiNW), their diverse surface chemistry, as well as associated process engineering, have provided a boon of sorts in terms of fundamental studies of relevance to its ultimate application in the field of biomedical devices. In this vein, SiNW have demonstrated some amazing properties in terms of biological functions to which it can contribute and analyze. These range from the transfection of individual cells (Kim *et al.*, 2007) to the detection of electrical signals within a cardiac cell (Tian *et al.*, 2010). Thus, it is the biomedical relevance of semiconducting SiNW that is the focus of this book, with a diverse range of experts from a number of institutions across the globe assembled to tackle the key themes of this area of research.

1.2 Origins of silicon nanowires

An overview of these issues will be highlighted momentarily, but let us begin with a brief historical perspective. The genesis of SiNW can be viewed as originating with seminal efforts regarding relatively larger diameter cylindrical structures of Si in the micrometer dimension, perhaps better known as Si whisker technology (Levitt, 1971). A key component of this effort was the fundamental discovery of the Vapor-Liquid-Solid (VLS) method, whereby the dimensions of a catalyst particle (such as Au) take advantage of the limited solubility of Si in a metal silicide liquid phase (e.g. Si(s) precipitating in $AuSi_x$(l)) at the proper temperature and reactant concentrations to form a crystalline Si microwire (Wagner and Ellis, 1964). It was Lieber and co-workers who had the prescient realization that proper reduction of the catalyst dimension from the macro- to the nanoscale could yield construction of the target well-defined cylindrical nanowire constructs (Morales *et al.*, 1998), thereby opening the door for the expansive number of papers on this nanomaterial that have subsequently appeared.

In that regard, let us look at the evolution of interest in the topic of silicon nanowires within the last 15 years. This is best exemplified by a search of the phrase 'silicon nanowire' appearing in the citations in the Web of Science database (Thomson Reuters) on an annual basis (up to, but not including, 2012) (see Fig. 1.1).

The number of citations containing 'silicon nanowire' in terms of content has clearly increased in an exponential manner, to a total of more than 7000 since 1997. From a biomedical context, inclusion of the term 'cell' along with 'silicon nanowire' results in a similar explosive growth curve, starting from a mere two references in 1999 to a value of 213 alone in 2011.

1.3 The structure of this book

This volume is broken down into a series of key concepts that are necessary building blocks to an understanding of how SiNW can be transformed into a bioactive platform and the unique advantages brought about as a consequence. A brief overview of the chapters is described below. The chapters may be loosely clustered into three main themes: 1 fabrication, characterization, and surface modification – with an associated impact on biocompatibility; 2 targeted use of SiNW in selected therapeutic approaches; 3 the ability of SiNW and associated composites to act as sensitive, selective biosensors.

It is logical to begin our detailed presentation in Chapter 2 with a discussion of common routes to silicon nanowire fabrication and their subsequent characterization by a broad range of experimental techniques. These key topics are presented by Gengfeng Zheng and co-workers at Fudan University, who carefully describe here the essence of the well-investigated VLS route to SiNW preparation, along with the increasingly popular metal-assisted top down etching

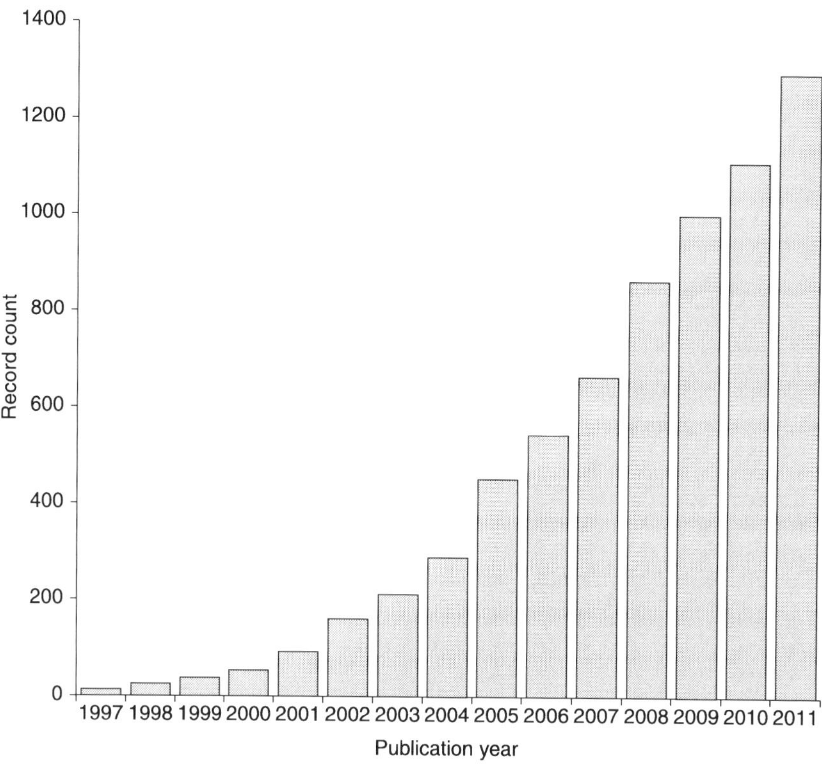

1.1 Silicon nanowire-related citations, 1997–2011 (Web of Science, Thomson-Reuters).

route to NW formation. In terms of characterization, the role of electron microscopy is discussed, along with the complementary technique of Raman vibrational spectroscopy and, very importantly, electrical measurements.

Chapter 3, by Rabah Boukerroub and his group at CNRS–Lille, presents general strategies critical to surface modification of nanowires and their significance in modulating function. Of course, a critical test of evaluating the success of such chemical transformation lies in an evaluation of the biocompatibility or bioactivity of such modified nanowires, a topic that is presented in Chapter 4 by the same group.

We then make the transition from the above topics to a series of chapters with a more explicit biological emphasis and long-term relevance to therapy. For example, in Chapter 5, Yang and co-workers describe the development of our understanding of SiNW interactions with biological systems mechanistically at the cellular level. This is exemplified by the case of folate covalently anchored to the surface of SiNW, and impact on uptake into specific cell lines (through the proper receptor).

It is also important to illustrate the relevance of these modified materials in more specific, targeted therapies. Thus, the ability of SiNW and polymeric composites to engage in high impact areas such as processes relevant to bone repair via a tissue engineering approach is described in Chapter 6.

A major area of related therapeutic research emphasis in regenerative medicine is stem cell-based therapies. Opportunities for SiNW to play a role in advances in this area are therefore described by Yen and co-workers from National Tsing-Hua University in Chapter 7. The emphasis here is on stereo-topographical cues that the nanowires can provide with respect to regulation of human mesenchymal stem cells, a source of great clinical potential.

Drug delivery is another area whereby nanoscale materials are making a significant impact. Thus, in Chapter 8, Chiappini, at the Imperial College London, describes a unique porous nanowire platform capable of engaging in sustained drug delivery. In this system, the porous morphology provides a convenient carrier for the release of novel therapeutics such as small interfering RNA (siRNA).

Finally, we focus on a series of chapters highlighting the role of SiNW in sensitive detection of biological molecules. Ji Wu of Georgia Southern University begins this series in Chapter 9 by providing a useful overview of some of the key general aspects of nanowire platform fabrication for high-throughput biosensing, a significant goal for long-term pragmatic implementation of these materials.

The logical similarities between the information processing roles of the brain in human systems along with that of Si in metal-oxide-semiconductor (MOSFET)-type architectures also makes use of SiNW in the field of neuroscience a natural fit. Thus, Chapter 10, by Villard and co-workers of CNRS–Grenoble, presents the challenges of controlling neuron position and axon polarity above SiNW field effect transistor (FET) arrays. In Chapter 11, Segal and Bussi of the Technion survey the utility of SiNW in associated composite formulations to serve as useful roles in biosensing. Finally, Carlen (University of Twente) follows this discussion with a detailed presentation of using as-prepared, probe-free nanowires to sensitively detect critical biomolecules (Chapter 12).

1.4 Conclusion

In this volume, the contributors seek to provide essential information for a rather novel form of such a ubiquitous material. The sensitivity brought about as a consequence of dimension of these nanowires, coupled with selectivity as a result of surface chemical functionality, has produced some impressive results that will be presented herein. Significant problems in drug delivery, neuroscience, tissue engineering, and other biomedical technologies have been addressed, and even more exciting research challenges clearly lie ahead.

1.5 References

Cui, Y., Wei, Q., Park, H., and Lieber, C.M. (2001), Nanowire nanosensors for highly sensitive and selective detection of biological and chemical species, *Science*, **293**, 1289–92. (doi: 10.1126/science.10672711)

Cui, Y. and Lieber, C.M. (2001), Functional nanoscale electronic devices assembled using silicon nanowire building blocks, *Science*, **291**, 851–3. (doi: 10.1126/science.291.5505.851)

Huang, Y., Duan, X., Cui, Y., Lauhon, L., Kim, K., and Lieber, C.M. (2001), Logic gates and computation from assembled nanowire building blocks, *Science*, **294**, 1313–17. (doi: 10.1126/science.1066192)

Kim, W., Ng, J.K., Kunitake, M.E., Conklin, B.R., and Yang, P. (2007), Interfacing silicon nanowires with mammalian cells. *J. Am. Chem. Soc.*, **129**, 7228–9. (doi: 10.1021/jr071456k)

Levitt, A.P. (1971) *Whisker Technology*. New York, John Wiley & Sons.

Mack, C.A., (2011), Fifty Years of Moore's Law, *IEEE Transactions On Semiconductor Manufacturing*, **24**, 202–7. (doi: 10.1109/TSM.2010.2096437)

Tian, B., Cohen-Karni, T., Qing, Q., Duan, X., Xie, P., and Lieber, C.M. (2010), Three-dimensional, flexible nanoscale field-effect transistors as localized bioprobes, *Science*, **329**, 830–4. (doi: 10.1126/science.1192033)

Wagner, R.S. and Ellis, W.C. (1964), Vapor-liquid-solid mechanism of single crystal growth, *Appl. Phys. Lett.*, **4**, 89–90. (doi: 10.1063/1.1753975)

2

Growth and characterization of semiconducting silicon nanowires for biomedical applications

GENGFENG ZHENG and MING XU, Fudan University, People's Republic of China

DOI: 10.1533/9780857097712.1.8

Abstract: One-dimensional silicon nanowires (SiNW) have attracted substantial interest in the study of interplays between these nanostructures and biology. This chapter briefly introduces several representative synthetic methods for the growth of SiNW, including the solution etching method and metal nanocluster-catalyzed chemical vapor deposition method. The latter will be extensively discussed to demonstrate the capability of controlled growth of a host of SiNW with modulated morphology, structure, and doping, including axial and radial heterostructures, kinked, branched, and/or modulated doped structures. Furthermore, this chapter will introduce several common characterization methods for SiNW, including electron microscopy, Raman spectroscopy, and electrical transport measurement.

Key words: silicon nanowire, chemical vapor deposition, vapor-liquid-solid mechanism, heterostructure, doping.

2.1 Introduction

Silicon nanowires (SiNW) represent an important class of one-dimensional (1D) nanostructures at the forefront of nanoscience and nanotechnology. They are promising building blocks for the assembly of nanoelectronic and nanophotonic systems because they can function both as nanoscale devices and interconnects (Lieber, 2011). The ability to control the electronic properties has been utilized for the reproducible assembly of field-effect transistors (Zheng *et al.*, 2004), integrated logic circuits (Yan *et al.*, 2011), photoelectric conversion (Tian *et al.*, 2007), and energy storage (Chan *et al.*, 2008). In addition, as the diameters of the structures are typically in the range of 10–100 nm, which are comparable to many of the chemical and biological targets of interest, SiNW are also capable of providing a unique and powerful platform for biomedical applications (Zheng *et al.*, 2005; Patolsky *et al.*, 2006). These nanoscale structures and devices serve as a 'bottom-up' approach paradigm and can offer advantages compared with lithographically patterned silicon devices as the physical and chemical characteristics of the NW, including diameter, surface composition, and electronic properties, can, in principle, be controlled during synthesis.

In this chapter, we will briefly summarize several synthetic methods of SiNW, including electroless metal-assisted solution etching as an example of a top-down approach, and nanocluster-catalyzed chemical vapor deposition and solution growth for the bottom-up approaches. The chemical vapor deposition method will be extensively illustrated to show the realization of several SiNW heterostructures, with the capability of tuning the physical dimension and chemical composition of the designed structures. Several common characterization methods, including electron microscopy, Raman spectroscopy, and electrical transport, will be briefly introduced. Finally, comments are made on the remaining challenges and future trends.

2.2 Synthesis methods for silicon nanowires (SiNWs)

The design and synthesis of semiconductor SiNW as building blocks with well-defined structures and physical properties are central to nanoscience and nanotechnology. Significant progress has been achieved in control of morphology, size, and composition on length scales ranging from the atomic and up. The synthesis of SiNW structures can be generally categorized into two distinct approaches: top-down and bottom-up. One well-known example of the solution approach is the metal nanoparticle-enhanced etching of a silicon wafer, which typically starts from an integral, homogeneous unit and uses a combination of lithography, deposition, and etching steps to create sub-structures with defined morphology and properties. Top-down approaches for the synthesis of Si nanostructures have been exceedingly successful in many fields such as microelectronics. However, the substantially increasing cost of the top-down approaches with the accompanying requirement of further reducing the feature resolution, as well as the need of developing multi-functional device building blocks, have motivated research efforts to search for new synthetic approaches. Bottom-up approaches start from molecular-level precursors to form individual SiNW, which waives the need for lithography steps and has the potential to develop multi-functionalities through direct synthesis and subsequent assembly. Gold nanocluster-catalyzed chemical vapor deposition (CVD) based on the vapor–liquid–solid (VLS) growth mechanism (Wagner and Ellis, 1964) represents the most important approach for bottom-up synthesis of SiNW. Central to the CVD synthesis is the precise control and tuning of the morphology, structure, and chemical composition of the catalyst targets. Several representative methods of top-down and bottom-up approaches are briefly summarized below.

2.2.1 Chemical etching of silicon wafers

Large-area, highly oriented 1D SiNW arrays can be formed by chemical etching of single-crystal silicon wafers, using a simple and rapid metal-nanoparticle-assisted solution approach (Peng *et al.*, 2005). The reaction mechanism includes both the electroless deposition of silver nanoparticles on silicon and the silver nanoparticle-catalyzed chemical etching of silicon in HF/$Fe(NO_3)_3$ solution, schematically shown in Fig. 2.1. First, a mixed solution of silver nitrate ($AgNO_3$) and hydrofluoric acid (HF) is placed on the surface of a single-crystal silicon wafer to galvanic reduction and deposition of Ag nanoparticles. In the meantime, silicon is oxidized into silicon dioxide, which is subsequently dissolved by HF. Both the cathodic (silver reduction and deposition) and the anodic reactions (silicon oxidation and dissolution) take place on the silicon surface simultaneously. After a uniform layer of Ag nanoparticles is formed and covered on the Si wafer surface, the wafer is transferred into a mixed solution of HF/$Fe(NO_3)_3$. As the redox level of Fe^{3+}/Fe^{2+} is located below the valence band of silicon, electrons are injected from silicon to reduce Fe^{3+} into Fe^{2+}, while silicon atoms are continuously oxidized and dissolved in HF. The Ag^+/Ag couple has a more positive redox potential than Fe^{3+}/Fe^{2+} and remains stable. However, Ag nanoparticles on silicon surface can attract electrons from silicon and transfer to Fe^{3+}, and thus serve as local catalysts to enhance the cathodic reaction. In addition, the Ag nanoparticles cannot move horizontally but sink into the pits where the oxidized silicon is

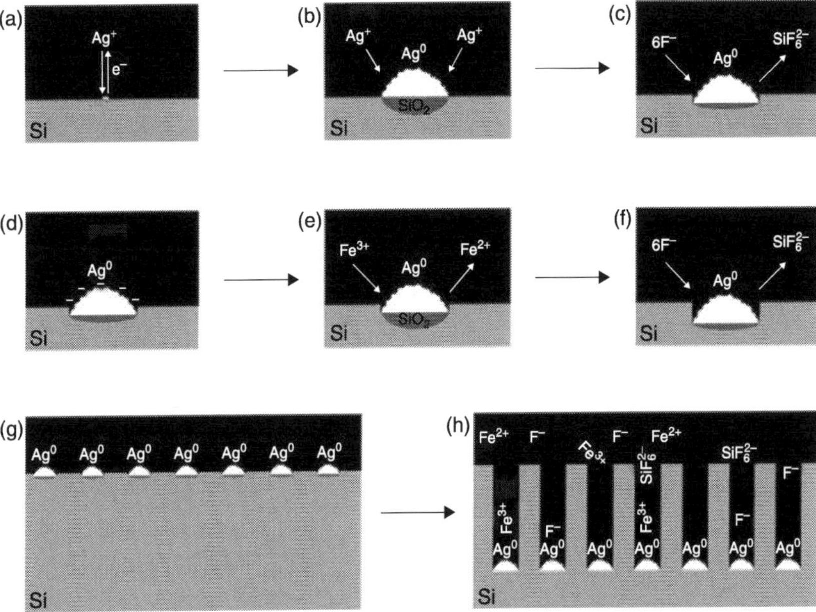

2.1 (a–h) A model illustrating the electroless deposition of silver on silicon in HF/$AgNO_3$ solution and the silver-nanoparticle-catalyzed chemical etching of silicon in HF/$Fe(NO_3)_3$ solution. Reproduced from Peng *et al.*, 2005, *Angewandte Chemie International Edition*, permitted by John Wiley and Sons.

dissolved, leading to selective etching of deeper pores. As a consequence, arrays of single-crystal SiNW are formed after continuous etching, with high orientation perpendicular to the original Si wafer surface. The diameter and length of the resultant SiNW are tuned by concentration of the electroplating solutions and the etching time, while the composition and doping are determined by the original Si wafers.

2.2.2 Chemical vapor deposition (CVD) for SiNW growth

The rational design and CVD synthesis of SiNW, with all key parameters including chemical composition, diameter, length and doping, can be realized with well-controlled growth conditions. For instance, the thinnest SiNW synthesized have diameters as small as a few nanometers (Wu *et al.*, 2004), and the longest SiNW can reach a length of several millimeters (Park *et al.*, 2008). In addition, the synthetic methods of SiNW with complex heterostructures, such as core-shell (Lauhon *et al.*, 2002), modulated doped (Yang *et al.*, 2005), branched (Jiang *et al.*, 2011) and kinked morphologies (Tian *et al.*, 2009), have been developed, enabling a variety of new electrical and optical properties that subsequently allow for many unconventional applications in biological and life sciences.

Growth of intrinsic (undoped) SiNWs

The CVD synthesis of SiNW with small diameter distribution is achieved using well-defined metal nanoclusters as catalysts in a vapor-liquid-solid (VLS) growth process (Lieber, 2003). The catalysts (usually gold nanoclusters) control the size of the initial nucleation event and eventually the SiNW diameters. The schematic of VLS growth process is shown in Fig. 2.2. Specifically, those gold nanoclusters are first deposited onto a flat substrate, usually a silicon wafer with thermal oxide layers. The substrate is then placed in a quartz tube reactor and heated to above the eutectic temperature of Au-Si (~363 °C). Silane (SiH_4) is used as the precursor for Si and introduced into the reactor and decomposed, where the Si atoms will dissolve into Au nanoclusters to form nanodroplets. When these nanodroplets become oversaturated, a nucleation event occurs, where Si atoms will precipitate from the liquid phase, and then by continuing feeding with the gas precursors, solid, crystalline SiNW are formed. The lengths of the nanowires are tuned by the reaction conditions including temperature, pressure, flow rate and growth time.

An important requirement of the VLS approach to the NW growth is that the SiNW grown should be predominantly controlled as in axial elongation. The homogeneous coating of amorphous silicon on the pre-formed SiNW surface, on the other hand, results in a NW diameter increase, which significantly affects its crystalline structure and the charge transport characteristics (Fig. 2.3). This phenomenon can become rather substantial in the preparation of modulated doped SiNW (*vide infra*). Unless designed intentionally, this homogenous deposition

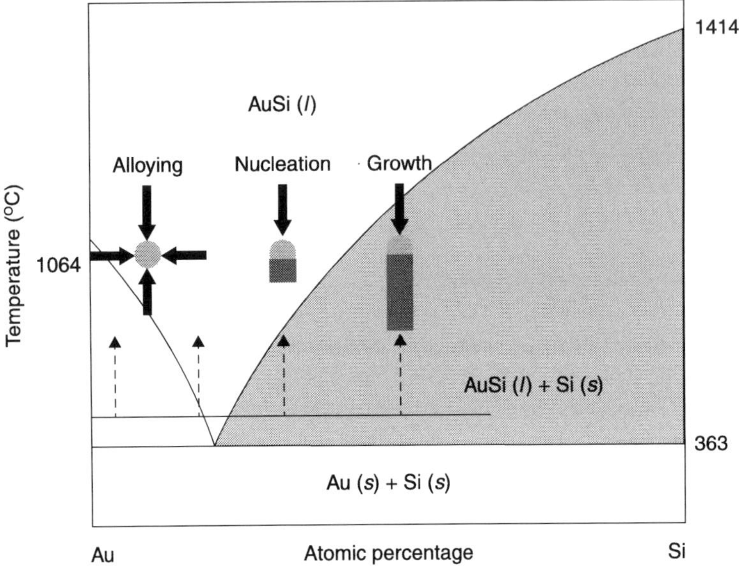

2.2 Au/Si bi-phase diagram illustrating the alloying, nucleation and growth steps of Au nanocluster-catalyzed SiNW synthesis by the chemical vapor deposition method.

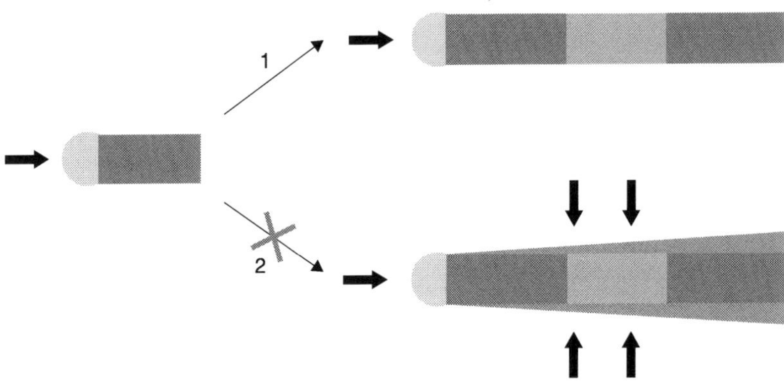

2.3 Illustrations of (1) pure axial and (2) simultaneous axial and radial growth occurring during gold nanocluster (at the tip) catalyzed silicon nanowire synthesis. Simultaneous radial growth (2) leads to undesirable deposition of amorphous silicon or other dopant materials over the entire nanowire. Reproduced from Yang *et al.*, 2005, *Science*, permitted by the American Association for the Advancement of Science.

should be avoided by tuning of the reaction conditions. Hydrogen gas (H_2) is often used as the carrier gas for the gas precursors, which can effectively inhibit homogeneous deposition.

Growth of p-type or n-type SiNWs

In the CVD process, a variety of dopant atoms can be readily incorporated into the intrinsic (undoped) SiNW to achieve desired functionalities. For instance, p-type (boron-doped) SiNW can be synthesized using mixed gas precursors of SiH_4 and diborane (B_2H_6), while n-type (phosphorus-doped) SiNW can be synthesized using SiH_4 and phosphine (PH_3) as reactants (see Section 2.4). The use of gas phase dopants and silicon reactants enables the dopant concentration to be readily controlled by varying the ratio of different gas reactants. Nonetheless, it should be pointed out that the atomic ratio in the obtained doped SiNW products is not necessarily equal to that of the gas phase, as the decomposition temperature profiles vary for different gas precursors. Furthermore, the addition of dopant gas reactants also affects the decomposition rate of SiH_4. For instance, the addition of B_2H_6 shows an enhancement (relative to the decomposition rate of SiH_4 alone), while adding PH_3 into the gas precursor mixture will reduce the rate.

Growth of millimeter-long SiNWs

Ultra-long (such as millimeter-long) SiNW can benefit device integration by facilitating the interconnection of individual SiNW arrays. Other techniques such as the high-temperature thermal evaporation of silicon monoxide and silicon powders have previously been reported for obtaining millimeter-long SiNW (Park *et al.*, 2008). However, this goal was challenging for the VLS-growth method, as most of the growth rates reported were predominantly in the order of 1–2 μm per minute. As the SiNW growth rate is strongly temperature-dependent, this phenomenon indicates that the kinetics of thermal decomposition of SiH_4 into atomic Si species is the rate-determining step, much more important than the gas-phase mass transport. Thus, the acceleration of the decomposition step can significantly enhance the overall growth rate. Because of the lower activation energy for cleavage of Si-Si versus Si-H bonds, disilane (Si_2H_6) is selected as the Si gas precursor for a higher catalytic decomposition rate. The growth rate of SiNW can be enhanced by almost 2 orders of magnitude, leading to SiNW tens of nanometers in diameter but ~2 mm in length, corresponding to an aspect ratio of close to 10^5 (Fig. 2.4).

Growth of axial SiNW heterostructures

The capability of controlling chemical composition and doping for SiNW offers a diverse set of building blocks for assembling nanodevices and biomedical

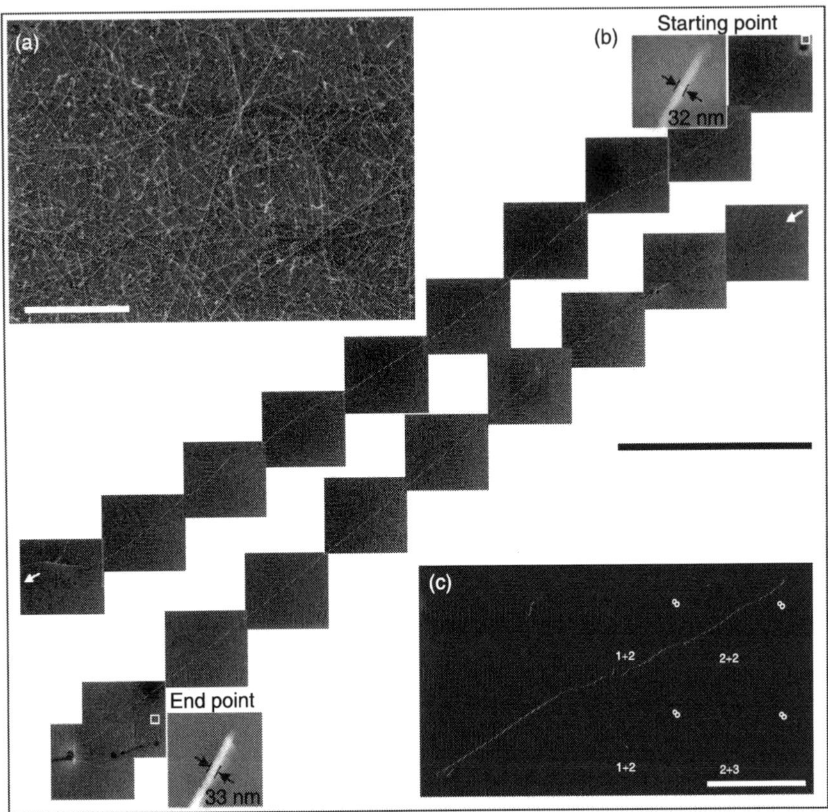

2.4 (a) Scanning electron microscopy (SEM) image of as-grown ultralong silicon nanowires (SiNW) synthesized by Si_2H_6 at 400 °C for 30 minutes. Scale bar: 20 μm. (b) A series of 20 SEM images of a 2.3 mm-long SiNW transferred on SiO_2/Si substrate. Scale bar: 200 μm. Insets: SEM images of starting and end segments of this SiNW. (c) Dark-field optical image of the same SiNW. Scale bar: 500 μm. Reproduced from Park *et al.*, 2008, *Nano Letters*, permitted by the American Chemical Society.

applications. Well-controlled variations in the composition and/or doping of SiNW heterostructures make possible the design and realization of unique electronic and photonic nanodevices via encoding functionality synthetically during growth. Conceptually, the axial SiNW heterostructures such as p-Si/n-Si can be realized by first deposition of a p-type SiNW segment for a period of time, followed by switching the reactant feedstock to deposit an n-type SiNW segment (Fig. 2.5). The exchange of the reactants can be repeated to produce superlattice structures, in which the doping profiles are modulated corresponding to the selected reactants. Nonetheless, this simplified scheme does not account for the possible homogenous coating on the pre-grown SiNW, especially for the case of

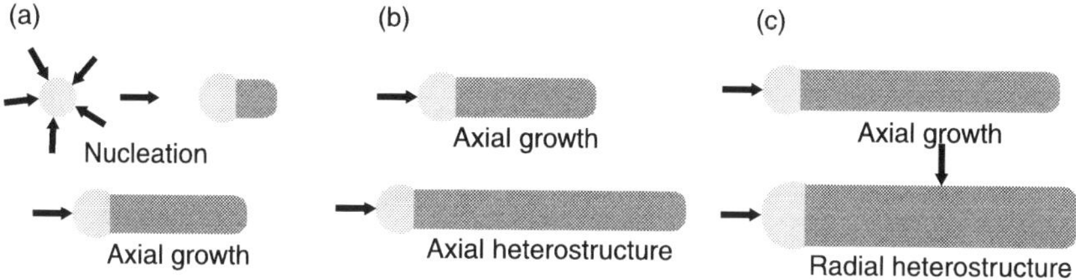

2.5 Growth and representative structures of (a) uniform single-crystal silicon nanowire (SiNW) nanowires, (b) axial SiNW heterostructures and (c) radial SiNW heterostructures. Reproduced from Lieber, 2003, *MRS Bulletin*, permitted by Cambridge University Press.

p-type SiNW deposition. Although almost no amorphous layer is observed by transmission electron microscopy (TEM) imaging, very tiny amounts of the deposition of boron atoms on the surface of an n-type SiNW segment can substantially affect the electronic properties achieved. A local heating technique is particularly useful in solving this problem, in which only the substrate covered with Au nanoclusters is heated, while the gas reactants are not heated and remain close to room temperature (Yang *et al.*, 2005). These gas reactants start to dissociate only when contacting the heated Au nanocluster catalyst, thus reducing the possibility of dissociation and homogenous deposition on the NW surface. By this means, the successful synthesis of modulation-doped SiNW was achieved with pure axial elongation without radial overcoating. The feature size of an individual segment can be less than 50 nm, surpassing the resolution achieved by conventional lithography-based top-down techniques.

Growth of radial SiNW heterostructures

For SiNW radial (or core-shell) heterostructures, the growth mode needs to be switched from the dominated VLS mechanism to homogenous deposition on the surface of existing nanowire core. This radial heterostructure can be achieved by changing the reactants and growth temperature, as schematically shown in Fig. 2.5(c). By sequentially modulating the reactants to form shells around a pre-grown SiNW core, it is, in principle, possible to create arbitrarily complex radial heterostructures, including crystalline Si/amorphous Si core-shell NWs (Lauhon *et al.*, 2002), and p-Si/n-Si core-shell structures (Tian *et al.*, 2007). A variety of other material composition combinations have also been demonstrated, such as different Si-Ge core-shell nanowire structures (Xiang *et al.*, 2006) and hollow SiNW (or Si nanotubes) (Ben-Ishai and Patolsky, 2012).

Growth of kinked or zigzag SiNWs

In addition to the straight SiNW structures where only one nucleation and growth cycle is involved, other hierarchical SiNW morphologies such as kinked and zigzag structures can be synthesized by iterative control over the nucleation and growth processes (Tian *et al.*, 2009). The structures of these kinked SiNW are analogous to metal-organic framework materials, having a secondary building unit consisting of two straight single-crystalline NW arms connected by one fixed 120° angle joint unit. Along these kinked SiNW, these kinks are introduced at defined positions during growth and are confined to a single plane, which involves three main steps during nanocluster-catalyzed growth (Fig. 2.6): 1 axial growth of a 1D SiNW arm segment, 2 purging of gas reactants to suspend SiNW elongation, and 3 re-introduction of gas reactants to supersaturate and nucleate the Au/Si catalyst for re-growth of SiNW to form kinked or zigzag structures. The concentration of Si species dissolved in the Au/Si nanocluster catalysts drops below the level for sustaining continuous growth during the purging of gas

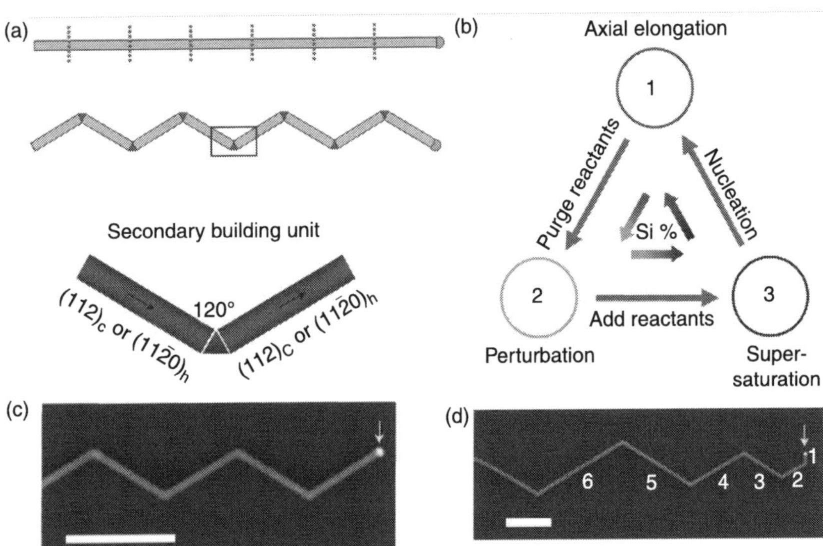

2.6 Design and controlled synthesis of multiply kinked nanowires. (a) Schematic of a coherently kinked nanowire and the secondary building unit (SBU), which contains two arms and one joint. The multiply kinked nanowires (middle panel) are derived from the corresponding one-dimensional nanowire by introducing the joints at the points indicated by the dashed lines in the upper panel. Subscripts c and h denote cubic and hexagonal structures, respectively. (b) Cycle for the introduction of a SBU by stepwise synthesis. (c, d) Scanning electron microscopy (SEM) images of multiply kinked two-dimensional silicon nanowires with (c) equal arm segment lengths and (d) decreasing arm segment lengths respectively. Scale bars are 1 μm. The arrows highlight the positions of the nanocluster catalysts. Reproduced from Tian *et al.*, 2009, *Nature Nanotechnology*, permitted by Nature Publishing Group.

reactants, and then increases again to reach a maximum upon supersaturation to re-nucleate when the gas reactants are re-introduced. This three-step cycle can be repeated to link a number of these SiNW secondary building units, leading to a kinked SiNW chain structure on a two-dimensional plane. The entire SiNW nanostructure is single crystalline, and no atomic scale twin defects or stacking faults are found across the complete arm-joint-arm junctions. The joint has a quasi-triangular structure with (111) top/bottom facets and two (112) side facets joining the adjacent arms. In addition, within each straight SiNW segment between kinks, a linear dependence of segment length on the axial growth time is identified, confirming the well-controlled VLS growth mode.

Growth of branched SiNWs

Branched or hyperbranched SiNW, in which one or more secondary NW branches grow in a radial direction from a primary NW backbone, offer another approach for increasing structural complexity and enabling more diverse and greater functions. The branches naturally provide access to higher dimensionality structures and the capability of achieving parallel connectivity and interconnection during synthesis (Jiang *et al.*, 2011). Analogous to the solution growth of branched nanostructures with morphology and size control, the approach exploits a multi-step nanocluster-catalyzed VLS growth process, in which the size and density of the Au nanoclusters can be independently tuned at each step.

One simple method to create secondary Au nanoclusters on the pre-grown SiNW backbones is via physical adsorption (Wang *et al.*, 2004), schematically exhibited in Fig. 2.7. First, a SiNW backbone of specific diameter and composition is prepared by Au nanocluster-mediated VLS growth. Then, new Au nanoclusters are deposited on the SiNW backbones as the catalysts for NW branch growth. This branch growth cycle can be repeated to yield higher order or hyperbranched SiNW structures. Another method for selective deposition of Au nanoparticles on the pre-grown SiNW backbones is carried out by galvanic surface reduction (Jiang *et al.*, 2011). First, the SiNW backbones are etched by a hydrogen fluoride (HF) solution to remove native oxide and produce hydrogen-terminated surface, which are then mixed with a $HAuCl_4$ solution for the reduction of Au nanoparticles. The density and size of the formed Au nanoparticles are controlled by the $HAuCl_4$ solution concentration and reaction time, with higher $HAuCl_4$ concentration and longer reaction time resulting in larger and denser Au nanoparticles. These galvanic deposited Au nanoparticles are further used as catalysts for the VLS-mode synthesis branches of SiNW or other material compositions.

2.2.3 Solution-liquid-solid growth of SiNWs

Colloidal methods are an important type of synthesis approach for semiconductor NW. Nonetheless, when compared with other semiconductor NW of group II-VI,

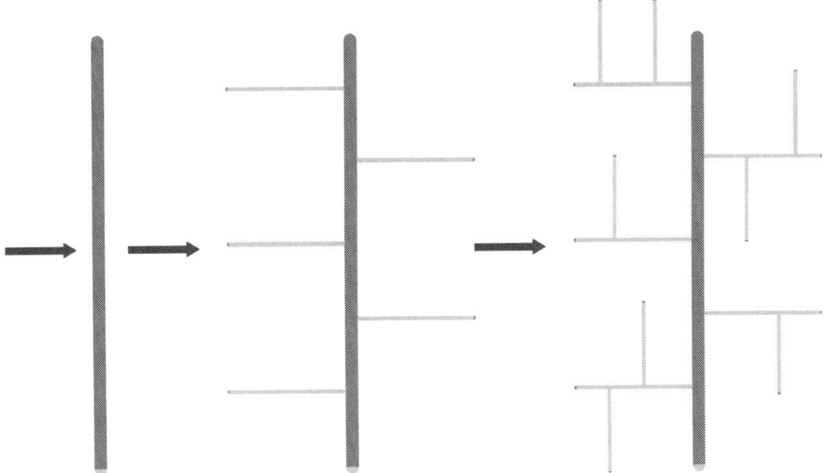

2.7 Schematic illustrating the multistep syntheses of branched and hyperbranched silicon nanowire (SiNW) structures, showing the nanowire trunks, the first-level branches, and the second-level branches. Reproduced from Wang *et al.*, 2004, *Nano Letters*, permitted by the American Chemical Society.

III-V and IV, Si is among the most challenging, partly because Si precursors, such as SiH_4, halogenerated silanes and organosilanes, are generally very stable and require high temperatures (>400 °C) to dissociate. Previously, the only solution phase route to produce large quantities of crystalline SiNW required extreme temperatures and pressures that exceed the critical point of the solvent (Holmes *et al.*, 2000). The growth of SiNW by the solution-liquid-solid (SLS) mechanism at atmospheric pressure, using trisilane (Si_3H_8) as a reactant in octacosane ($C_{28}H_{58}$) or squalane ($C_{30}H_{62}$) and either gold (Au) or bismuth (Bi) nanocrystals as seeds, was first reported in 2008 (Heitsch *et al.*, 2008). Au or Bi can form a eutectic with Si at 363 °C and 264 °C, respectively, well below the boiling temperatures of $C_{28}H_{58}$ (T_b=430 °C) and $C_{30}H_{62}$ (T_b=423 °C). During the growth, trisilane (Si_3H_8) decomposes to generate Si atoms, which are consumed by the Au (or Bi) seeds to form a Au/Si (or Bi/Si) eutectic that promotes the SiNW growth (Fig. 2.8).

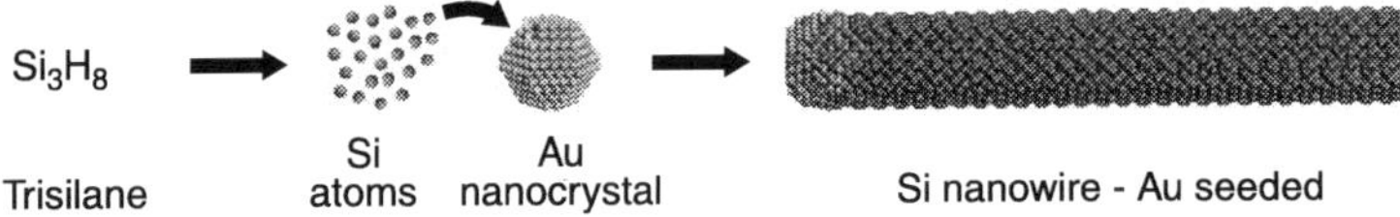

2.8 Schematic illustrating the solution-liquid-solid (SLS) growth of a silicon nanowire (SiNW) using an Au nanocrystal catalyst and Si_3H_8 precursors. Reproduced from Heitsch *et al.*, 2008, *Journal of the American Chemical Society*, permitted by the American Chemical Society.

2.3 Characterization methods

2.3.1 Electron microscopy techniques

Electron microscopy techniques such as scanning electron microscopy (SEM) and TEM are important analytical tools for characterizing the morphology, composition and structure of grown SiNW. Electron microscopy provides a feedback to rationally grow the desired SiNW, with compositional information obtained using energy dispersive X-ray analysis (EDS). General structural information of SiNW, such as SiNW thickness and whether the SiNW are single crystals or amorphous, can readily be determined using TEM. Once the desired SiNW structures are achieved, more advanced techniques can be applied to investigate the structure and properties of SiNW. A thorough characterization includes the determination of the NW growth direction, cross-section, surface morphology, dislocations and stacking faults, which provide insights into the material properties that affect the electronic or optical response of a SiNW device. Thus, these electron microscopy techniques provide necessary inputs to tune the synthesis parameters to produce the desired SiNW structures, and are also critical for the future development of new NW synthetic methods.

2.3.2 Raman spectroscopy

Vibrational spectroscopic techniques such as Raman are non-destructive and relatively high-throughput characterization techniques for NWs. The Raman spectra of bulk single-crystal Si and SiNW can be well differentiated, where the first-order Raman peak of SiNW is at $516\,cm^{-1}$. Qualitatively, the smaller the crystalline grain, the larger the frequency shifts and the more asymmetric and the broader the Raman peak becomes. In addition, it has been reported that strong and stable third-order nonlinear optical (NLO) signals, including four-wave mixing and third harmonic generation, can be observed from SiNW with diameters as small as 5 nm (Jung *et al.*, 2009).

2.3.3 Electrical transport measurement

Electrical transport measurements are important techniques for characterizing the electronic structure and property of SiNW. SiNW can be configured as field-effect transistors (FETs), where two metal electrodes, designated as drain and source, are contacting to both leads of NW with well-defined spacing in between, designated as the channel. A third electrode, known as gate, is capacitively coupled with the SiNW channel through a thin dielectric layer. When a bias voltage is applied across the drain and source electrodes, a current is injected and collected, and the conductance of the SiNW FET can be controlled by the voltage applied over the gate electrode. For example, in the case of p-type (e.g. boron-doped) SiNW, applying a positive gate voltage depletes carriers (holes) and reduces the conductance, whereas applying a negative gate voltage leads to an accumulation of carriers (holes) and an increase in conductance. Thus, the SiNW-based FET devices can exhibit a conductivity change in response to variation in the electric field or potential at the surface. A variety of electronic and optoelectronic structural information, including the type and concentration of charge carriers, conductivity, carrier mobility and temperature-dependent modulation, can be obtained by the designed electrical measurement.

2.4 Synthesis of semiconductor SiNWs by the chemical vapor deposition (CVD) method

In this section, an experimental procedure is provided as an example of synthesizing semiconductor silicon nanowires by the chemical vapor deposition method. More detailed growth and characterization steps can be referred to a published protocol (Patolsky, et al. *Nature Protocols*, 2006):

1. Use a diamond scriber to cut a silicon <100> wafer with 600 nm thermal oxide into small chips of desired size (e.g. 2 cm × 2 cm). These chips will be used as growth substrates for SiNW. Sonicate these chips in acetone and then ethanol, for 10 minutes each. Then place one growth chip on a clean surface with the polished side facing up, and cover its surface with 200 μl poly-L-lysine (0.1% in deionized (DI) water) for 2 minutes. Afterwards, rinse the chip surface with DI water for 5–10 seconds, and then dry the chip with a N_2 gas stream.
2. Cover the growth chip surface with 200 μl gold colloid solution (20 nm colloid diameter, 1:4 v/v dilution in DI water) for 10 seconds. Rinse the chip surface with DI water for 5–10 seconds, and then dry the chip with a N_2 gas stream.
3. Place the growth chip inside an oxygen plasma cleaner, and clean the chip surface using 100 W plasma power and 50 sccm (standard cubic centimeters per minute; 1 sccm = 1.7×10^{-8} m^3/s) O_2 flow, for 5 minutes.
4. Insert the growth chip into the middle of a quartz tube (inner diameter ~1 inch), and place the quartz tube in the tube furnace of a CVD system, with one end connected to reactant gas lines and mass flow controllers, and the other end connected to a control valve and a dry pump. A schematic of the CVD setup is shown in Fig. 2.9.
5. Close all the gas lines and completely open the valve to fully evacuate the quartz tube (to pressure less than 3 mTorr). Then start to flow 10 sccm Ar and increase the tube furnace temperature to the designated temperature for SiNW growth.
6. At the growth temperature, start to flow all reactant gases and control the valve opening degree to achieve certain pressure inside the quartz tube, for growing different designed nanowires. A typical recipe for p-type SiNW is: 460 °C, 10 sccm Ar, 6 sccm SiH_4, 7.5 sccm B_2H_6 (100 ppm in H_2), total chamber

pressure is 25 Torr. A typical p-type SiNW growth rate at this condition is ~1.2–1.5 μm/min. A typical recipe for n-type SiNW is: 460 °C, 30 sccm H_2, 8 sccm SiH_4, 2 sccm PH_3 (1000 ppm in H_2), total chamber pressure is 40 Torr. A typical n-type SiNW growth rate at this condition is ~0.8–1.2 μm/min.

7. When the growth is finished, shut down the furnace heating; turn off all the reactant gas lines and completely open the valve to fully evacuate the quartz tube. When the quartz tube drops to room temperature, close the pumping valve and start to flow Ar gas to vent the quartz tube, and then take out the chip. The growth chip should now be fully covered by a layer of grown SiNW, which can be observed by a high-magnification optical microscope (Olympus, Model BX51) under the dark-field mode, or by a scanning electron microscope.
8. It is suggested to store the as-grown SiNW on growth chips in a dessicator to reduce the degradation rate. Before it is used for device fabrication, gently sonicate the chip in 1–2 ml ethanol for 5–10 seconds. The SiNW will be transferred from the growth chip to the ethanol solution.

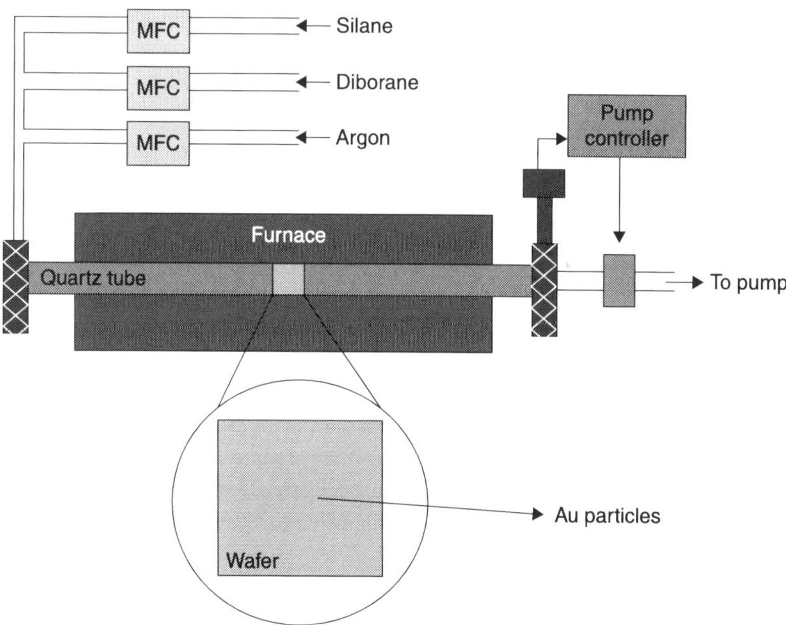

2.9 Schematic showing a chemical vapor deposition (CVD) setup for synthesis of silicon nanowires (SiNW). Silane and diborane are used as Si and B precursors, respectively, and Ar is used as the carrier gas. Reproduced from Patolsky *et al.*, 2006, *Nature Protocols*, permitted by Nature Publishing Group.

2.5 Conclusion

In this chapter, typical growth approaches for semiconductor NW were discussed, which fall into two main categories: vapor-phase and wet-chemical approaches. We focused on the vapor-liquid-solid growth mode, and reviewed recent research progress in the synthesis of conventional and complex SiNW building blocks, including different electronic doping, millimeter-long, axial and radial modulation, kinked and branched structures. The capability to create new SiNW and their assemblies with tunable composition and structures on many length scales is important for and drives the scientific breakthroughs that can enable revolutionary advances and future technologies. The rational design and controlled synthesis have enabled, and will continue to inspire, exploration of physical limits of nanostructures, investigating a broad range of scientific problems, discovering new concepts and ultimately driving technologies of the future.

2.6 Future trends

Over the last decade, remarkable research progress has been achieved on the synthesis of semiconductor SiNW with superior control. The direct growth of semiconductor SiNW with a variety of key structural parameters during synthesis allows for many opportunities of new properties and functionalities. The bottom-up paradigm of SiNW growth by the vapor-liquid-solid CVD methods provides a technical opportunity for realizing the rational control of physical dimension and morphology, chemical composition, electronic structure and doping, which are the central elements that determine predictable device functions. The examples described in this chapter illustrate how it is possible to achieve increasing control over key parameters of the basic SiNW building blocks from homogeneous doped materials to increasingly complex axial and radial heterostructures. Nonetheless, in order to use these SiNW as building blocks to construct complex architectures and integrated systems with novel functions such as interfacing with biomaterials, several main tasks remain to be accomplished.

One major challenge to the ultimate control of SiNW is to achieve controlled modulation of chemical composition and doping with atomic accuracy. The scaling of material dimensions down to the molecular regime presents fundamental and technological challenges for fabricating well-defined structures with controlled atomic composition. In particular, the desired uniformity in miniaturized material structures with nanoscale abruptness has not been achieved by conventional synthetic methods and technologies. One possible route for realizing such fine control is the integration of self-limiting and self-assembly processes where surface and chemical phenomena guide the synthesis of the designed SiNW structures.

In addition, the capability of patterning SiNW assemblies into arrays over multiple length scales is critical to the realization of integrated electronic and photonic systems. Conventionally, patterning of nanomaterials is accompanied by progress in different templating techniques, where nearly all existing technologies, from conventional photolithography to nano-imprint lithography, have been employed for the spatial control of the positioning of nanowires. However, for the nanopatterned growth of semiconductor NW on a macroscopic scale, substantial improvement of current techniques is still needed and/or an innovative technology should be developed. Ideally, new integration strategies will be developed to remove the constraints of lithography facing conventional top-down technologies today. Such a new technique should give high processing resolution and efficiency, and allow for epitaxial NW growth with good crystallinity.

Looking into the future, continued efforts to achieve the capability in controlling the structural/compositional complexity of SiNW during growth, which correspondingly determines the functional complexity of building blocks, together with advances in organizing them into larger integrated arrays and interfacing with biomaterials, will lead to many new exciting opportunities for the biomedical applications.

2.7 Sources of further information and advice

Below is additional information on several research groups working on the synthesis of SiNW:

- Professor Jeffery L. Coffer, Department of Chemistry, Texas Christian University, USA. http://www.chm.tcu.edu/faculty/coffer/
- Professor Yi Cui, Department of Materials Science and Engineering, Stanford University, USA. http://www.stanford.edu/group/cui_group/
- Professor James R. Heath, Department of Chemistry, California Institute of Technology, USA. http://www.its.caltech.edu/~heathgrp/
- Professor Brian A. Korgel, Department of Chemical Engineering, University of Texas at Austin, USA. http://www.che.utexas.edu/korgel-group/korgel.htm
- Professor Ali Javey, Department of Electrical Engineering and Computer Sciences, University of California at Berkeley. http://nano.eecs.berkeley.edu/
- Professor Lincoln J. Lauhon, Department of Materials Science and Engineering, Northwestern University, USA. http://lauhon.mccormick.northwestern.edu/
- Professor Shui-Tong Lee, Institute of Functional Nano & Soft Materials, Soochow University, China. http://funsom.suda.edu.cn
- Professor Nathan S. Lewis, Division of Chemistry and Chemical Engineering, California Institute of Technology, USA. http://nsl.caltech.edu/nslewis
- Professor Charles M. Lieber, Department of Chemistry and Chemical Biology, Harvard University, USA. http://cml.harvard.edu/
- Professor Fernando Patolsky, School of Chemistry, Tel-Aviv University, Israel. http://chemistry.tau.ac.il/patolsky/index.php
- Professor Mark A. Reed, Department of Electrical Engineering, Yale University, USA. http://www.eng.yale.edu/reedlab/index.html

- Professor John A. Rogers, Department of Material Sciences and Engineering, University of Illinois at Urbana-Champaign, USA. http://rogers.matse.illinois.edu/
- Professor Peidong Yang, Department of Chemistry, University of California at Berkeley, USA. http://nanowires.berkeley.edu/

2.8 References

Ben-Ishai M and Patolsky F. (2012). 'From crystalline germanium-silicon axial heterostructures to silicon nanowire-nanotubes.' *Nano Letters* **12**(3): 1121–8.

Chan CK, Peng H, Liu G, McIlwrath K, Zhang XF, *et al.* (2008). 'High-performance lithium battery anodes using silicon nanowires.' *Nature Nanotechnology* **3**(1): 31–5.

Heitsch AT, Fanfair DD, Tuan HY and Korgel BA. (2008). 'Solution-liquid-solid (SLS) growth of silicon nanowires.' *Journal of the American Chemical Society* **130**(16): 5436–7.

Holmes JD, Johnston KP, Doty RC and Korgel BA. (2000). 'Control of thickness and orientation of solution-grown silicon nanowires.' *Science* **287**(5457): 1471–3.

Jiang X, Tian B, Xiang J, Qian F, Zheng G, *et al.* (2011). 'Rational growth of branched nanowire heterostructures with synthetically encoded properties and function.' *Proceedings of the National Academy of Sciences of the United States of America* **108**(30): 12212–6.

Jung Y, Tong L, Tanaudommongkon A, Cheng JX and Yang C. (2009). 'In vitro and in vivo nonlinear optical imaging of silicon nanowires.' *Nano Letters* **9**(6): 2440–4.

Lauhon LJ, Gudiksen MS, Wang D and Lieber CM. (2002). 'Epitaxial core-shell and core-multishell nanowire heterostructures.' *Nature* **420**(6911): 57–61.

Lieber CM. (2003). 'Nanoscale science and technology: Building a big future from small things.' *MRS Bulletin* **28**(7): 486–91.

Lieber CM. (2011). 'Semiconductor nanowires: A platform for nanoscience and nanotechnology.' *MRS Bulletin* **36**(12): 1052–63.

Park WI, Zheng G, Jiang X, Tian B and Lieber CM. (2008). 'Controlled synthesis of millimeter-long silicon nanowires with uniform electronic properties.' *Nano Letters* **8**(9): 3004–9.

Patolsky F, Zheng GF and Lieber CM. (2006). 'Fabrication of silicon nanowire devices for ultrasensitive, label-free, real-time detection of biological and chemical species.' *Nature Protocols* **1**(4): 1711–24.

Peng K, Wu Y, Fang H, Zhong X, Xu Y and Zhu J. (2005). 'Uniform, axial-orientation alignment of one-dimensional single-crystal silicon nanostructure arrays.' *Angewandte Chemie-International Edition* **44**(18): 2737–42.

Tian B, Xie P, Kempa TJ, Bell DC and Lieber CM. (2009). 'Single-crystalline kinked semiconductor nanowire superstructures.' *Nature Nanotechnology* **4**(12): 824–9.

Tian B, Zheng X, Kempa TJ, Fang Y, Yu N, *et al.* (2007). 'Coaxial silicon nanowires as solar cells and nanoelectronic power sources.' *Nature* **449**(7164): 885–8.

Wagner RS and Ellis WC. (1964). 'Vapor-liquid-solid mechanism of single crystal growth (new method growth catalysis from impurity whisker epitaxial + large crystals Si E).' *Applied Physics Letters* **4**(5): 89.

Wang D, Qian F, Yang C, Zhong Z and Lieber CM. (2004). 'Rational growth of branched and hyperbranched nanowire structures.' *Nano Letters* **4**(5): 871–4.

Wu Y, Cui Y, Huynh L, Barrelet CJ, Bell DC and Lieber CM. (2004). 'Controlled growth and structures of molecular-scale silicon nanowires.' *Nano Letters* **4**(3): 433–6.
Xiang J, Lu W, Hu Y, Wu Y, Yan H and Lieber CM (2006). 'Ge/Si nanowire heterostructures as high-performance field-effect transistors.' *Nature* **441**(7092): 489–93.
Yan H, Choe HS, Nam SW, Hu Y, Das S, *et al.* (2011). 'Programmable nanowire circuits for nanoprocessors.' *Nature* **470**(7333): 240–4.
Yang C, Zhong ZH and Lieber CM. (2005). 'Encoding electronic properties by synthesis of axial modulation-doped silicon nanowires.' *Science* **310**(5752): 1304–7.
Zheng GF, Lu W, Jin S and Lieber CM. (2004). 'Synthesis and fabrication of high-performance n-type silicon nanowire transistors.' *Advanced Materials* **16**(21): 1890–3.
Zheng GF, Patolsky F, Cui Y, Wang WU and Lieber CM. (2005). 'Multiplexed electrical detection of cancer markers with nanowire sensor arrays.' *Nature Biotechnology* **23**(10): 1294–301.

3

Surface modification of semiconducting silicon nanowires for biosensing applications

Y. COFFINIER and R. BOUKHERROUB,
CNRS and University of Lille1, France

DOI: 10.1533/9780857097712.1.26

Abstract: Silicon nanowires (SiNW) are one of the most important 1-D semiconductors, partly because of their ready implementation in modern technology. As sensing applications using SiNW increase, it is necessary to have well-defined chemical attachment schemes to provide the desired biomolecular recognition properties, chemical stability and interfacial electrical properties. Such surface functionalization allows controlled immobilization of biomolecules on SiNW, important in designing selective and sensitive biosensors. Moreover, such SiNW modification is not only dedicated to sensing, but also is useful for many other applications (not described in this chapter) such as device integration, controlled cell micro-patterning, SiNW internalization and analyte confinement.

Key words: silicon nanowires, surface functionalization, biological applications.

3.1 Introduction

The integration of nanotechnology with biology has received increasing attention in recent years[1] largely because of the desire to use biomolecular recognition to aid nanoscale assembly[2–7] or to create sensitive biosensors based on nanoscale devices.[8,9] Silicon nanostructures (nanowires, nanoribbons, rods or tubes) are particularly attractive in biology, because of their morphology, their high surface to volume ratio and their semiconducting properties. Silicon nanowires (SiNW) are an important class of 1-D objects that have attracted a big deal of interest recently. SiNW-based devices represent an attractive technology for future miniaturized and multiplexed biosensing platforms, but can also be extended to high-throughput functional assays (e.g. drug screening), a topic also addressed in a later chapter of this book. Their sensitivity depends on several parameters: the intrinsic properties of the device, and the experimental conditions such as the ionic strength of the solution defining the Debye length and the preserved function of the probes (proteins, DNA, peptides, aptamers . . .) used as recognition motifs. A large variety of biomolecular interactions, including oligonucleotide hybridization,[10–12] protein–protein interactions,[9] protein–ligand binding[13,14] and immunodetection[15,16] have been demonstrated using SiNW field-effect transistor-based sensors. Probe immobilization in these applications is mostly achieved either via physisorption or chemical cross-linking of the recognition motif to the SiNW surface.

In this chapter, we will focus on the chemical modification (chemisorption) of SiNW surface for essentially biosensing applications. Indeed, it is necessary to have well-defined chemical attachment schemes that will provide the desired biomolecular recognition properties, chemical stability and good interfacial electrical properties.[9,17,18]

3.2 Methods for fabricating silicon nanowires (SiNWs)

Numerous methods have been developed for the preparation of Si nanostructures using top-down or bottom-up approaches, such as vapour-liquid-solid (VLS) growth,[19] reactive ion etching (RIE),[20] electrochemical etching[21] or metal-assisted chemical etching,[22] all of which aim to control various parameters of the Si nanostructures. These techniques are nicely described in Chapter 2 of this book as well as in several review articles cited therein.

3.3 Chemical activation/passivation of SiNWs

Usually, the immobilization of biomolecules on surfaces first requires the modification of the solid substrate with an appropriate functional layer. The surface chemistries of silicon and silicon dioxide surfaces are well known and fully described in the literature.[23,24]

The chemical modification of SiNW can be performed either on native oxide-surrounding SiNW or on hydrogen-terminated SiNW. The choice is often related to the targeted application. Most of SiNW-based devices for biosensing use electrical detection to assess the change in the nanowire electrical properties induced when a target binds to the recognition receptor immobilized on the nanowire surface. The 1–2 nm thick native oxide layer on the nanowire surface stabilizes the nanowire against corrosion in aqueous biological environments. However, like any gate oxide, it screens the silicon core of the nanowire from the charged analyte species captured by the bioaffinity layer on the oxide surface. Thus, there is a good reason to replace the native oxide layer with a thinner, covalently linked molecular layer that passivates the silicon surface. Moreover, it was shown that a non-oxidized SiNW substrate presented better electrical properties (conductance) when compared with an oxidized one.[25] Bunimovich *et al.* found that non-oxidized SiNW gave a higher sensitivity of DNA detection with an improvement in magnitude of 2 orders compared with oxidized SiNW.[26] However, most of the chemical modifications performed on SiNW for biosensing devices are still achieved via silanization. In fact, modification of SiNW via hydrosilylation requires a de-oxidation step by dipping SiNW in HF (manipulate with caution!), before the reaction with alkene or alkyne molecules. Secondly, only few functional alkene reagents are commercially available. Most of the time, the chemical synthesis of compounds is necessary. However, when semiconductive properties of SiNW are not needed, modification of native oxide will be preferred.

Chemical modification of SiNW surfaces can be achieved depending on their surface termination:

1. **Modification of native oxide SiO_x/SiNW**
   - Silanization.
   - Reaction with organophosphonates.
2. **Modification of hydrogen-terminated SiNW**
   - Hydrosilylation.
   - Halogenation/alkylation (Grignard reaction).
   - Arylation via aryldiazonium salts.

All these reactions permit the activation/passivation of SiNW by introducing a desired chemical function or tail group on the SiNW surface and confer to the SiNW surface resistance against corrosion, oxidation and stability of probe immobilization under experimental conditions over time.

The tail group is introduced via chemical compounds composed by an anchoring moiety (silane (silanization) or alkene/alkyne (hydrosilylation)), a spacer (alkyl, aryl, ethylene glycol chains . . .) and the tail group. Furthermore, by controlling SiNW surface functionalization, it is also possible to inhibit non-specific interactions, improve blood circulation (PEG-spacer) and increase electrical properties (defect passivation).[16,25,27–31]

3.4 Modification of native oxide layer

3.4.1 Silanization reaction

The silanization reaction involves linking molecules through the intermediate oxide sheath that typically surrounds air-exposed SiNW surface,[17,18] with the molecule being anchored via siloxane bonds (Si-O-Si) (Fig. 3.1). Silanes are more commonly used on oxide surfaces, where they can covalently bind to the surface by the transfer of a proton from the surface hydroxyl group to a silane leaving group, eliminating an alcohol (in the case of methoxy or ethoxysilanes) or HCl (in case of chlorosilanes).

This method is widely used to provide SiNW with biomolecular recognition capability.[14,17,18] However, care must be taken to limit formation of 3-D silane networks by siloxane cross-linking that can predominate over surface attachment.[32] Indeed, the degree of siloxane cross-condensation depends critically on the water content of the deposition solvent. One method to overcome organosilane condensation on the silicon surface is to perform the reaction in vapour phase.[33–35] Indeed, a good comparison between vapour and liquid phase silanization was made by Hunt *et al.* for the functionalization of flat SiO_x/Si with

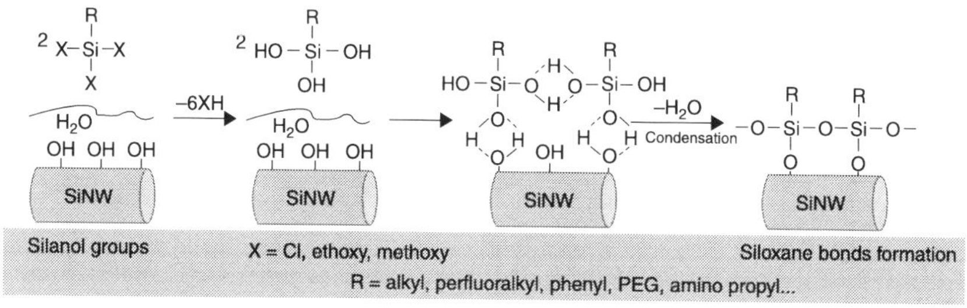

3.1 Silanization reaction.

aminopropyltriethoxysilane (APTES) and found that chemical vapour deposition (CVD) provides more ordered monolayers.[36] However, such a protocol cannot be applied to all silane compounds as it is dependent on the vapour pressure of the silane.

In addition, if the number of available surface OH groups is limited, low yields of direct surface attachment can result.[37] For this reason and immediately before silanization, SiNW should be cleaned with oxidant to remove organic pollutants and to increase the hydroxyl density on the surface ($\approx 10^{15}\ cm^{-2}$).[38] A cleaning process to generate reactive hydroxyl groups is critical for the effective immobilization of silanes. There are several types of Si-OH groups that can be formed on silica surfaces. Some (germinal and isolated silanols) are reactive, whereas others (vicinal silanols and siloxane groups) are not. The most widely used oxidants are oxygen plasma[39] and piranha solution[36] (consisting of a concentrated sulfuric acid/hydrogen peroxide mixture at different ratios). This treatment is well performed at room temperature or by heating, but usually for only a few minutes. The literature also describes other oxidants and cleaning agents comprising UV-ozone,[40] sodium hydroxide,[41] ammonia/hydrogen peroxide mixture,[42,43] nitric acid, hydrochloric acid,[44] sulfuric acid,[45] chromic acid[46] or mineral acids with hydrogen peroxide.[43] Sometimes more than one of these treatments are combined and sequentially applied to the surface.[40,43,44]

In addition, some silane films have been shown to be hydrolytically unstable in aqueous base and in biological media.[47] This can be a challenge for applications that involve ambient conditions or a biological environment. The choice of the tail group of the organic layer depends on the targeted applications.

Control of wetting properties by introduction of alkyl or perfluoroalkyl chains on SiNWs

The combination of the surface roughness, provided by silicon oxide nanowires or SiNW obtained either by CVD growth or by metal-assisted chemical etching, with chemical modification (silanization) by low surface energy molecules such as perfluoroalkyl or alkyl chains, leads to the formation of superhydrophobic (SH) or omniphobic (SO) surfaces (Fig. 3.2).[48–54]

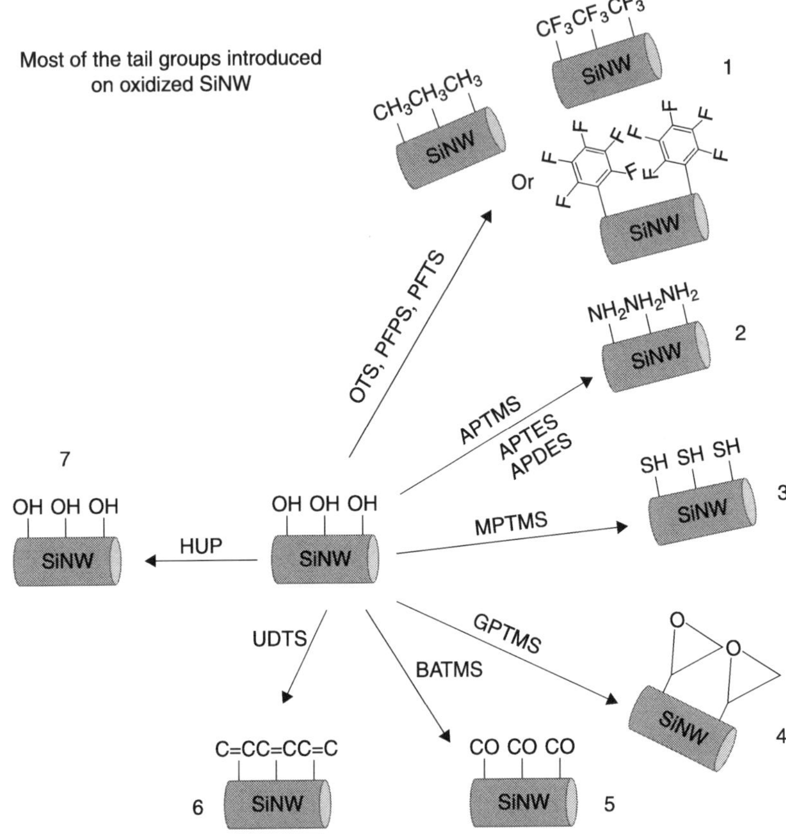

3.2 Main tail groups grafted on SiOx/SiNW. HUP, 11-hydroxyundecylphosphonate; OTS, octadecyltrichlorosilane; PFTS, 1H, 1H, 2H, 2H, perfluorodecyltrichlorosilane; PFPS, perfluorophenyltrichlorosilane; APTMS, aminopropyltrimethoxysilane; APTES, aminopropyltriethoxysilane; APDES, aminopropyldiethoxymethylsilane; MPTMS, 3-mercaptopropyltrimethoxysilane; GPTMS, 3-glycidoxypropyltrimethoxysilane; BATMS, 3-(trimethoxysilyl) butyl aldehyde; UDTS, 10-undecenyltrichlorosilane.

Such superhydrophobic surfaces have received tremendous attention in the past few years and present a non-wetting behaviour with high contact angles (>150°) and facile sliding of drops, a 'rolling ball effect', corresponding to a low contact angle hysteresis. Such non-wetting properties are desirable for many industrial and biological applications such as anti-biofouling paints for boats, anti-sticking of snow for antennas and windows, self-cleaning windshields for automobiles, anticorrosion coatings for metals and microfluidics.[53] However, these liquid-repellent surfaces are effective only for high-surface-tension (γ) liquids such as

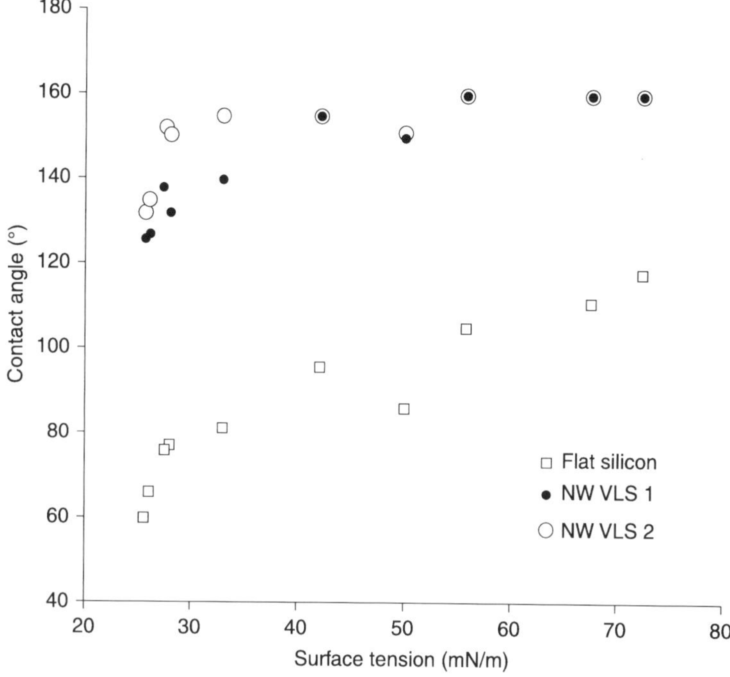

3.3 Static contact angles measured on silicon nanowires for liquids of various surface tensions. Reproduced with permission from reference 53. Copyright (2010) American Chemical Society.

water. One of the current challenges in lab-on-a-chip devices based on discrete microfluidics, is to design surfaces that repel all types of liquids of either high or low surface tension to make easier the liquid droplet motion.

Recently, we have demonstrated that surfaces composed by SiNW grown by CVD and coated with perfluorodecyl molecules were able to repel not only water but also liquids of low surface tension such as hexadecane (γ=27 mN/m) or a mixture of ethanol/water (65/35, γ=25.6 mN/m) with high robustness, that is with high resistance to impalement of liquid droplets inside the texturation (Fig. 3.3).[53]

Amine-terminated silicon nanowires (NH_2-SiNW)

Among the vast variety of commercially available organosilanes, only a few have been used to functionalize SiNW.[23] Thus, NH_2, epoxy or aldehyde functionalities are mainly employed and the amino group is one of the most used for biofunctionalization of SiNW.[55] Usually, aminopropyltrimethoxy (APTMS) or aminopropyltriethoxy (APTES) silanes are reacted with the oxide layer to introduce the amino group that can subsequently react with aldehyde, carboxylic acid or epoxy groups (Fig. 3.2).

However, the conditions employed differ for silane concentration, solvent and incubation time. Moreover, a curing process is often performed after silanization. The trimethoxy compound is more reactive and can be deposited on a substrate in pure organic solvents. The advantage of this process is the control over the film thickness and the density of aminopropyl groups on the surface. For triethoxysilane, the reaction must occur in the presence of water; otherwise the ethoxy groups are not reactive enough to spontaneously couple to the hydroxyl groups on the surface. Given the possibility of hydrogen bond formation between the amine of APTES and the SiO_x surface, both the head and tail groups in the organosilane can be oriented toward the surface, which can result in a disordered layer.[56] Additionally, cross-linking among alkoxysilane units may yield oligomerized silane structures, resulting in multilayers.

In most biosensing applications, NH_2-SiNW surface requires post-modification for probe immobilization (see Section 3.4.2). However, there are some examples where NH_2-SiNW can either be used to favour protein physisorption prior to specific interaction with the target from sample, or directly used for pH sensing.[9] This latter example was the first case demonstrating the ability of SiNW field-effect devices to detect, in liquid, hydrogen ion concentration or pH sensing. In that case, a *p*-doped nanowire device surrounded by oxide native layer was chemically modified with APTES to yield amino groups at the nanowire surface along with silanol groups (Si-OH) from the oxide. Both chemical groups were used as receptors of hydrogen ions, which undergo protonation/deprotonation reactions, thereby changing the net nanowire surface charge provoking conductance variation. In that case, the authors demonstrated that the sensing mechanism was the result of a field-effect phenomenon analogous to an applied voltage using a metallic gate electrode.[9]

More recently, Shalek *et al.* have demonstrated that vertical NH_2-SiNW can also be used for non-covalent binding of various compounds such as DNA, RNA, peptides, proteins, anti-apoptosis agents, etc., for their release and delivery into cells.[57]

Finally, NH_2-SiNW bearing a positive charge can be assembled in devices driven by electrostatic interactions.[58–63] In such cases, the negative charge of silicon oxide at pH 7 will attract positively charged nanowires.

Thiol-terminated silicon nanowires (HS-SiNW)

Incorporation of thiol functional groups on the SiNW surface involves employment of a thiolated silane (Fig. 3.2). This functionalization can serve for the attachment of thiolated oligonucleotides to the surface via disulfide bond linkage(s) or through their amine groups using the heterobifunctional cross-linker *m*-maleimidobenzoyl-N-hydroxysuccinimide ester (MBS) (see Section 3.4.3), as demonstrated by Xu *et al.*[46] In addition, employment of disulfide bonds to attach thiolated oligonucleotides on silanized surfaces can offer the advantage of reusability. Disulfide bond formation is reversible; the surface can be regenerated, for instance, by treatment with dithiothreitol (DTT). Another interesting approach is the reaction between thiol and alkene/alkyne moieties, the so-called thiol-e(y)ne click reaction, which takes place at close-to-visible wavelength ($\lambda = 365$ nm) using short reaction times (~10 min).[64,65] Very importantly, this procedure is compatible with aqueous media, which is crucial for its bioavailability. In the last few years, thiol-ene chemistry has also found important applications in the area of surface derivatization, but has yet to be applied to SiNW.[66]

However, HS-terminated organosilanes and especially 3-mercapto-propyltriethoxysilane (MPTS) have not been widely used on SiNW. To the best of our knowledge, there is one known example using HS-terminated SiNW that has been described by Salhi *et al.*[67] They used HS-SiNW for self-assembly and integration in devices assisted by protocollagen, a low-cost, soluble, long-fibre protein and precursor of collagen fibrils. Firstly, the collagen was combed on an octadecyl-terminated flat SiO_x/Si surface bearing gold electrodes. Then, the combed surface was exposed to an aqueous suspension of chemically modified SiNW. Indeed, in order to increase electrostatic interactions between the positively charged collagen (under experimental conditions) and the nanowires, SiNW were chemically modified with negatively charged sulfonate groups. The interaction of collagen with the sulfonated nanowires, which mimics the native collagen/heparin sulfate interaction, induced self-assembly of the nanowires localized between gold electrodes. This proof of concept for the formation of spontaneous electrode–nanowire–electrode junctions using collagen as a template was supported by current–voltage measurements.[67]

Epoxy-terminated SiNWs

Epoxide chemistry is an alternative coupling system for biomolecule immobilization given its stability under aqueous conditions and its reactivity with several nucleophiles such as amine and sulfhydryl groups.[68,69] Thus, SiNW that are covalently coated with 3-glycidoxypropyltrimethoxysilane (GPTMS) can be used to conjugate thiol-, amine- or hydroxyl-containing ligands (Fig. 3.2). GPTMS can be employed to covalently attach antibodies and aminated oligonucleotides through epoxide ring opening.[41] Ingebrandt *et al.* have developed SiNW-based devices for label-free biomedical applications.[70] For that, they fabricated SiNW arrays on a wafer-scale, combining nanoimprint lithography and wet chemical etching. Then, the devices were cleaned and activated as described above in section 3.1. For chemical activation, chips were cleaned for 10 minutes with a piranha solution at 60°C and then rinsed with ultrapure water. For the covalent attachment of the probe amino-DNA molecules, a gas phase silanization protocol with GPTMS was used.[70]

Aldehyde-terminated SiNWs

Another functionality that can be introduced onto SiNW surface is aldehyde function (Fig. 3.2). An example of direct immobilization of biomolecules was shown by Gao *et al.*[11] They used freshly etched nanowires bearing a native oxide layer which was reacted with 3-(trimethoxysilyl) butyl aldehyde (BATMS) and subsequently reacted with amine-terminated peptide nucleic acids (PNA). Capture probe 21-mer PNA are DNA analogues in which a 2-aminoethyl-glycine linkage generally replaces the normal phosphodiester backbone. These synthetic molecules are non-ionic, achiral and not susceptible to hydrolytic (by enzyme) cleavage. PNA are capable of sequence-specific binding to DNA and RNA leading to a complex with high thermal stability, and so can be used as capture probes. Arrays of an unspecified number of these nanowires, with radii ranging from 5 to 50 nm, allowed a limit of detection of 10 fM for the detection of fully complementary ssDNA 21-mers in a 40-mM tris buffer maintained at 50 °C.[11]

Vinyl-terminated SiNWs

Cho *et al.* have depicted the chemical modification of SiNW with 10-undecenyltrichlorosilane (UDTS) to yield vinyl-terminated SiNW (Fig. 3.2).[71] This termination was used for subsequent modification to generate terminal carboxylic acid groups (see Section 3.4.2 for further detail).

Modification with carboxylic acid/organosilane reagents

Although the introduction of carboxylic acid groups on oxidized silicon surfaces can be directly achieved using molecules such as N-(trimethoxysilylpropyl) ethylene-diaminetriacetic acid or carboxyethylsilanetriol sodium salt, only a few examples have been described in the literature.[44,72] This can be explained because the product is expensive or needs to be chemically synthesized and/or leads to organic layers of poor quality. In fact, most of the time the carboxylic acid function is introduced via post-chemical modification of a chemical group or functionality previously covalently grafted on SiNW, such as an ester, $-NH_2$ or vinyl species (see Section 3.4.2) allowing subsequent conjugation with amine-containing molecules.

3.4.2 Post-functionalization

Although direct immobilization of the probe can be performed on activated SiNW surfaces (oxide and oxide-free) (Fig. 3.4(a)), the ω-functionality (tail group) of the organic layer is often required to be chemically modified for the desired chemistry necessary for attachment of biomolecules/ligands/probes (Fig. 3.4(b)). Such chemical modification should not destroy the existing organic layer underneath and can be used either on oxidized or non-oxidized chemically modified SiNW.

Ester and amide linkages are widely used groups for surface modification and biomolecule immobilization. Carboxylic acid-terminated SiNW can be suitably activated by conversion into anhydride, acyl fluoride or active ester. The activated

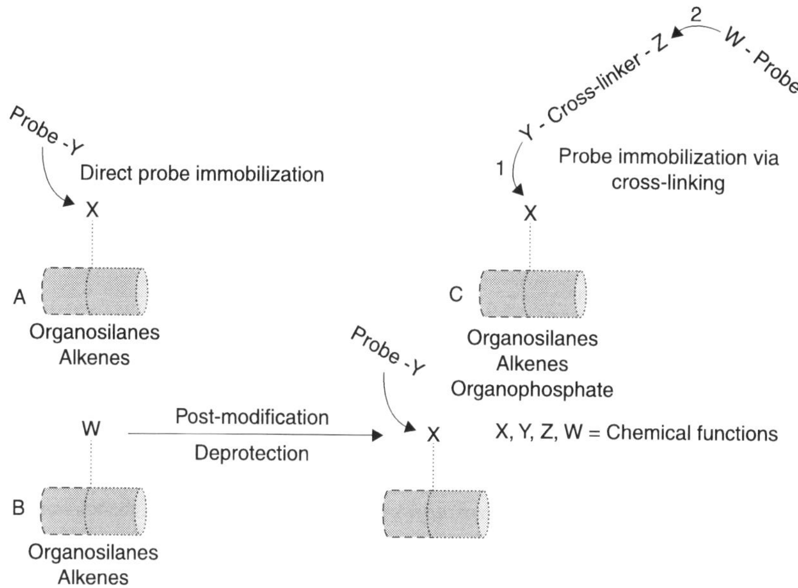

3.4 Main strategies for probe immobilization on oxidized (dashed line) or oxide-free (dotted line) silicon nanowires.

acid derivatives can then react with alcohols or amines to form esters or amides, respectively. One of the most used strategies is the *N*-hydroxysuccinimide ester (NHS)/carbodiimide (EDC or DCC) procedure allowing biomolecule attachment through amide-bond formation. However, this chemistry leads to random immobilization of peptides that usually present several primary amino groups and also other competing nucleophiles such as thiols or hydroxyl groups.[73] Lee *et al.* used this strategy to immobilize an anti-vascular endothelial growth factor (VEGF) RNA aptamer on SiNW. For this, SiNW were first reacted with 3-aminopropyldiethoxysilane (APDES) in ethanol solution under an inert N_2 environment. The resulting amino-terminated SiNW were immersed in a solution containing succinic anhydride to convert the terminal amine groups to carboxyl groups. Then, carboxylated SiNW were activated by the addition of EDC and sulfo-NHS. After that, an anti-VEGF RNA aptamer was immobilized onto SiNW through amide bond formation (Fig. 3.5).[74]

However, it must be noted that any NHS/EDC-based strategy for biomolecule immobilization should be carried out carefully because reproducibility is highly dependent on the experimental conditions. Indeed, NHS esters typically undergo rapid hydrolysis under aqueous conditions, and functional activity is compromised over time.[75] Hence, the quality of the probe immobilization process is highly dependent on the experimental conditions. So, besides taking into account a risk of hydrolysis of the active ester, control assays to demonstrate the covalent nature

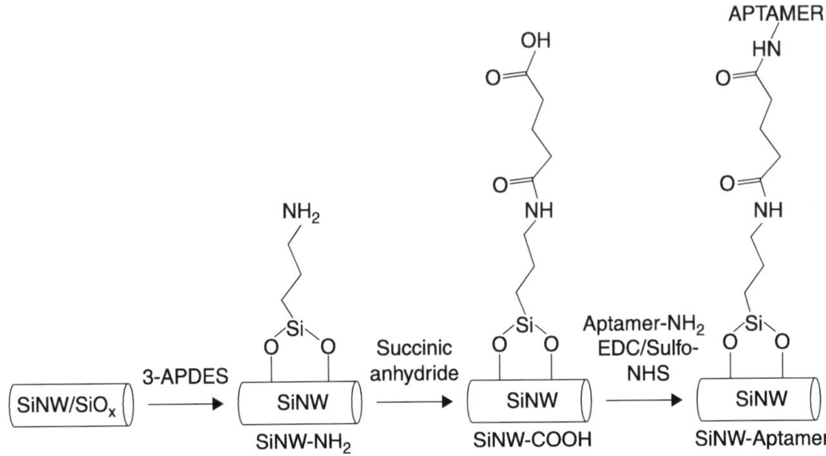

3.5 Anti-vascular endothelial growth factor (VEGF) RNA aptamer immobilization strategy on silicon nanowires (SiNW). 3-APDES, 3-aminopropyldiethoxysilane; EDC, carbodiimide; NHS, *N*-hydroxysuccinimide ester.

of the link between the protein and the surface are recommended. In fact, the biomolecule can remain on the surface through electrostatic interactions between the amine and the carboxylate moieties, without having the advantages of the covalent link.

Duan *et al.* showed that SiNW field-effect transistors can be used as affinity biosensors to effectively determine the affinities and kinetics of two representative protein–receptor binding pairs: 1 the high mobility group box 1 (HMGB1) proteins/DNA and 2 biotin/streptavidin (Fig. 3.6).[76] HMGB1 has been covalently

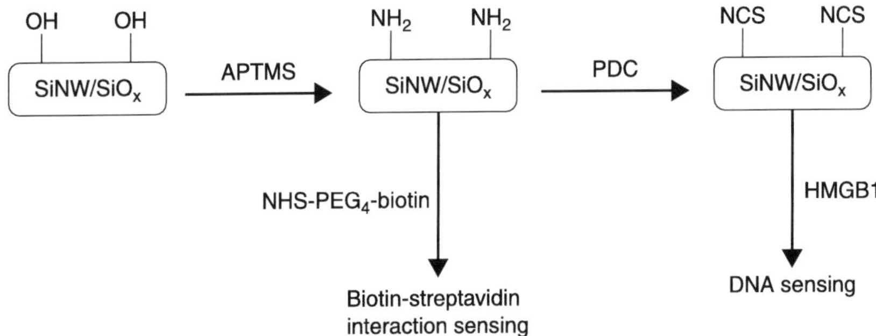

3.6 Surface chemistry strategies for DNA and streptavidin sensing on functionalized silicon nanowires (SiNW). APTMS, aminopropyltrimethoxysilane; HMGB1, high mobility group box 1; NHS, *N*-hydroxysuccinimide ester; PDC, 1,4-phenylene diisothiocyanate.

bound, as a ligand, through amine coupling onto isothiocyanate-functionalized SiNW. The conversion of the amino-functionalized device to an isothiocyanate-bearing layer was accomplished by exposure to 1,4-phenylene diisothiocyanate (PDC) in ethanol. HMGB1 was immobilized for 30 minutes in MES buffer (pH 5.6). DNA solutions were then injected for interaction.[76]

Biotin was immobilized by a NHS–PEG_4–biotin linker (NHS-biotin, with a 2.9 nm PEG arm) using succinimidyl ester chemistry onto an amine-functionalized SiNW. After biotinylation, streptavidin was allowed to bind and the sensor responses were recorded.[76]

Another example is the use of terminal vinyl groups on the surface of SiNW that have been oxidized by $KMnO_4$, K_2CO_3 and $NaIO_3$ with a loss of one carbon (CO/CO_2) to form carboxylic acid groups. The carboxylic acid groups were then activated with EDC and pentafluorophenol (PFP). The formation of the organic layers on SiNW and the successive reactions were confirmed by polarized infrared external reflectance spectroscopy (PIERS). PFP linked to SiNW via ester bond was finally reacted with biotin-amine for further interaction with streptavidin (Fig. 3.7).

Vinyl termination could also be used for thiol-ene (thiol-yne) reaction (see Section *Thiol-terminated silicon nanowires (HS-SiNW)*). Indeed, a huge variety

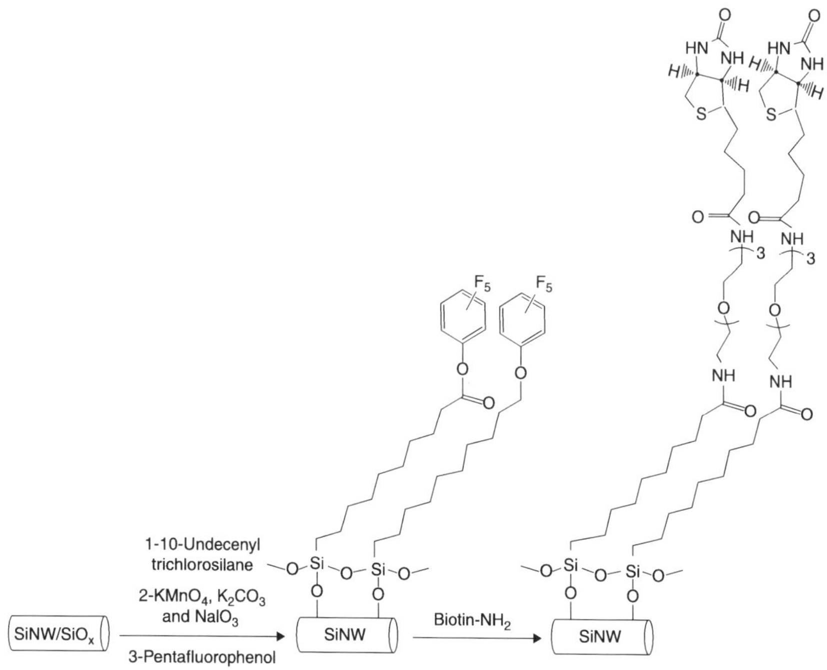

3.7 Biotin immobilization on silicon nanowires (SiNW).

of thiol compounds are commercially available and could be easily immobilized on vinyl-terminated SiNW, offering new probe immobilization strategies.

3.4.3 Heterobifunctional cross-linkers

The strategies presented above are often used when small compounds have to be immobilized on SiNW. Large molecules such as oligonucleotides/proteins can also be attached directly to the nanowire surface and require often the use of a bifunctional cross-linker (Fig. 3.4(c)). The cross-linkers presented can be used either on oxidized or non-oxidized chemically modified SiNW.

Glutaraldehyde (GA) is often used to link amine species onto amine-terminated nanowires, as illustrated in Fig. 3.8(a). Maleimide-activated surfaces are important for further reactivity with thiol groups, for example thiol-tagged DNA strains. Hydroxyl-terminated SiNW react with 3-maleimidopropionic-acid-*N*-hydroxy-succinimide ester (BMPS),[77] whereas amine-terminated SiNW react with sulfo-succinimidyl 4-(*N*-maleimidomethyl) cyclohexene-1-carboxylate (SSMCC),[102] as shown in Fig. 3.8(b) and (c), respectively.

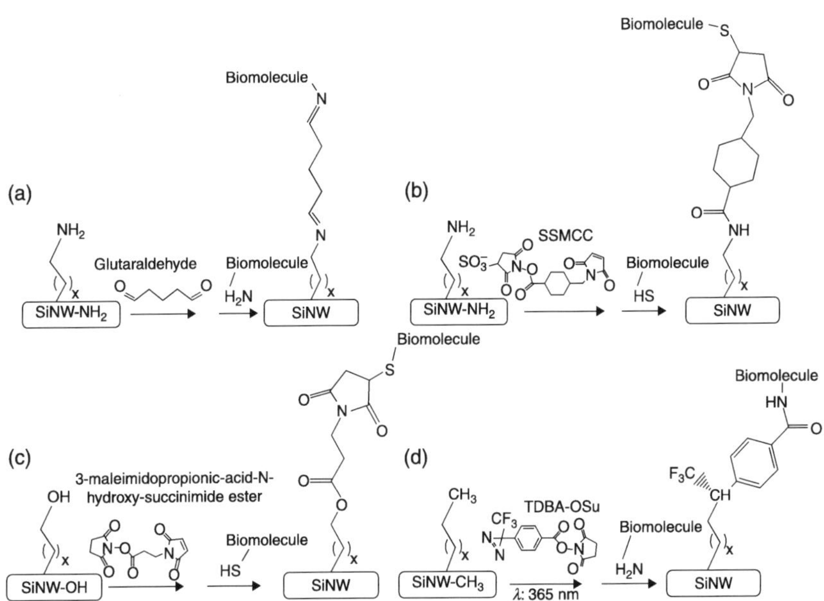

3.8 Bifunctional cross-linker molecules used for SiNW functionalization with (a) glutaraldehyde (GA), (b) sulfo-succinimidyl 4-(*N*-maleimidomethyl) cyclohexene-1-carboxylate (SSMCC), (c) 3-maleimidoproprionic-acid-*N*-hydroxy-succinimide ester (BMPS) and (d) 4′-(3-trifluoromethyl-3*H*-diazirin-3-yl)-benzoic acid *N*-hydroxy-succinimide ester (TDBA-OSu).

Recently, Zhang *et al.* have studied cellular binding and internalization of folate-modified SiNW.[78] SiNW were grown by CVD and subsequently modified by aminopropyltrimethoxysilane (APTMS) leading to amino-terminated-SiNW ($SiNW-NH_2$). The amino-modified samples were reacted with N-[β-maleimidopropyloxy] succinimide ester (BMPS, 3 mg/ml in anhydrous DMSO) in the presence of triethylamine (TEA), leading to maleimide-terminated-SiNW. The unreacted NH_2 groups were blocked by adding excess acetic anhydride. Following rinsing and drying, folate-cysteine (1 mM in MES buffer, pH 6.5) was added onto the substrates and reacted for 2–3 hours. Excess folate-cysteine was washed away with phosphate-buffered saline (PBS). This second chemical modification was denoted as 'SiNW-Folate', representing a folate-functionalized surface (Fig. 3.9).

Then, using non-linear optical signal of SiNW, they visualized the interaction between the folate- and amine-modified SiNW and cells by monitoring the

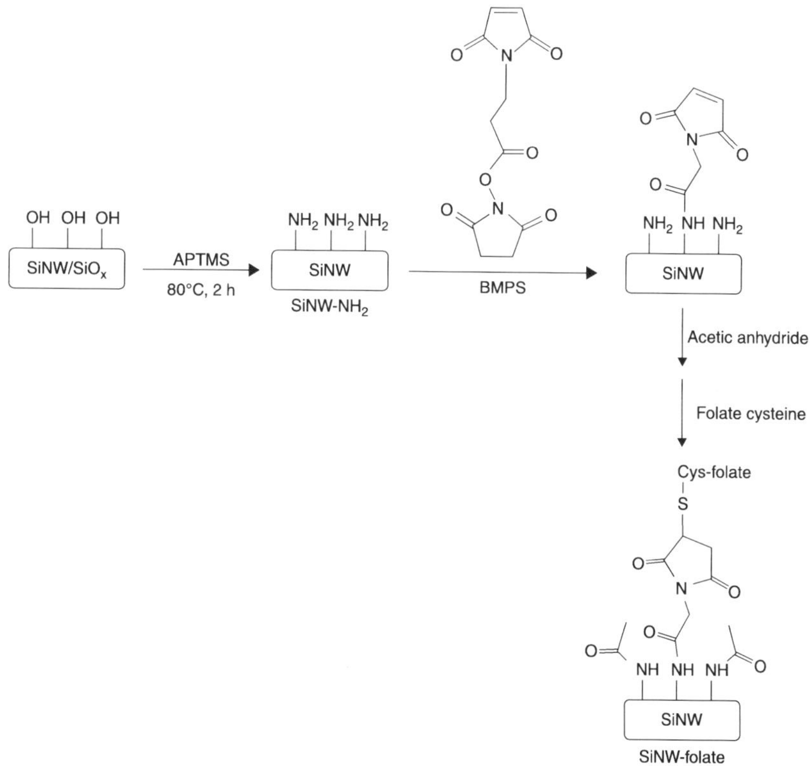

3.9 Folate immobilization on silicon nanowires (SiNW). APTMS, aminopropyltrimethoxysilane; BMPS, 3-maleimidoproprionic-acid-*N*-hydroxy-succinimide ester.

cellular binding and uptake of SiNW in real time. They demonstrated that the strong specific ligand–receptor interaction between folate on NW and folate receptors on CHO-β cell membranes (negatively charged) expedited agglomeration of folate-modified SiNW on cells and internalization of NW. Such specific targeting was further confirmed through control experiments performed with normal CHO cells without folate receptors. This type of approach to *in vitro* processes probed by SiNW is examined in further detail in Chapter 5 of this book.

Assad *et al.* used a photoactive aryldiazirine cross-linker, 4′-(3-trifluoromethyl-3*H*-diazirin-3-yl)-benzoic acid *N*-hydroxysuccinimide ester (TDBA-OSu), shown in Fig. 3.8(d), to cross-link methyl-terminated SiNW with amine groups.[79] The use of this linker molecule is particularly useful as it can covalently attach to methyl-terminated SiNW with no need for the presence of a terminal heteroatom functional group at the nanowire surface. Furthermore, the photochemically induced reaction (365 nm) was complete after 15 minutes and an oxide-free SiNW surface was preserved.

3.4.4 Reaction with organophosphates

SiNW chemical modification with organophosphonates provides stable and elegant systems that can be used to bond biological systems to native silicon oxide surfaces and thus obviates many disadvantages of silanization such as limited hydrolytic stability, critical dependence on available hydroxyl binding sites on the SiO_2 and the intrinsic risk of multilayer formation (see Section 3.4.1 and Fig. 3.2).[32,80]

Phosphonate organic layer formation involves two steps: first, the phosphonic acid is adsorbed on the oxide surface and then converted to a phosphonate organic layer by heating. In contrast to silanization, where only surface OH groups react, both surface OH and bridging surface oxide groups can react during this process. Phosphonate layers adhere strongly to the substrate surface and are homogeneous and versatile for further chemical modification. They are resistant to removal by moisture and oxidation and are stable in electronically active environments.[81]

In 2008, Cattani-Scholz *et al.*, used such a chemical strategy to introduce hydroxyl groups onto the nanowire surface using 11-hydroxyundecylphosphonate (HUP). Then, thiolated PNA molecules were immobilized using a maleimide heterobifunctional cross-linker (see Section 3.4.3) for label-free detection of DNA via electrical measurements (Fig. 3.10). They showed that such modification can be an interesting alternative to native oxide modification through silanization.[77]

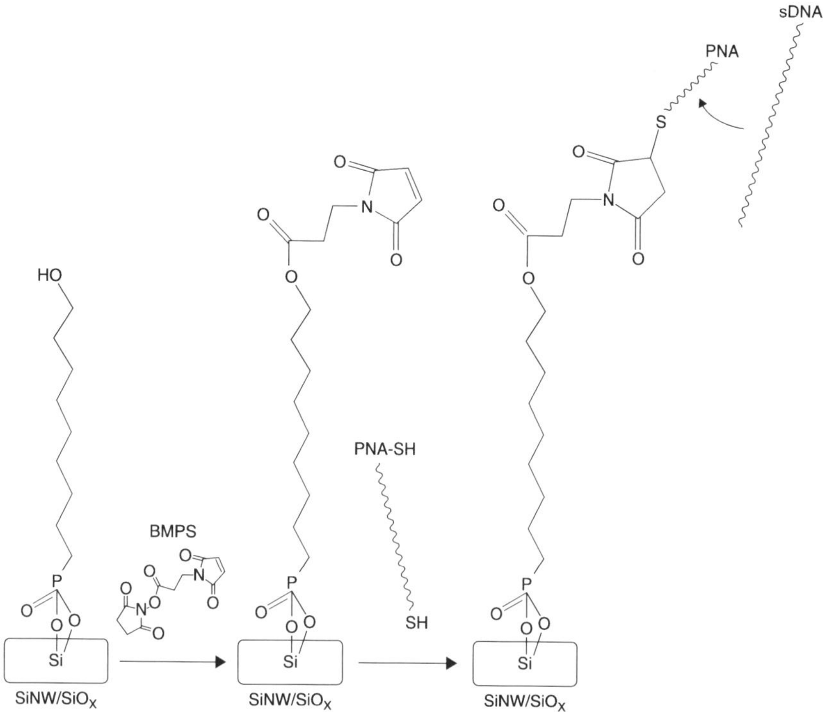

3.10 11-hydroxyundecylphosphonate (HUP) grafting and peptide nucleic acid (PNA) immobilization on oxidized silicon nanowires (SiNW). BMPS, 3-maleimidoproprionic-acid-*N*-hydroxy-succinimide ester.

3.5 Modification of hydrogen-terminated silicon nanowires (H-SiNW)

3.5.1 Hydrosilylation reaction

The hydrosilylation reaction consists of the use of organic molecular layers bearing unsaturated alkene or alkyne bonds that can be linked directly to hydrogen-terminated SiNW under photochemical (UV[82–87] or visible[88,89]), thermal (150–200 °C),[90–93] peroxide activation (radical initiator)[90,94] or Lewis acid catalysts,[95,96] leading to the formation of strong Si–C bonds without an intervening oxide (Fig. 3.11). The method yields surfaces with improved stability and higher reproducibility of modification.[24,82,84,92,97–101]

Both strategies, initially developed for flat silicon surface modification, can be easily transferred to silicon nanostructures as shown by Streifer *et al.*[102] Indeed,

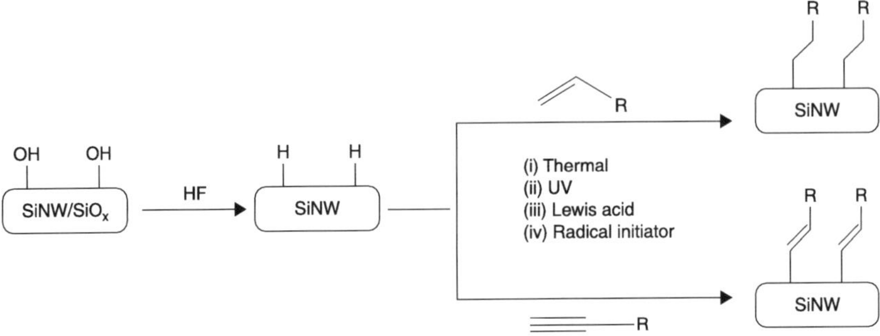

3.11 Hydrosilylation reactions on silicon nanowires (SiNW).

they used hydrosilylation under photochemical conditions to immobilize DNA covalently on H-SiNW in order to achieve biomolecular recognition. The group of Haick has demonstrated that alkylation of SiNW, when compared with Si wafer, led to higher surface coverage and that a shorter reaction time is needed to get the same level of coverage as that of a surface.[103,104]

3.5.2 Deprotection

However, when bifunctional molecules such as ω-amino alkenes are reacted with a hydrogenated silicon surface, both functional groups – alkene and amino – will competitively react with the Si-H bonds, resulting in disordered monolayers.[105] Thus, the introduction of the required functionality can be achieved using protecting groups. Figure 3.12 illustrates preparation of commonly employed protecting groups and corresponding deprotection reactions. Amino or

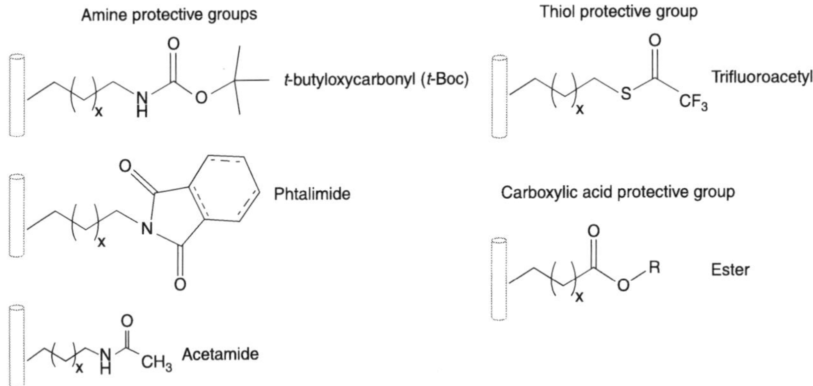

3.12 Protective groups for Si surface functionalization.

semicarbazide groups can be introduced using a *t*-butyloxycarbonyl (*t*-Boc)-protected amine.[16,87,106] Alternatively, phthalimide or acetamide moieties also serve as effective protecting groups for the introduction of well-ordered amino functionalities.[106] Ester-terminated monolayers can also be easily modified to obtain a variety of functional groups, reduction with $NaBH_4$ or $LiAlH_4$ results in alcohol termination, acid hydrolysis leads to carboxylic acid formation, and reaction with alkyl Grignard reagents produces a tertiary alcohol.[84,91,107] Trifluoroacetyl groups are effective protecting agents for thiol groups.[108]

3.5.3 Post-modification/cross-linking

As described above for functionalization of oxidized SiNW, immobilization often needs to proceed post-modification to promote biomolecule/probe attachment. The same strategies or cross-linkers can be used as the same chemical functionalities will be present on SiNW.

Recently, Coffinier *et al.* used NHS ester-activated SiNW to immobilize amino-NTA ligands and performed His*6-Tag-peptides enrichment prior to their laser-assisted desorption/ionization and subsequent detection by mass spectrometry.[109] In this process, H-SiNW (1) surfaces were reacted with undecylenic acid (UA) via hydrosilylation reaction initiated under thermal conditions (~150 °C) to yield carboxylic acid-terminated SiNW. Then, the terminal carboxylic group was converted into an amino-reactive linker, NHS-ester (2), allowing the immobilization of N-(5-amino-1-carboxylpentyl)iminodiacetic acid (NH_2–NTA) via amide bond formation (3). Finally, the NTA–Ni^{2+} complex was formed by nickel loading (4) allowing capture of the His-tag-peptide (5) (Fig. 3.13).

Streifer *et al.* functionalized H-SiNW with a *t*-BOC-protected amine by UV-initiated hydrosilylation.[102] Following deprotection, the amine-terminated nanowires were reacted with the bifunctional linker sulfo-succinimidyl 4-(*N*-maleimidomethyl) cyclohexene-1-carboxylate (SSMCC). Then, the maleimide group was used to immobilize thiol-terminated DNA olignucleotides onto SiNW, subsequently hybridated with fluorescently labelled complementary DNA targets (Fig. 3.14).[102]

Bunimovich *et al.* used the same strategy, but instead of using SMCC as cross-linker, they used glutaraldehyde (GA) for the attachment of uncharged PNA 16-mer strands.[26]

3.5.4 Halogenation/alkylation followed by Grignard reaction

An alternative approach to formation of alkylated Si surfaces is via alkyl Grignard (R-MgX) reagents.[27,79,84] First of all, a halogenation step is performed to obtain (Cl, Br, I)-terminated Si surfaces that can be prepared via a two-step process: on initial removal of the surface oxide using aqueous HF, followed by a treatment with an appropriate halogenation reagent. The preparation of chlorinated SiNW

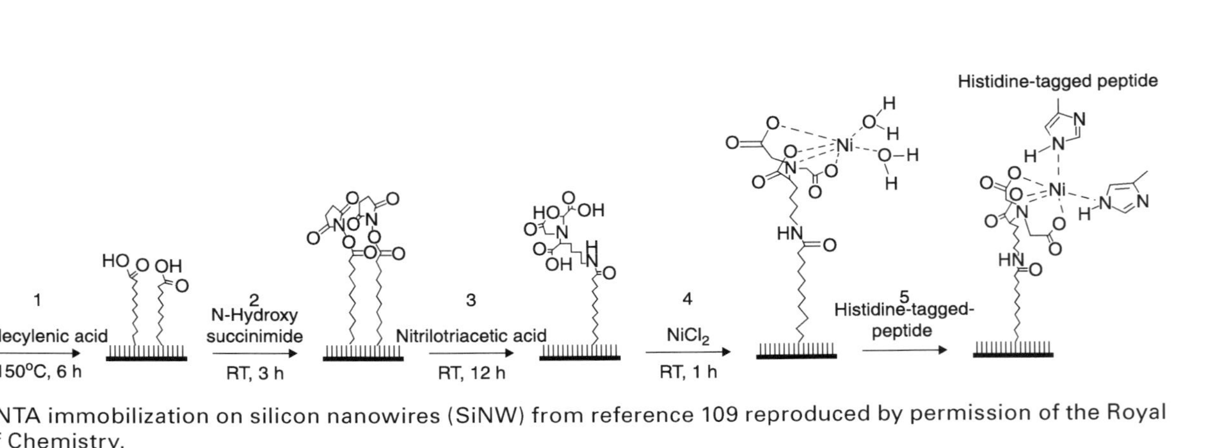

3.13 NH_2-NTA immobilization on silicon nanowires (SiNW) from reference 109 reproduced by permission of the Royal Society of Chemistry.

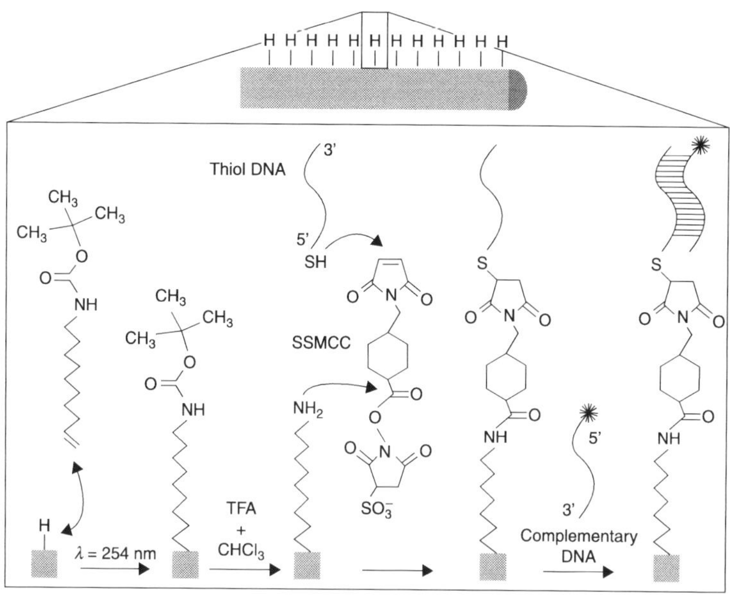

3.14 Schematic illustration of the functionalization of silicon nanowires (SiNW) with DNA strands from reference 102. Reproduced by permission of IOP Publishing. All rights reserved. SSMCC, sulfo-succinimidyl 4-(*N*-maleimidomethyl) cyclohexene-1-carboxylate.

surfaces, often required for further organic functionalization, can be achieved using saturated solutions of PCl_5 in chlorobenzene heated to temperatures between 80 °C and 100 °C (Fig. 3.15).[27] The reaction can be mediated by a benzoyl peroxide radical initiator or by UV irradiation.[110,111]

Grignard reagents can then react with halogenated-SiNW surfaces, typically Cl-terminated. It has been shown that Grignard reagents can also react with H-Si planar surfaces.[95] The functionalization reaction is carried out at elevated temperatures (60–80 °C) and requires long reaction times (up to 8 days), particularly for Grignard reagents consisting of long alkyl chains. Bashouti *et al.* functionalized Au-seeded SiNW with alkyl chains ranging from C_1–C_6 through a chlorination/alkylation route and found that the chain length influenced the surface coverage, saturation time and oxidation resistance of the functionalized nanowires.[103] The smaller van der Waals radius of C_1 groups (2.5 Å) compared

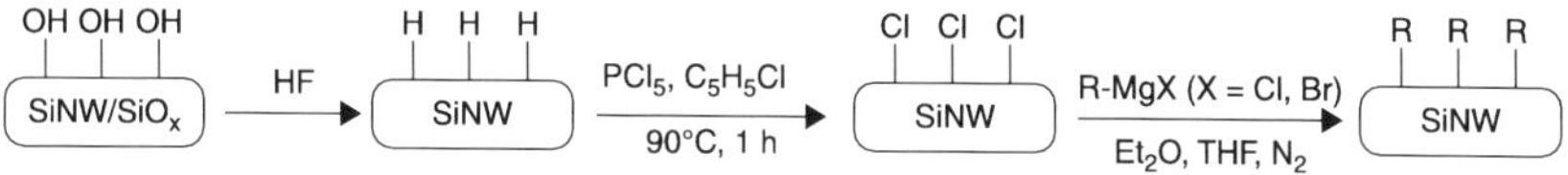

3.15 Halogenation/alkylation reaction on silicon nanowires (SiNW).

with the Si-Si separation distance (3.8 Å), facilitated a nearly full surface coverage for SiNW modified with methyl groups. However as the chain length increased, the van der Waals radius restricted the packing of alkyl groups and limited maximum surface coverage to ~50–55%. Assad *et al.* compared the stability of methyl (CH_3)-, propenyl- and propynyl-terminated SiNW, prepared by a chlorination/alkylation route using the corresponding Grignard reagents.[79] Both alkenyl and alkynyl functionalization layers exhibited nearly full coverage of the Si atop sites with 103±5% and 97±5%, respectively, and relative to the full coverage obtained for CH_3-SiNW surface.

Stability studies of the modified nanowires showed that CH_3- and CH_3–C≡C-functionalized surfaces were oxide-free up to 100 hours and 300 hours of ambient exposure, respectively. Then, the rate of oxidation of CH_3-terminated nanowires increased continuously with time. In contrast, CH_3–CH=CH–Si nanowires displayed an initial oxidation equivalent to ~0.12 monolayer, but then stabilized on further ambient exposure. SiNW functionalized with CH_3–C≡C- and CH_3-exhibited the fastest rates of re-oxidation.[79] The greater stability of the CH_3-CH=CH–Si passivation layers can be attributed to the $\pi-\pi$ interactions of adjacent chains, which inhibit oxidation of the underlying Si atoms.[112]

3.5.5 Electrografting on H-SiNWs

Electrografting provides another method for the direct (Si–C) covalent functionalization of silicon surfaces. The formation of alkane and alkene organic layers via electrochemical grafting has been demonstrated on H-terminated SiNW.[113] Scheibal *et al.* used cathodic electrografting of hexynoic acid to H-SiNW surfaces (dense arrays of silicon nanowires approximately 50–100 nm in diameter and 10 μm in length). Then, using NHS/EDC strategy, bovine serum albumin (BSA) was immobilized. Protein immobilization was achieved via an amidation reaction with carboxylic moieties of electrografted hexynoic acids. Such protein immobilization was proved using attenuated total reflectance Fourier transform infrared spectroscopy (FTIR) and fluorescence microscopy.

3.5.6 Arylation via aryldiazonium salt

Electrografting of arenediazonium salts has been used to prepare phenyl layers on H-terminated planar Si surfaces.[114–116] Stewart *et al.* demonstrated the spontaneous grafting of organic ligands via arenediazonium salts on planar semiconductor surfaces such as Si and GaAs (Fig. 3.16).[117] The functionalization procedure was carried out in anhydrous acetonitrile at room temperature, with reaction times of 1 hour resulting in successful covalent attachment of organic ligands to the Si surface. Haight and co-workers observed that surface modification of SiNW could be achieved by exposing H-SiNW to a fine mist of phenylterpyridinediazonium solution.[118]

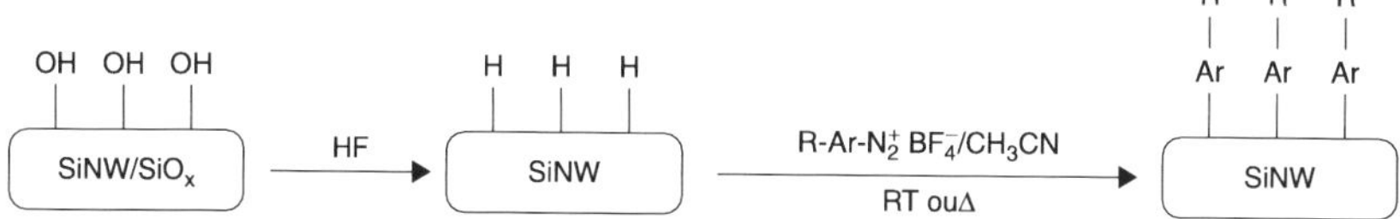

3.16 Arylation reaction on silicon nanowires (SiNW) via reduction of diazonium salts ($R\text{-}Ar\text{-}N_2^+BF_4^-$).

3.6 Site-specific immobilization strategy of biomolecules on SiNWs

In all previous cases, the active sites of a substantial population of immobilized molecules are not accessible to targets in the solution phase.[119] To ensure a specific orientation on the surface, various covalent chemoselective and site-specific immobilization strategies have been developed on flat surfaces. Indeed, it has been demonstrated that the direct interface between solid surfaces and probes can affect the quality of the molecular interactions (e.g. the catalytic efficiency of an enzyme towards its substrate, or antibody–antigen recognition)[120,121] and that site-specific and oriented immobilization can improve the detection.[119,122,123]

Among them, we can mention Schiff-base site-specific ligation methods such as oxime[124–127] or α-oxo-semicarbazone ligations,[87,128–130] the Staudinger ligation,[131,132] the Diels-Alder reaction,[133] 'click' chemistry[134] and native chemical ligation (NCL).[135–139] However, to the best of our knowledge, there are only a few examples of site-specific ligation which have been performed on oxide-free SiNW.

3.6.1 Native chemical ligation (NCL)

Dendane *et al.* have described site-specific and chemoselective immobilization of peptides on H-SiNW, grown via a VLS mechanism, using NCL.[140] First, the direct reaction of the H-SiNW surface with undecylenic acid (UA) under thermal conditions led to the formation of an organic layer covalently attached to the surface through Si–C bonds (1). Then, the carboxylic acid terminal group was converted to a benzylthiol ester group, allowing the immobilization of model Cys peptides using NCL (Fig. 3.17). For this, carboxylic acid groups were activated with dicyclohexylcarbodiimide in the presence of benzylmercaptan and 4-dimethylaminopyridine (DMAP) used as an acylation catalyst (2).

NCL is the reaction between a free cysteine residue and a thioester group. It proceeds through a transthioesterification step and a formation of a transient thioester-linked intermediate. This reaction is followed by a rapid, spontaneous intramolecular *S,N*-acyl shift resulting in the formation of an amide bond. The formation of the thioester-linked intermediate is reversible, but subsequent

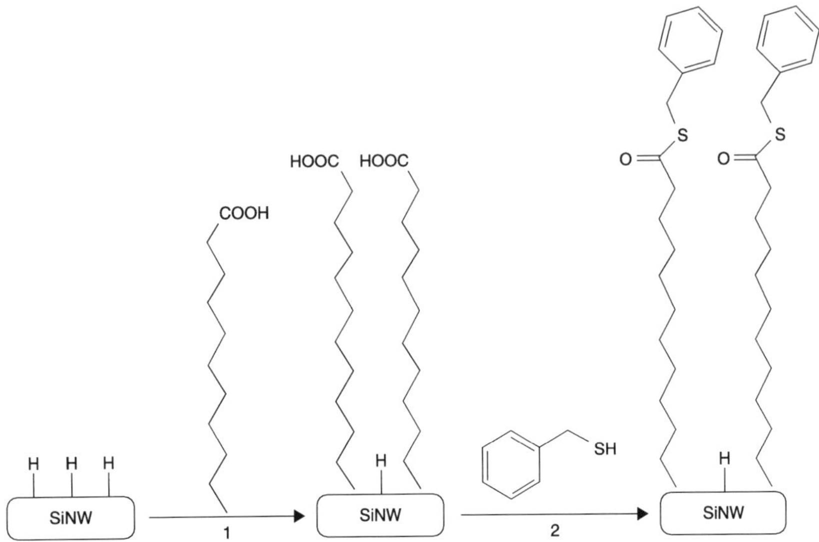

3.17 Schematic representation of the chemical steps for silicon nanowire (SiNW) functionalization with benzylthiol groups.

rearrangement to the amide is irreversible, thereby allowing displacement of the equilibrium toward the target amide bond ligated product (Fig. 3.18). NCL has several features that make it uniquely attractive as a potential methodology for grafting molecules to surfaces. Indeed, NCL generates a stable amide bond in water under physiological conditions and is chemo- and regioselective. Moreover, free cysteine residue can easily be incorporated into peptides or other non-peptidic biomolecules using solution- or solid-phase synthesis methods, whereas *N*-terminal Cys proteins can be produced using recombinant techniques.

3.6.2 'Click' chemistry

The workhorse of 'click' chemistry is the copper-catalyzed azide-alkyne cycloaddition (CuAAC) process, which has proven its versatility and efficiency not only for solution phase reactions but also for surface modifications of a broad range of solid substrates including flat silicon[141] and porous silicon.[142]

For certain applications, however, the use of a metal catalyst is precluded, for example, in biological environments because of the cytotoxic properties of copper[143] or for surface modifications of electronic materials, where traces of copper are retained on the surface and alter the electronic properties dramatically.[144] Copper ions can also cause degradation of DNA molecules, induce protein denaturation, and inhibit the luminescence of quantum dots attached via 'click' chemistry to biomolecules for *in vivo* imaging purposes.[145]

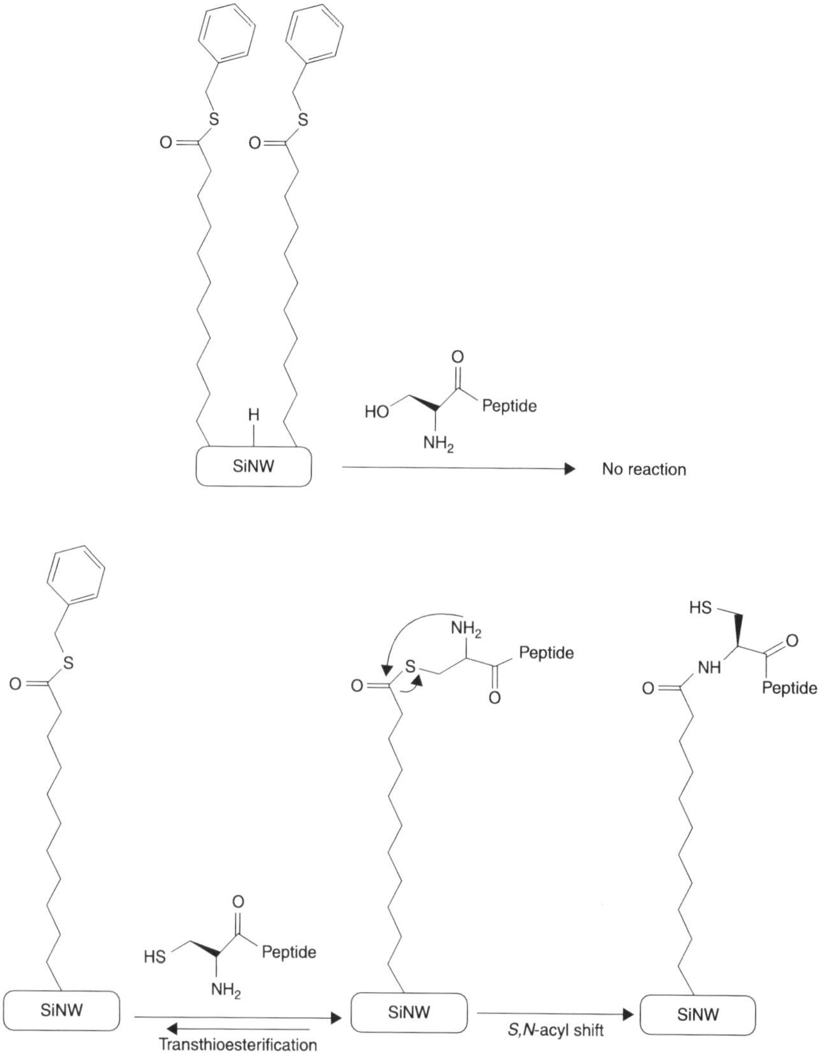

3.18 Native chemical ligation (NCL) on silicon nanowires (SiNW).

Henrikson *et al.* have described a process based on copper-free 'click' chemistry, by which the surface of SiNW can be functionalized with specific organic substituents.[145] A hydrogen-terminated SiNW surface was first primed with a monolayer of an R,ω-diyne and thereby turned into an alkyne-terminated, clickable platform. Then, an azide, carrying the desired terminal functionality, was subsequently coupled (Fig. 3.19). They demonstrated that a reactive, but air-stable primer layer can be first attached to the SiNW, onto which the desired

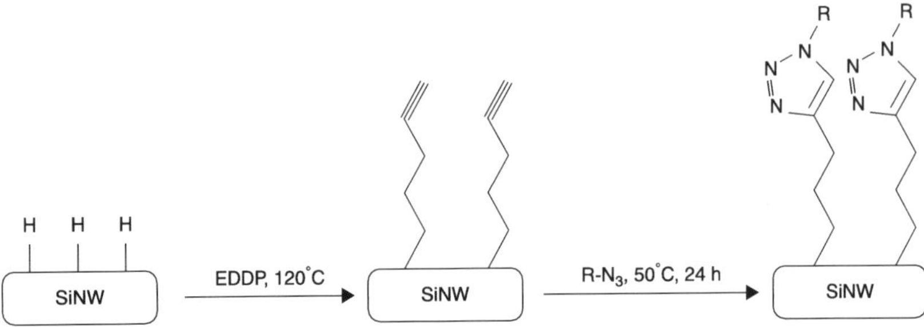

3.19 'Click-chemistry' on silicon nanowires (SiNW) and formation of a triazole group.

functionalities can subsequently be clicked via alkyne-azide coupling. The mild, catalyst-free reaction conditions for this process in combination with the extensive pool of available click transformations offer a promising approach for fine-tuning the electronic properties of SiNW through surface functionalization and present a biocompatible route for fabrication of SiNW-based bioanalytical devices. It has to be noted that such strategies can be easily adapted to oxidized SiNW.

3.7 Control of non-specific interactions

Non-specific binding depends mainly on surface modification, but also on the fluidics and buffers employed during a given biorecognition process. Thus, optimization of such variables will strongly influence the specificity of the analyte recognition on the developed 2-D or 1-D devices. After bioreceptor attachment, a blocking step is often performed to avoid non-specific binding by using a blocking agent after probe attachment. When 2-D devices are considered, bovine serum albumin (BSA) or ovalbumin (OVA) can be used, especially when proteins or antibodies are employed as probes.[146] Generally in oligonucleotide probes, no blocking step with protein is required, although the chemical blocking of the remaining active sites is necessary. Thus, reducing agents or ethanolamine are used to block aldehyde, isocyanate and epoxy surfaces after bioreceptor attachment.[146]

However, in place of or in addition to these processes, surface functionalization can strongly avoid undesired biofouling, notably using poly(ethylene glycol) derivatives. Indeed, organic layers containing oligo, poly (ethylene oxide) (OEG) or PEG moieties have been extensively studied and have been shown to resist protein adsorption and limit the non-specific interaction, an important parameter to take into account when biosensing devices are developed.[16,147,148]

Grafting OEG or PEG onto silicon oxide surfaces has been mostly based on siloxane chemistry using trichloro- or trialkoxylsilane derivatives.[149] However, hydrosilylation of PEG containing alkenes on silicon has been also reported

(Fig. 3.20(a)).[150] The ability of Si-C linked mixed monolayers formed from PEG and amine-terminated alkenes to protein adsorption has been demonstrated by assaying the specific versus the non-specific binding of common serum proteins and avidin on these surfaces. Such protein resistance was also demonstrated by Stern *et al.* on a PEG-functionalized SiNW device that yielded no response when 1 nM streptavidin solution was injected.[16] Post grafting of PEG bearing specific function can also be performed on surfaces terminated with amino, epoxy or hydroxyl groups, as shown by Voicu *et al.* They formed undecanoic acid-terminated monolayers on Si(111) followed by coupling of amino-terminated tetra- (ethylene oxide) derivatives (Fig. 3.20(b)).[151] The grafted PEG molecule can also carry specific function that can react in a post-modification process for biomolecule ligation (Fig. 3.20(c)).[147,148] Finally, heterobifunctional PEGs can also be used, that is PEG molecules carrying thiol and carboxylic acid moieties or containing two amine functionalities, one with a protecting group. The use of such reagents allowed introduction of reactive carboxyl and amino groups onto the surface for further biomolecule immobilization (Fig. 3.20(d)).[39]

In addition, PEG moieties allow 1) steric stabilization of particles/wires/tubes and prevent binding of plasma proteins (opsonization), thereby prolonging half-life in circulation, 2) a reduction of immunogenicity, and 3) are also non-toxic, non-immunogenic, non-antigenic, highly soluble in water and FDA-approved.[152–155]

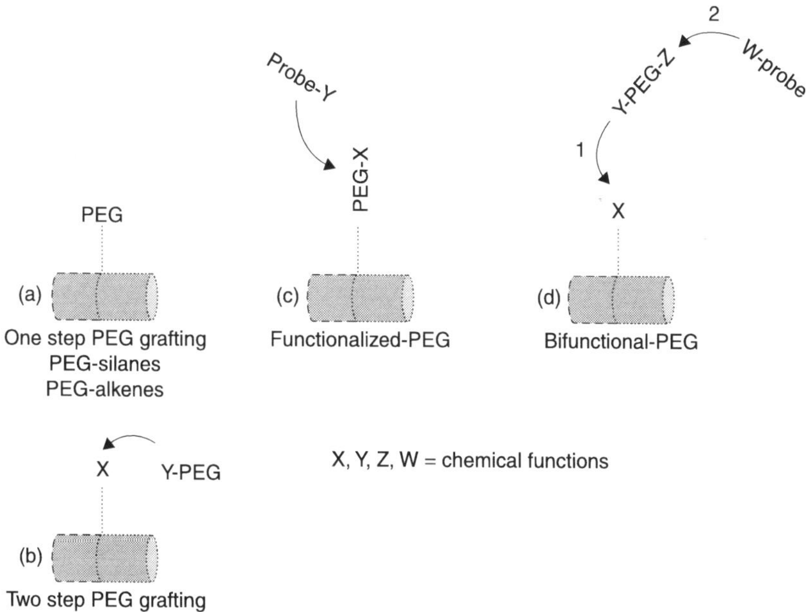

3.20 (a–d) Poly(ethylene glycol) (PEG) introduction strategies on oxidized (dashed line) and oxide-free (dotted line) silicon nanowires (SiNW). See text for explanation.

Jung *et al.* studied *in vivo* and *in vitro* optical imaging of silicon nanowires (5–40 nm in diameter) using strong and stable third-order non-linear optical (NLO) signals, including four-wave mixing (FWM) and third-harmonic generation (THG). They also performed *in vivo* monitoring of SiNW circulating in the peripheral blood of a live mouse and to map the organ distribution of systemically administrated SiNW (liver, spleen . . .), opening up further opportunities to investigate *in vivo* cellular response to nanomaterials as a function of size, aspect ratio and surface chemistry.[28] To do that, they chemically modified SiNW with PEG molecules to prolong blood circulation time.

3.8 Conclusion

In this chapter, we present some of the chemical strategies commonly used for modifying SiNW and related nano-objects. Surface chemistries on SiNW described in the literature are mostly dedicated to sensing and especially to electrical sensing. The surface chemistry on either flat or nanostructured surfaces is still the key step in bio- or chemical sensing device fabrication. Indeed, surface modification should be reproducible, robust and stable over time, which is not often reported. Moreover, chemical strategies for probe immobilization should permit the limitation of non-specific interactions, to promote the probe–target interaction by unambiguously controlling the probe orientation. Although a large number of chemical strategies have been developed on 2-D-silicon surfaces (oxide or oxide-free) for probe immobilization in a well controlled manner, only a few have been transferred to 1-D-silicon objects, most notably the NHS/EDC strategy.

In addition, and perhaps one of the most important challenges is that surface chemistry on SiNW should be accessible to non-chemists, easy to scale up, avoid any complicated chemical steps or the use of very specific equipment and dangerous solvents (i.e. environmentally benign processes), in order to get a high degree of reproducibility and chemical stability.

A possible answer to the above challenges could be site-specific ligation strategies such as NCL, ‘click’ chemistry or thiol-yne (ene) reactions, all of which are very promising but still not widely applied to SiNW.

With a global biosensors market that reached US$6.1 billion in 2012, chemical modifications of SiNW can also be achieved for many other industrial or research level applications, such as semiconductor-based technology, energy conversion and storage technologies, and medical technology.

3.9 References

1. Cui, D. and Gao, H. (2003) ‘Advance and prospect of bionanomaterials’, *Biotechnol. Prog.*, 19, 683–92.
2. Wang, Y., Tang, Z., Tan, S. and Kotov, N. A. (2005) ‘Biological assembly of nanocircuit prototypes from protein-modified CdTe nanowires’, *Nano Lett.*, 5, 243–8.

3. Mbindyo, J. K. N., Reiss, B. D., Martin, B. R., Keating, C. D., Natan, M. J. and Mallouk, T. E. (2001) 'DNA-directed assembly of gold nanowires on complementary surfaces', *Adv. Mater.*, 13, 249–54.
4. Shim, M., Kam, N. W. S., Chen, R. J., Li, Y. and Dai, H. (2002) 'Functionalization of carbon nanotubes for biocompatibility and biomolecular recognition', *Nano Lett.*, 2, 285–8.
5. Chen, R. J., Zhang, Y., Dunwei, W. and Dai, H. (2001) 'Non covalent sidewall functionalization of single-walled carbon nanotubes for protein immobilization', *J. Am. Chem. Soc.*, 123, 3838–9.
6. Monson, C. F. and Woolley, A. T. (2003) 'DNA-templated construction of copper nanowires', *Nano Lett.*, 3, 359–63.
7. Niemeyer, C. M. (2000) 'Self-assembled nanostructures based on DNA: towards the development of nanobiotechnology', *Curr. Opin. Chem. Biol.*, 4, 609–18.
8. Jianrong, C., Yuqing, M., Nongyue, H., Xiaohua, W. and Sijiao, L. (2004) 'Nanotechnology and biosensors', *Biotechnol. Adv.*, 22, 505–18.
9. Cui, Y., Wei, Q., Park, H. and Lieber, C. M. (2001) 'Nanowire nanosensors for highly sensitive and selective detection of biological and chemical species', *Science*, 293, 1289–92.
10. Maki, W. C., Mishra, N. N., Cameron E. G., Filanoski B., Rastogi S. K. and Maki G. K. (2008) 'Nanowire-transistor based ultra-sensitive DNA methylation detection', *Biosens. Bioelectron.*, 23, 780–7.
11. Gao, Z. Q., Agarwal, A., Trigg, A. D., Singh, N., Fang, C., *et al.* (2007) 'Silicon nanowire arrays for label-free detection of DNA', *Anal. Chem.*, 79, 3291–7.
12. Zhang, G. J., Zhang, G., Chua, J. H., Chee, R. E., Wong, E. H., *et al.* (2008) 'DNA sensing by silicon nanowire: charge layer distance dependence', *Nano Lett.*, 8, 1066–70.
13. Patolsky, F., Zheng, G. and Lieber C. M. (2006) 'Nanowire sensors for medicine and the life sciences', *Nanomedicine*, 1, 51–65.
14. Wang, W. U., Chen, C., Lin, K. H., Fang, Y. and Lieber C. M. (2005) 'Label-free detection of small molecule–protein interactions by using nanowire nanosensors', *Proc. Natl. Acad. Sci. USA*, 102, 3208–12.
15. Kim, A., Ah, C. S., Yu, H. Y., Yang, J. H., Baek, I. B., *et al.* (2007) 'Ultrasensitive label-free and real-time immunodetection using silicon field-effect transistors', *Appl. Phys. Lett.*, 91, 103901–4.
16. Stern, E., Klemic, J. F., Routenberg, D. A., Wyrembak, P. N., Turner-Evans, D. B., *et al.* (2007) 'Label-free immunodetection with CMOS-compatible semiconducting nanowires', *Nature*, 445, 519–22.
17. Li, Z., Chen, Y., Li, X., Kamins, T. I., Nauka, K. and Williams, R. S. (2004) 'Sequence-specific label-free DNA sensors based on silicon nanowires', *Nano Lett.*, 4, 245–7.
18. Hahm, J. and Lieber, C. M. (2004) 'Direct ultrasensitive electrical detection of DNA and DNA sequence variations using nanowire nanosensors', *Nano Lett.*, 4, 51–4.
19. Schmidt, V., Wittemann, J. V., Senz, S. and Gösele, U. (2009) 'Silicon nanowires: A review on aspects of their growth and their electrical properties', *Adv. Mater.*, 21, 2681–702.
20. Seo, K., Wober, M., Steinvurzel, P., Schonbrun, E., Dan, Y., *et al.* (2011) 'Multicolored vertical silicon nanowires', *Nano Lett.*, 11, 1851–6.
21. Kang, Z., Tsang, C. H. A., Zhang, Z., Zhang, M., Wong, N. B., *et al.* (2007) 'A polyoxometalate-assisted electrochemical method for silicon nanostructures preparation: From quantum dots to nanowires', *J. Am. Chem. Soc.*, 129, 5326–7.

22. Huang, Z., Geyer, N., Werner, P., de Boor, J. and Gösele, U. (2011) 'Metal-assisted chemical etching of silicon: A review', *Adv. Mater.*, 23, 285–308.
23. Haensch, C., Hoeppener, S. and Schubert, U. S. (2010) 'Chemical modification of self-assembled silane based monolayers by surface reactions', *Chem. Soc. Rev.*, 39, 2323–34.
24. Buriak, J. M. (2002) 'Organometallic chemistry on silicon and germanium surfaces', *Chem. Rev.*, 102, 1271–308.
25. Haick, H., Hurley, P. T., Hochbaum, A. I., Yang, P. D. and Lewis, N. S. (2006) 'Electrical characteristics and chemical stability of non-oxidized, methyl-terminated silicon nanowires', *J. Am. Chem. Soc.*, 128, 8990–1.
26. Bunimovich, Y. L., Shin, Y. S., Yeo, W. Y., Amori, M., Kwong, G. and Heath, J. R. (2006) 'Quantitative real-time measurements of DNA hybridization with alkylated nonoxidized silicon nanowires in electrolyte solution', *J. Am. Chem. Soc.*, 128, 16323–31.
27. Bashouti, M. Y., Stelzner, T., Christiansen, S. and Haick, H. (2009) 'Surface chemistry of non-oxidized, molecule-terminated silicon nanowires: monolayer formation, kinetics, and oxidation resistance', *J. Phys. Chem. C*, 113, 14823–8.
28. Jung, Y., Tong, L., Tanaudommongkon, A., Cheng, J. X. and Yang, C. (2009) '*In vitro* and *in vivo* non linear optical imaging of silicon nanowires', *Nano Lett.*, 9, 2440–4.
29. Rochdi, N., Tonneau, D., Jandard, F., Dallaporta, H. and Safarov, V. (2008) 'Fabrication and electrical properties of ultra-thin silicon nanowires', *Phys. Stat. Solidi A*, 205, 1157–61.
30. Bashouti, M. Y., Tung, R. T. and Haick, H. (2009) 'Tuning the electrical properties of Si nanowire field-effect transistors by molecular engineering', *Small*, 5, 2761–9.
31. Cui, Y., Zhong, Z. H., Wang, D. L., Wang, W. U. and Lieber, C. M. (2003) 'High performance silicon nanowire field effect transistor', *Nano Lett.*, 3, 149–52.
32. Silverman, B. M., Wieghaus, K. A. and Schwartz, J. (2005) 'Comparative properties of siloxane *vs* phosphonate monolayers on a key titanium alloy', *Langmuir*, 21, 225–8.
33. Densmore, A., Vachon, M., Xu, D. X., Janz, S., Ma, R., *et al.* (2009) 'Silicon photonic wire biosensor array for multiplexed real-time and label-free molecular detection', *Opt. Lett.*, 34, 3598–600.
34. Shi, C., Mehrabani, S. and Armani, A. M. (2012) 'Leveraging bimodal kinetics to improve detection specificity', *Opt. Lett.*, 37, 1643–5.
35. García-Rupérez, J., Toccafondo, V., Bãnuls, M. J., García Castelló, J., Griol, A., *et al.* (2010) 'Label-free antibody detection using band edge fringes in SOI planar photonic crystal waveguides in the slow-light regime', *Opt. Express*, 18, 24276–86.
36. Hunt, H. K., Soteropolos, C. and Armani, A. M. (2010) 'Bioconjugation strategies for microtoroidal optical resonators', *Sensors*, 10, 9317–36.
37. Gawalt, E. S., Avaltroni, M. J., Danahy, M. P., Silverman, B. M., Hanson, E. L., *et al.* (2003) 'Bonding organics to Ti alloys: Facilitating human osteoblast attachment and spreading on surgical implant materials', *Langmuir*, 19, 200–4.
38. Aswal, D. K., Lenfant, S., Guerin, D., Yakhami, J. V. and Vuillaume, D. (2006) 'Self assembled monolayers on silicon for molecular electronics', *Anal. Chim. Acta*, 568, 84–108.
39. De Vos, K., Girones, J., Popelka, S., Schacht, E., Baets, R. and Bienstman, P. (2009) 'SOI optical microring resonator with poly(ethylene glycol) polymer brush for label-free biosensor applications', *Biosens. Bioelectron.*, 24, 2528–33.
40. Piret, G., Drobecq H., Coffinier, Y., Melnyk, O. and Boukherroub, R. (2010) 'Desorption/ionization mass spectrometry on silicon nanowire arrays prepared by chemical etching of crystalline silicon', *Langmuir*, 26, 1354–61.

41. Ramachandran, A., Wang, S., Clarke, J., Ja, S. J., Goad, D., *et al.* (2008) 'A universal biosensing platform based on optical micro-ring resonators', *Biosens. Bioelectron.*, 23, 939–44.
42. Lee, S., Eom, S. C., Chang, J. S., Huh, C., Sung, G. Y. and Shin, J. H. (2010) 'Label-free optical biosensing using a horizontal air-slot SiN_x microdisk resonator', *Opt. Express*, 18, 20638–44.
43. Sharma, S., Johnson, R. W. and Desai, T. A. (2004) 'XPS and AFM analysis of antifouling PEG interfaces for microfabricated silicon biosensors', *Biosens. Bioelectron.*, 20, 227–39.
44. Zlatanovic, S., Mirkarimi, L. W., Sigalas, M. M., Bynum, M. A., Chow, E., *et al.* (2009) 'Photonic crystal microcavity sensor for ultracompact monitoring of reaction kinetics and protein concentration', *Sens. Actuators B*, 141, 13–19.
45. Schneider, B. H., Dickinson, E. L., Vach, M. D., Hoijer, J. V. and Howard, L. V. (2000) 'Highly sensitive optical chip immunoassays in human serum', *Biosens. Bioelectron.*, 15, 13–22.
46. Xu, J., Suarez, D. and Gottfried, D. S. (2007) 'Detection of avian influenza virus using an interferometric biosensor', *Anal. Bioanal. Chem.*, 389, 1193–9.
47. Xiao, S. J., Textor, M., Spencer, N. D. and Sigrist, H. (1998) 'Covalent attachment of cell-adhesive, (Arg-Gly-Asp)-containing peptide to titanium surfaces', *Langmuir*, 14, 5507–16.
48. Coffinier, Y., Janel, S., Addad, A., Blossey, R., Gengembre, L., *et al.* (2007) 'Preparation of superhydrophobic silicon oxide nanowire surfaces', *Langmuir*, 23, 1608–11.
49. Verplanck, N., Galopin, E., Camart, J. C., Thomy, V., Coffinier, Y. and Boukherroub, R. (2007) 'Reversible electrowetting on superhydrophobic silicon nanowires', *Nano Lett.*, 7, 813–17.
50. Brunet, P., Lapierre, F., Thomy, V., Coffinier, Y. and Boukherroub, R. (2008) 'Extreme resistance of superhydrophobic surfaces to impalement: reversible electrowetting related to the impacting/bouncing drop test', *Langmuir*, 24, 11202–8.
51. Lapierre, F., Thomy, V., Coffinier, Y. and Boukherroub, R. (2009) 'Reversible electrowetting on superhydrophobic double-nanotextured surfaces', *Langmuir*, 25, 6551–8.
52. Lapierre, F., Brunet, P., Coffinier, Y., Thomy, V., Blossey, R. and Boukherroub, R. (2010) 'Electrowetting and droplet impalement experiments on superhydrophobic multiscale structures', *Faraday Discuss.*, 146, 125–39.
53. Nguyen, T. P. N., Brunet, P., Coffinier, Y. and Boukherroub, R. (2010) 'Quantitative testing of robustness on superomniphobic surfaces by drop impact', *Langmuir*, 26, 18369–73.
54. Daniels, R. H., Dikler, S., Li, E. and Stacey, C. (2008) 'Break free of the matrix: sensitive and rapid analysis of small molecules using nanostructured surfaces and LDI-TOF mass spectrometry', *J. Assoc. Lab. Autom.*, 13, 314–21.
55. Hermanson, G.T. (2008), *Bioconjugate Techniques*, 2nd ed., Academic Press, London, UK, pp. 562–81.
56. Vanderberg, E. T., Bertilsson, L., Liedberg, B., Uvdal, K., Erlandson, R., *et al.* (1991) 'Structure of 3-aminopropyltriethoxysilane on silicon oxide', *J. Colloid Interface Sci.*, 147, 103–18.
57. Shalek, A. K., Robinson, J. T., Karp, E. S., Lee, J. S., Ahn D. R., *et al.* (2010) 'Vertical silicon nanowires as a universal platform for delivering biomolecules into living cells', *Proc. Natl. Acad. Sci. USA,* 107, 1870–5.

58. Heo, K., Cho, E., Yang, J. E., Kim, M. H., Lee, M., *et al.* (2008) 'Large-scale assembly of silicon nanowire network-based devices using conventional microfabrication facilities', *Nano Lett.*, 8, 4523–7.
59. Kang, J., Myung, S., Kim, B., Oh, D., Kim, G. T. and Hong, S. (2008) 'Massive assembly of ZnO nanowire-based integrated devices', *Nanotechnology*, 19, 095303–9.
60. Myung, S., Lee, M., Kim, G. T., Ha, J. S. and Hong, S. (2005) 'Large-scale surface-programmed assembly of pristine vanadium oxide nanowire-based device', *Adv. Mater.*, 17, 2361–4.
61. Myung, S., Im, J., Huang, L., Rao, S. G., Kim, T., *et al.* (2006) 'Lens effect in directed assembly of nanowires on gradient molecular patterns', *J. Phys. Chem. B*, 110, 10217–19.
62. Kim, Y. K., Park, S. J., Koo, J. P., Kim, G. T., Hong, S. and Ha, J. S. (2007) 'Control of adsorption and alignment of V_2O_5 nanowires via chemically functionalized patterns', *Nanotechnology*, 18, 015304–10.
63. Myung, S., Heo, K., Lee, M., Choi, Y. H., Hong, S. H. and Hong, S. (2007) 'Focused assembly of V_2O_5 nanowire masks for the fabrication of metallic nanowire sensors', *Nanotechnology*, 18, 205034–8.
64. Hoyle, C. E. and Bowman, C. N. (2010) 'Thiol-ene click chemistry', *Angew. Chem. Int. Ed.*, 49, 1540–73.
65. Meziane, D., Barras, A., Kromka, A., Houdkova, J., Boukherroub, R. and Szunerits, S. (2012) 'Thiol-yne reaction on boron-doped diamond electrodes: application for the electrochemical detection of DNA-DNA hybridization events', *Anal. Chem.*, 84, 194–200.
66. Caipa Campos, M. A., Paulasse, J. M. J. and Zuilhof, H. (2010) 'Functional monolayers on oxide-free silicon surfaces via thiol–ene click chemistry', *Chem. Commun.*, 46, 5512–14.
67. Salhi, B., Vaurette, F., Grandidier, B., Stiévenard, D., Melnyk, O., *et al.* (2009) 'The collagen assisted self-assembly of silicon nanowires', *Nanotechnology*, 20, 235601–7.
68. Mateo, C., Abian, O., Fernandez-Lorente, G., Pedroche, J., Fernández-Lafuente, R. and Guisán, J. M. (2002) 'Epoxy sepabeads: a novel epoxy support for stabilization of industrial enzymes via very intense multipoint covalent attachment', *Biotechnol. Prog.*, 18, 629–34.
69. Thierry, B., Jasienak, M., de Smet, L. C. P., Vasilev, K. and Griesser, H. J. (2008) 'Reactive epoxy-functionalized thin films by a pulsed plasma polymerization process', *Langmuir*, 24, 10187–95.
70. Ingebrandt, S., Vu, X. T., Eschermann, J. F., Stockmann, R. and Offenhausser, A. (2011) 'Top-Down processed SOI nanowire device for biomedical applications', *ECS Trans*, 35, 3–15.
71. Cho, W. K., Jung, Y. H., Lee, K. B., Park, H. J., Kim, Y., *et al.* (2006) 'Chemical modification of Si nanowires for bioconjugation', *Bull. Korean Chem. Soc.*, 27, 111–14.
72. Duval, D., Gonzales-Guerrero, A. B., Dante, S., Osmond, J., Monge, R., *et al.* (2012) 'Nanophotonic lab-on-a-chip platforms including novel bimodal interferometers, microfluidics and grating couplers', *Lab Chip*, 12, 1987–94.
73. Abad, S., Nolis, P., Gispert, J. D., Spengler, J., Albericio, F., *et al.* (2012) 'Rapid and high-yielding cysteine labelling of peptides with N-succinimidyl 4-[18F] fluorobenzoate', *Chem. Commun.*, 48, 6118–20.
74. Lee, H. S., Kim, K. S., Kim, C. J., Hahn, S. K. and Jo, M. H. (2009) 'Electrical detection of VEGFs for cancer diagnoses using anti-vascular endothelial growth factor aptamer-modified Si nanowire FETs', *Biosens. Bioelectron.*, 24, 1801–5.

75. Wong, S. Y. and Putnam, D. (2007) 'Overcoming limiting side reactions associated with an NHS-activated precursor of polymethacrylamide-based polymers', *Bioconjugate Chem.*, 18, 970–82.
76. Duan, X., Li, Y., Rajan, N. K., Routenberg, D. A., Modis, Y. and Reed, M. A. (2012) 'Quantification of the affinities and kinetics of protein interactions using silicon nanowire biosensors', *Nature Nanotechnol.*, 7, 401–7.
77. Cattani-Scholz, A., Pedone, D., Dubey, M., Neppl, S., Nickel, B., *et al.* (2008) 'Organophosphonate-based PNA-functionalization of silicon nanowires for label-free DNA detection', *ACS Nano*, 2, 1653–60.
78. Zhang, W., Tong, L. and Yang, C. (2012) 'Cellular binding and internalization of functionalized silicon nanowires', *Nano Lett.*, 12, 1002–6.
79. Assad, O., Puniredd, S. R., Stelzner, T., Christiansen, S. and Haick, S. (2008) 'Stable scaffolds for reacting Si nanowires with further organic functionalities while preserving Si-C passivation of surface sites', *J. Am. Chem. Soc.*, 130, 17670–1.
80. Hanson, E. L., Schwartz, J., Nickel, B., Koch, N. and Danisman, M. F. (2003) 'Bonding self-assembled, compact organophosphonate monolayers to the native oxide surface of silicon', *J. Am. Chem. Soc.*, 125, 16074–80.
81. McDermott, J. F., McDowell, M., Hill, I. G., Hwang, J., Kahn, A., *et al.* (2007) 'Organophosphonate self-assembled monolayers for gate dielectric surface modification of pentacene-based organic thin-film transistors: A comparative study', *J. Phys. Chem. A*, 111, 12333–8.
82. Cicero, R. L., Linford, M. R., and Chidsey, C. E. D. (2000) 'Photoreactivity of unsaturated compounds with hydrogen-terminated silicon(111)', *Langmuir*, 16, 5688–95.
83. Effenberger, F., Gotz, G., Bidlingmaier, B., and Wezstein, M. (1998) 'Photoactivated preparation and patterning of self-assembled monolayers with 1-alkenes and aldehydes on silicon hydride surfaces', *Angew. Chem. Int. Ed.*, 37, 2462–4.
84. Boukherroub, R. and Wayner, D. D. M. (1999) 'Controlled functionalization and multistep chemical manipulation of covalently modified Si(111) surfaces', *J. Am. Chem. Soc.*, 121, 11513–15.
85. Langner, A., Panarello, A., Rivillon, S., Vassylyev, O., Khinast, J. G. and Chabal, Y. J. (2005) 'Controlled silicon surface functionalization by alkene hydrosilylation', *J. Am. Chem. Soc.*, 127, 12798–9.
86. Wang, X. Y., Ruther, R. E., Streifer, J. A. and Hamers, R. J. (2010) 'UV-induced grafting of alkenes to silicon surfaces: Photoemission versus excitons', *J. Am. Chem. Soc.*, 132, 4048–9.
87. Coffinier, Y., Olivier, C., Perzyna, A., Grandidier, B., Wallart, X. and Stievenard, D. (2005) 'Semicarbazide-functionalized Si(111) surfaces for the site-specific immobilization of peptides', *Langmuir*, 21, 1489–96.
88. Scheres, L., Achten, R., Giesbers, M., de Smet, L., Arafat, A., *et al.* (2009) 'Covalent attachment of Bent-core mesogens to silicon surfaces', *Langmuir*, 25, 1529–33.
89. Sun, Q. Y., de Smet, L., van Lagen, B., Giesbers, M., Thune, P. C., *et al.* (2005) 'Covalently attached monolayers on crystalline hydrogen-terminated silicon: extremely mild attachment by visible light', *J. Am. Chem. Soc.*, 127, 2514–23.
90. Linford, M. R., Fenter, P., Eisenberger, P. M. and Chidsey, C. E. D. (1995) 'Alkyl monolayer on silicon prepared from 1-alkene and hydrogen terminated silicon', *J. Am. Chem. Soc.*, 117, 3145–55.

91. Sieval, A. B., Demirel, A. L., Nissink, J. W. M., Linford, M. R., van der Maas, J. H., *et al.* (1998) 'Highly stable Si-C linked functionalized monolayers on the silicon (100) surface', *Langmuir*, 14, 1759–68.
92. Sung, M. M., Kluth, G. J., Yauw, O. W. and Maboudian, R. (1997) 'Thermal behavior of alkyl monolayers on the Si(100) surface', *Langmuir*, 13, 6164–8.
93. Sieval, A. B., Vleeming, V., Zuilhof, H. and Sudholter, E. J. R. (1999) 'An improved method for the preparation of organic monolayers of 1-alkenes on hydrogen-terminated silicon surfaces', *Langmuir*, 15, 8288–91.
94. Linford, M. R. and Chidsey, C. E. D. (1993) 'Alkyl monolayers covalently bonded to silicon surfaces', *J. Am. Chem. Soc.*, 115, 12631–2.
95. Boukherroub, R., Morin, S., Bensebaa, F. and Wayner, D. D. M. (1999) 'New synthetic routes to alkyl monolayers on the Si(111) surface', *Langmuir*, 15, 3831–5.
96. Buriak, J. M., Stewart, M. P., Geders, T. W., Allen, M. J., Choi, H. C., *et al.* (1999) 'Lewis acid mediated hydrosilylation on porous silicon surfaces', *J. Am. Chem. Soc.*, 121, 11491–502.
97. Stewart, M. P. and Buriak, J. M. (1998) 'Photopatterned hydrosilylation on porous silicon', *Angew. Chem. Int. Ed.*, 37, 3257–60.
98. Strother, T., Cai, W., Zhao, X., Hamers, R. J. and Smith, L. M. (2000) 'Synthesis and characterization of DNA-modified silicon (111) surfaces', *J. Am. Chem. Soc.*, 122, 1205–9.
99. Yang, W., Auciello, O., Butler, J. E., Cai, W., Carlisle, J. A., *et al.* (2002) 'DNA-modified nanocrystalline diamond thin-films as stable, biologically active substrates', *Nature Mater.*, 1, 253–7.
100. Strother, T., Hamers, R. J. and Smith, L. M. (2000) 'Covalent attachment of oligodeoxyribonucleotides to amine-modified Si (001) surfaces', *Nucleic Acids Res.*, 28, 3535–41.
101. Lin, Z., Strother, T., Cai, W., Cao, X., Smith, L. M. and Hamers, R. J. (2002) 'DNA attachment and hybridization at the silicon (100) surface', *Langmuir*, 18, 788–96.
102. Streifer, J. A., Kim, H., Nichols, B. M. and Hamers, R. J. (2005) 'Covalent functionalization and biomolecular recognition properties of DNA-modified silicon nanowires', *Nanotechnology*, 16, 1868–73.
103. Bashouti, M. Y., Stelzner, T., Berger, A., Christiansen, S., and Haick, H. (2008) 'Chemical passivation of silicon nanowires with C1-C6 alkyl chains through covalent Si-C bonds', *J. Phys. Chem. C*, 112, 19168–72.
104. Bashouti, M. Y., Paska, Y., Puniredd, S. R., Stelzner, T., Christiansen, S. and Haick, H. (2009) 'Silicon nanowires terminated with methyl functionalities exhibit stronger Si-C bonds than equivalent 2D surfaces', *Phys. Chem. Chem. Phys.*, 11, 3845–8.
105. Sieval, A. B., Linke, R., Heij, G., Meijer, G., Zuilhof, H. and Sudholter, E. J. R. (2001) 'Amino-terminated organic monolayers on hydrogen-terminated silicon surfaces', *Langmuir*, 17, 7554–9.
106. Haensch, C., Erdmenger, T., Fijten, M. W. M., Hoeppener, S. and Schubert, U. S. (2009) 'Fast surface modification by microwave assisted click reactions on silicon substrates', *Langmuir*, 25, 8019–24.
107. Shao, M. W., Wang, H., Fu, Y., Hua, J. and Ma, D. D. D. (2009) 'Surface functionalization of HF-treated silicon nanowires', *J. Chem. Sci.*, 121, 323–7.
108. Böcking, T., Salomon, A., Cahen, D. and Gooding, J. J. (2007) 'Thiol-terminated monolayers on oxide-free Si: assembly of semiconductor–alkyl–S–metal junctions', *Langmuir*, 23, 3236–41.

109. Coffinier, Y., Nguyen, N., Drobecq, H., Melnyk, O., Thomy, V. and Boukherroub, R. (2012) 'Affinity surface-assisted laser desorption/ionization mass spectrometry for peptide enrichment', *Analyst*, 137, 5527–32.
110. Rivillon, S., Chabal, Y. J., Webb, L. J., Michalak, D. L., Lewis, N. S., *et al.* (2005) 'Chlorination of hydrogen-terminated silicon (111) surfaces', *J. Vac. Sci. Technol. A*, 23, 1100–7.
111. Bansal, A., Li, X. L., Lauermann, I., Lewis, N. S., Yi, S. I. and Weinberg, W. H. (1996) 'Alkylation of Si surfaces using a two-step halogenation/Grignard route', *J. Am. Chem. Soc.*, 118, 7225–6.
112. Scheres, L., Giesbers, M. and Zuilhof, H. (2010) 'Self-assembly of organic monolayers onto hydrogen-terminated silicon: 1-Alkynes are better than 1-alkenes', *Langmuir*, 26, 10924–9.
113. Scheibal, Z. R., Xu, W., Audiffred, J. F., Henry, J. E. and Flake, J. C. (2008) 'Protein immobilization onto silicon nanowires via electrografting of hexynoic acid', *Electrochem. Solid-State Lett.*, 11, 81–4.
114. De Villeneuve, C. H., Pinson, J., Bernard, M. C. and Allongue, P. (1997) 'Electrochemical formation of close-packed phenyl layers on Si(111)', *J. Phys. Chem. B*, 101, 2415–20.
115. Allongue, P., de Villeneuve, C. H., Pinson, J., Ozanam, F., Chazalviel, J. N. and Wallart, X. (1998) 'Organic monolayers on Si(111) by electrochemical method', *Electrochim. Acta*, 43, 2791–8.
116. Allongue, P., de Villeneuve, C. H. and Pinson, J. (2000) 'Structural characterization of organic monolayers on Si(111) from capacitance measurements', *Electrochim. Acta*, 45, 3241–8.
117. Stewart, M. P., Maya, F., Kosynkin, D. V., Dirk, S. M., Stapleton, J. J., *et al.* (2004), 'Direct covalent grafting of conjugated molecules onto Si, GaAs, and Pd surfaces from aryldiazonium salts', *J. Am. Chem. Soc.*, 126, 370–8.
118. Haight, R., Sekaric, L., Afzali, A. and Newns, D. (2009) 'Controlling the electronic properties of silicon nanowires with functional molecular groups', *Nano Lett.*, 9, 3165–70.
119. Cha, T. W., Guo, A. and Zhu, X. Y. (2005) 'Enzymatic activity on a chip: the critical role of protein orientation', *Proteomics*, 5, 416–19.
120. Wacker, R., Schroder, H. and Niemeyer, C. M. (2004) 'Performance of antibody microarrays fabricated by either DNA-directed immobilization, direct spotting, or streptavidin–biotin attachment: a comparative study', *Anal. Biochem.*, 330, 281–7.
121. Zhu, H., Bilgin, M., Bangham, R., Hall, D., Casamayor, A., *et al.* (2001) 'Global analysis of protein activities using proteome chips', *Science*, 293, 2101–5.
122. Alonso, J. M., Reichel, A., Piehler, J. and del Campo, A. (2008) 'Photopatterned surfaces for site-specific and functional immobilization of proteins', *Langmuir*, 24, 448–57.
123. Bonroy, K., Frederix, F., Reekmans, G., Dewolf, E., De Palma, R., *et al.* (2006) 'Comparison of random and oriented immobilization of antibody fragments on mixed self-assembled monolayers', *J. Immunol. Methods*, 312, 167–81.
124. Scheibler, L., Dumy, P., Boncheva, M., Leufgen, K., Mathieu, H. J., *et al.* (1999) 'Functional molecular thin films: topological templates for the chemoselective ligation of antigenic peptides to self-assembled monolayers', *Angew. Chem. Int. Ed.*, 38, 696–9.

125. Dendane, N., Hoang, A., Defrancq, E., Vinet, F. and Dumy, P. (2008) 'Use of gamma-aminopropyl-coated glass surface for the patterning of oligonucleotides through oxime bond formation', *Bioorg. Med. Chem. Lett.*, 18, 2540–3.
126. Dendane, N., Hoang, A., Guillard, L., Defrancq, E., Vinet, F. and Dumy, P. (2007) 'Efficient surface patterning of oligonucleotides inside a glass capillary through oxime bond formation', *Bioconjugate Chem.*, 18, 671–6.
127. Dendane, N., Hoang, A., Renaudet, O., Vinet, F., Dumy, P. and Defrancq, E. (2008) 'Surface patterning of (bio)molecules onto the inner wall of fused-silica capillary tubes', *Lab Chip*, 8, 2161–3.
128. Melnyk, O., Duburcq, X., Olivier, C., Urbès, F., Auriault, C. and Gras-Masse, H. (2002) 'Peptide arrays for highly sensitive and specific antibody-binding fluorescence assays', *Bioconjugate Chem.*, 13, 713–20.
129. Duburcq, X., Olivier, C., Desmet, R., Halasa, M., Carion, O., *et al.* (2004) 'Polypeptide semicarbazide glass slide microarrays: characterization and comparison with amine slides in serodetection studies', *Bioconjugate Chem.*, 15, 317–25.
130. Olivier, C., Perzyna, A., Coffinier, Y., Grandidier, B., Stiévenard, D., *et al.* (2006) 'Detecting the chemoselective ligation of peptides to silicon with the use of cobalt-carbonyl labels', *Langmuir*, 22, 7059–65.
131. Kohn, M., Wacker, R., Peters, C., Schroder, H., Soulère, L., *et al.* (2003) 'Staudinger ligation: a new immobilization strategy for the preparation of small-molecule arrays', *Angew. Chem. Int. Ed.*, 42, 5830–4.
132. Soellner, M. B., Dickson, K. A., Nilsson, B. L. and Raines, R. T. (2003) 'Site specific protein immobilization by Staudinger ligation', *J. Am. Chem. Soc.*, 125, 11790–1.
133. Houseman, B. T., Huh, J. H., Kron, S. J. and Mrksich, M. (2002) 'Peptide chips for the evaluation of protein kinase activity', *Nature Biotechnol.*, 20, 270–4.
134. Devaraj, N. K., Dinolfo, P. H., Chidsey, C. E. D. and Collman, J. P. (2006) 'Selective functionalization of independently addressable microelectrodes by electrochemical activation and deactivation of a coupling catalyst', *J. Am. Chem. Soc.*, 128, 1794–5.
135. Lesaicherre, M. L., Uttamchandani, M., Chen, G. Y. J. and Yao, S. Q. (2002) 'Developing site-specific immobilization strategies of peptides in a microarray', *Bioorg. Med. Chem. Lett.*, 12, 2079–83.
136. Uttamchandani, M., Chan, E. W. S., Chen, G. Y. J. and Yao, S. Q. (2003) 'Combinatorial peptide microarrays for the rapid determination of kinase specificity', *Bioorg. Med. Chem. Lett.*, 13, 2997–3000.
137. Girish, A., Sun, H., Yeo, D. S. Y., Chen, G. Y. J., Chua, T.-K. and Yao, S. Q. (2005) 'Site-specific immobilization of proteins in a microarray using intein-mediated protein splicing', *Bioorg. Med. Chem. Lett.*, 15, 2447–51.
138. Camarero, J. A., Kwon, Y. and Coleman, M. A. (2004) 'Chemoselective attachment of biologically active proteins to surfaces by expressed protein ligation and its application for 'protein chip' fabrication', *J. Am. Chem. Soc.*, 126, 14730–1.
139. Wojtyk, J. T. C., Morin, K. A., Boukherroub, R. and Wayner, D. D. M. (2002) 'Modification of porous silicon surfaces with activated ester monolayers', *Langmuir*, 18, 6081–7.
140. Dendane, N., Melnyk, O., Xu, T., Grandidier, B., Boukherroub, R., *et al.* (2012) 'Direct characterization of native chemical ligation of peptides on silicon nanowires', *Langmuir*, 28, 13336–44.
141. Ciampi, S., Böcking, T., Kilian, K. A., James, M., Harper, J. B. and Gooding, J. J. (2007) 'Functionalization of acetylene-terminated monolayers on Si(100) surfaces: A click chemistry approach', *Langmuir*, 23, 9320–9.

142. Ciampi, S., Böcking, T., Kilian, K. A., Harper, J. B. and Gooding, J. J. (2008) 'Click chemistry in mesoporous materials: functionalization of porous silicon rugate filters', *Langmuir*, 24, 5888–92.
143. Jewett, J. C. and Bertozzi, C. R. (2010) 'Cu-free click cycloaddition reactions in chemical biology', *Chem. Soc. Rev.*, 39, 1272–9.
144. Zhou, C. and Walker, A. V. (2010) 'Formation of multilayer ultrathin assemblies using chemical lithography', *Langmuir*, 26, 8441–9.
145. Henriksson, A., Friedbacher, G. and Hoffmann, H. (2011) 'Surface modification of silicon nanowires via copper-free click chemistry', *Langmuir*, 27, 7345–8.
146. Bãnuls, M. J., Puchades, R. and Maquiera Á. (2013) 'Chemical surface modifications for the development of silicon-based label-free integrated optical (IO) biosensors: A review', *Anal. Chim. Acta*, 777, 1–16.
147. Böcking, T., Kilian, K. A., Hanley, T., Ilyas, S., Gaus, K., *et al.* (2005) 'Formation of tetra(ethylene oxide) terminated Si-C linked monolayers and their derivatization with glycine: An example of a generic strategy for the immobilization of biomolecules on silicon', *Langmuir*, 21, 10522–9.
148. Böcking, T., Kilian, K. A., Gaus, K. and Gooding, J. J. (2006) 'Single-step DNA immobilization on antifouling self-assembled monolayers covalently bound to silicon (111)', *Langmuir*, 22, 3494–6.
149. Lee S. W. and Laibinis, P.E. (1998) 'Protein-resistant coatings for glass and metal oxide surfaces derived from oligo(ethylene glycol)-terminated alkyltrichlorosilanes', *Biomaterials*, 19, 1669–75.
150. Lasseter, T. L., Clare, B. H., Abbott, N. L. and Hamers, R. J. (2004) 'Covalently modified silicon and diamond surfaces: resistance to nonspecific protein adsorption and optimization for biosensing', *J. Am. Chem. Soc.*, 126, 10220–1.
151. Voicu, R., Boukherroub, R., Bartzoka, V., Ward, T., Wojtyk, J. T. C. and Wayner, D. D. M. (2004) 'Formation, characterization, and chemistry of undecanoic acid-terminated silicon surfaces: patterning and immobilization of DNA', *Langmuir*, 20, 11713–20.
152. Lasic, D. D. and Needham, D. (1995) 'The stealth liposome: a prototypical biomaterial', *Chem. Rev.*, 2601–28.
153. Niidome, T., Yamagata, M., Okamoto, Y., Akiyama, Y., Takahashi, H., *et al.* (2006) 'PEG-modified gold nanorods with a stealth character for in vivo applications', *J. Controlled Release*, 114, 343–7.
154. Liu, Z., Cai, W., He, L., Nakayama, N., Chen, K., *et al.* (2007) 'In vivo biodistribution and highly efficient tumour targeting of carbon nanotubes in mice', *Nature Nanotechnol.*, 2, 47–52.
155. Uskokovi, V., Lee, P. P., Walsh, L. A., Fischer, K. E. and Desai, T. A. (2012) 'The effect of PEGylation on particle-epithelium interactions', *Biomaterials*, 33, 1663–72.

4

Biocompatibility of semiconducting silicon nanowires

L. MARCON and R. BOUKHERROUB,
CNRS and University of Lille1, France

DOI: 10.1533/9780857097712.1.62

Abstract: Silicon nanowires (SiNW) have been shown to be potential candidates for biological applications such as drug delivery systems, *in vivo* imaging agents and biosensors. However, concerns have been raised over adverse effects that SiNW may exert on biological systems. The first objective of this chapter is to offer a general review of the studies on the biocompatibility of SiNW *in vitro* and *in vivo*. The second objective is then to discuss the relevance of published data and present trends in the toxicity studies of SiNW in relation to their possible future applications.

Key words: silicon nanowires, biological applications, *in vitro* and *in vivo* models, biocompatibility assessment.

4.1 Introduction

In the late 1990s, the field of silicon nanowires (SiNW) underwent a significant expansion as developments moved toward biological applications. Several subfields emerged, including interfacing SiNW within living cells, *in vivo* nanotoxicity studies, drug delivery and synthetic bone coatings. In contrast to the efforts aimed at exploiting remarkable properties of SiNW for biomedical applications are also the attempts to evaluate potential adverse effects of SiNW on biological systems. With the ongoing development of nanotechnology-based products, concerns are growing as to whether unintentional exposure to nanoparticles, nanotubes and other nanoscale components during manufacture might have unpredicted impacts when exposed to humans (Nel *et al.*, 2006). Do nanowires induce side effects in live cells after short and long exposure? How do different cell types or organs deal with nanowire administration? These are still pending questions.

Currently, there is a common assumption that the small size of nanostructures allows them to easily cross tissues and cell organelles as their actual physical size is similar to that of many biomolecules. Previous studies suggested that nanoparticles are not inherently benign and that they affect biological phenomena at the cellular, subcellular and protein levels. For instance, inhaled or instilled ultrafine particles (<100 nm) can induce pulmonary inflammation, oxidative stress and distal organ involvement (Oberdörster, 2010). In a similar fashion, asbestos fibres, with structure similar to the structure of SiNW, induce oxidative injury, inflammation, fibrosis, cytotoxicity and mediator release from lung cells (Oberdörster, 2010).

In parallel, there have been numerous investigations of the cytotoxicity of carbon-based nanowires and nanotubes. Oxidative stress, reduced respiratory activity and enzyme inhibition are among the main risks mentioned. These data have raised questions about the occupational health and safety risks of nanowires as human exposure to nanotechnology-based materials dramatically increases. An evaluation of their potential toxicity is thus needed urgently.

The main characteristic of SiNW is precisely their size, which falls in the transitional zone between nanoparticles and the corresponding bulk materials (Oberdörster, 2010). Toxicity may be greater than a similar bulk material because the surface area-to-volume ratio is much higher. This, along other physicochemical parameters, can facilitate movement across cellular and intracellular barriers, interact with subcellular structures and finally increase pathological and physiological responses including inflammation, fibrosis and genotoxicity. For example, when taken up via the respiratory tract, nanoparticles below 100 nm are expected to be translocated to secondary target organs and to affect the genetic material integrity of cells, whereas larger particles above 500 nm should not. Additionally, the degradation and/or breakdown of these structures could lead to unique toxic effects that are difficult to predict.

This specific intermediary state raises several questions: if SiNW can be degraded and broken up, what is their persistence within the body? What could possibly be the behaviour of SiNW in the presence of cultured cells or *in vivo*? What is, in sum, their biocompatibility, that is the extent to which the material does not have toxic or injurious effects on biological systems?

In the past decade, several studies have tried to answer this question by investigating the cytotoxicity of SiNW and their potential adverse effects on biological systems. Typically, biocompatibility studies require both *in vitro* and *in vivo* experiments in order to test the local and systemic effects of a biomaterial on culture cells, tissue sections and the whole body. In this chapter, we summarize the studies conducted so far and discuss the essential data obtained. Finally, an analysis of the possible future trends in the field of SiNW biocompatibility is outlined.

Prior to any discussion, it is essential to determine which type of structures will be encompassed in this chapter. Some researchers consider that nanomaterials are structures with at least one dimension of 100 nm or less. Other researchers fixed that, to be truly nano, the relevant length scale must be small enough for its properties and behaviour to be different from those observed in the bulk. In our case, the specific term of nanowire will be restrained to structures having a diameter below 400 nm, as such dimensions are typically obtained using standard techniques (focused ion beam, chemical vapour deposition and chemical etching) and represent 90% of the SiNW produced in laboratories.

4.2 *In vitro* biocompatibility of silicon nanowires (SiNWs)

In vitro model systems provide a fast and reliable means to evaluate SiNW for a variety of toxicological endpoints. Because of the expense of *in vivo* experiments, along with public and governmental incentives to develop alternatives to animal models, *in vitro* assays are often more attractive for the preliminary testing of nanomaterials. Advantages of *in vitro* systems include identification of primary response of target cells, determination of the mechanisms of toxicity, cost-effectiveness, reduced inter-experiment variability and efficiency, and definition of starting points for the rational design of *in vivo* studies. Other advantages are detailed in a review by Arora *et al.* (2012).

In this regard, there are no clear guidelines by the regulatory agencies on the evaluation of nanomaterials. For instance, the US Food and Drug Administration (FDA) relies on International Organization for Standardization (ISO) 10993 standards for biological evaluation of medical devices prior to clinical testing, although it recommends additional steps (Gad, 2005). Regulatory aspects on the evaluation of SiNW for biomedical applications will be omitted hereafter as this is beyond the scope of this chapter.

To perform these tests, mammalian cells are typically used, usually of mouse or human origins, and cultured in flasks using nutrient culture media until they reach approximately 80% confluence. Cells are then exposed to the biomaterial for at least 24 hours before proceeding to the determination of biocompatibility parameters. Two experimental parameters of biocompatibility considered as paramount are (1) cytocompatibility, to measure the qualitative and quantitative aspect of the impact of the biomaterial with regard to the viability of cultured cells, and (2) genotoxicity, that is deleterious action on a given cell's genetic material. A relatively large number of *in vitro* studies have been performed since 2005. As presented below, all have stressed the cytocompatibility aspects of SiNW, either in a general manner (see Section 4.2.1) or by using focused models (sections 4.2.2 and 4.2.3).

4.2.1 Cytotoxicity

Cytotoxicity is determined by qualitative means, such as a microscopic monitoring to observe morphological changes, cell lysis and detachment, and quantitative means, including analysis of cell death, cell proliferation and cell viability.

Traditional assays measure cytotoxicity using either an end-stage event (e.g. permeability of cytoplasmic membranes) or metabolic parameters such as cell division or enzymatic reactions. Cell viability assays based on cellular metabolism include proliferative assays (tetrazolium MTT test, Alamar blue), apoptosis assays (caspase activity and Annexin V assays), and death assays (trypan blue assay, lactate dehydrogenase LDH assay). These general tests may be accompanied by more focused tests such as oxidative stress assays, for example the

2,7-dichlorofluorescein diacetate (DCFH-DA) assay, or inflammatory assays via the enzyme-linked immunosorbent assay (ELISA) to test inflammatory markers (e.g. interleukin-8 or TNF-α). Although fragmentary, *in vitro* cytotoxicity studies performed hitherto mostly focused on vertical SiNW bound to a substrate in the presence of a wide range of cell lines (Table 4.1).

The first attempt to interface living cells with an array of vertically aligned SiNW was demonstrated by Han *et al.* (2005). Si nanoneedles of 200 nm in diameter were constructed by focused ion beam etching of Si AFM cantilevers. DNA was then attached to the Si tip using a biotin-streptavidin linker, and the assembly was inserted into and retracted from a single human embryonic kidney HEK293 cell up to ten times. Microscopy observations revealed that the cell apparently suffered no obvious ill effects and was reported to divide normally.

Kim *et al.* (2007) provided more insight by culturing mouse embryonic stem (mES) cells and human embryonic kidney HEK293T (expressing the SV40 large T antigen) cells on vertically aligned SiNW (diameter ~ 90 nm, length ~ 6 μm). The cells were penetrated by SiNW within 1 hour, with an average of two to three nanowires per cell. After 48 hours, propidium iodide (PI) was used for identifying dead cells. PI is a membrane impermeant DNA intercalating agent that is excluded from viable cells. This test showed that 78% of cells remained viable. However, cell viability was strongly affected by the diameter of the wires. Cell death occurred within 1 day on 400 nm wires and after 5 days on 30 nm wires.

In contrast, Yang's group (Qi *et al.*, 2009) observed a higher viability (92%) of human hepatocellular carcinoma HepG2 cells and human hepatic LX-2 cells with longer SiNW (L ~ 20 μm) of similar diameter. Experimentally, the authors used an Alamar blue viability assay in which the reagent works as a cell health indicator by using the reducing environment in the cytosol of living cells. Cells mostly attached to the top of the NW with cell filopodia holding the NW underneath, and exhibited a compacted shape (Fig. 4.1), although change in the NW density did not affect the cell behaviour. Using a cell centrifugation method, they demonstrated that the adhesion forces between cells and SiNW were stronger than between cells and a reference silicon wafer. Interestingly, the authors attempted to gain insight into the mechanisms governing the NW/cell interaction. They quantified via RT-PCR the expression level of genes known to be related to cell adhesion process and cellular components, such as integrin, focal adhesion kinase (FAK), type I collagen (Col I) and α-actin. Col I and α-actin were down-regulated, possibly explaining the restricted spreading behaviour of cells.

Park and co-workers (Shalek *et al.*, 2010) complemented these data by demonstrating spatially localized delivery of biomolecules into human cervical cancer HeLa cells and primary mammalian cells, such as human fibroblasts, rat neural progenitor cells (NPCs) and rat hippocampal neurons, using aminated vertical SiNW (diameter ~ 100 nm, length ~ 1 μm). A kinetic study showed that cell penetration by NWs was complete after 1 hour, regardless of the NW density. The authors reported that some cell dysfunctions, such as slowed cell division and

Table 4.1 Summary of *in vitro* biocompatibility studies of silicon nanowires

SiNW formulation	Techniques	Cell line	Observations	Application	Author
Single SiNW coated with plasmid on a Si AFM tip Ø 200 nm L 6 μm	Confocal microscopy	Human embryonic kidney (HEK293) cells	Normal proliferation	Plasmid delivery	Han (2005)
Vertical SiNW on Si(111) wafer Ø 90–400 nm L 6 μm	Confocal microscopy and SEM PI	Mouse embryonic stem (mES) cells HEK293	Ø 90 nm: 22% of necrotic cells after 48 hours Ø 400 nm: death <24 hours Ø 30 nm: alive>5 days	Plasmid delivery	Kim (2007)
Vertical SiNW on P(100) Si wafer Ø 100 nm L 20 μm	Alamar blue assay RT-PCR for expression profile of adhesion-specific genes	HepG2 human hepatocellular carcinoma LX-2 human hepatic cells	92% viability after 18 hours and 48 hours	Cell adhesion and spreading study	Qi (2009)
Aminated vertical NW coated with biomolecules on Si(111) wafer Ø 90 nm L 1 μm	Confocal microscopy Q-PCR for analysis of 5 housekeeping genes Annexin V binding assay Trypan blue	Rat hippocampal neurons Hela cells Human fibroblasts Rat neural progenitor cells	Growth over weeks 10–20% apoptotic cells after 7 hours Initial cell perturbation within a few hours	Biomolecule delivery (siRNA, DNA, peptides, proteins)	Shalek (2010)
Vertical *n*-type and undoped SiNW on Si(100) wafer No dimensions provided, no control over the diameter and orientation	MTT assay SEM	L929 mouse fibroblasts	80% viability after 48 hours Proliferation similar to a Si wafer Flattened shape on doped NW	Biocompatibility study	Garipcan (2010)

Vertical SiNW on Si(100) wafer with superhydrophilic/ superhydrophobic contrasts Ø 20–80 nm L 2.5 μm	SEM and TEM PI	Chinese hamster ovary (CHO) cells	2% of necrotic cells after 48 hours	Biocompatibility study	Piret (2011)
Porous vertical SiNW with antibiotic loading on Si(111) wafer Ø 10–40 nm L 1–3 μm	SEM	MC3T3-E1 mouse fibroblasts	Inhibition of cell adhesion and spreading	Antibiotic delivery	Brammer (2009)
Free-standing SiNW 0–100 μg/mL Ø 20–30 nm L 100 nm–1 μm	Alamar blue assay Cell cycle phase analysis RT-PCR for expression profile of adhesion-specific genes	HepG2 human hepatocellular carcinoma Cancer and normal cells	Dose-dependent cytotoxicity 90% viability at 25 μg/mL after 48 hours	Cell adhesion and spreading study	Qi (2007)
Free-standing PEGylated SiNW 0–60 μg/mL decorated with 500 AuNP of 9 nmØ per nanowire Ø 150 nm L 2.6 μm	MTT	Human oral squamous carcinoma KB cells	Dose-dependence cytotoxicity 60% cell viability at 60 μg/mL after 24 hours Increased viability with the PEG coating	SiNW-based hyperthermia agent	Su (2012)

Abbreviations: Ø, nanowire diameter; L, nanowire length; AFM, atomic force microscopy; SEM, scanning electron microscopy; PI, propidium iodide; RT-PCR, reverse transcription polymerase chain reaction; Q-PCR, quantitative PCR; MTT, 3-(4,5-dimethylthiazol-2-yl)-2,5-diphenyltetrazolium bromide; TEM, transmission electron microscopy; PEG, poly(ethylene glycol).

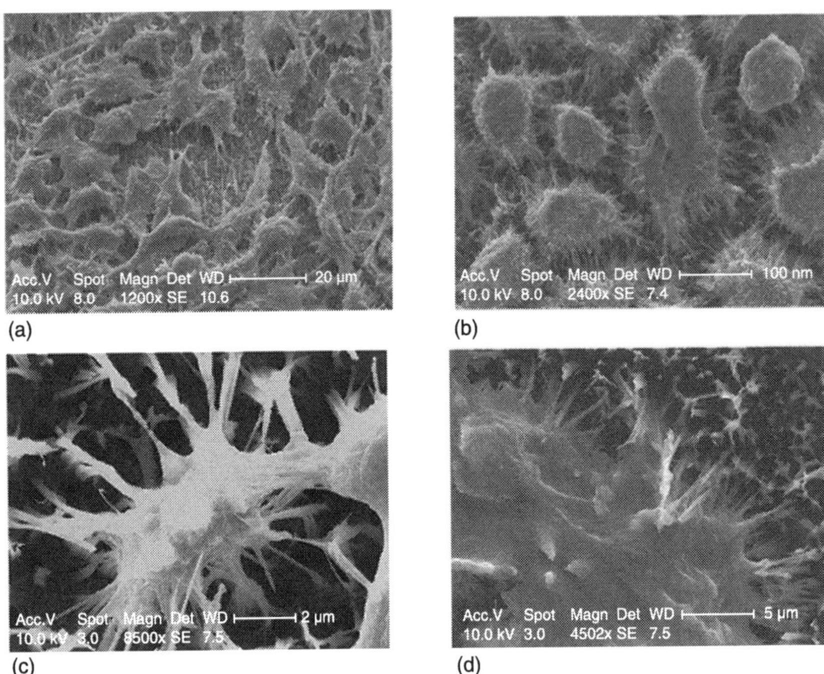

4.1 Scanning electron microscope images of (a, c) LX-2 cells and (b, d) HepG2 cells cultured on SiNW arrays. Reprinted from Qi *et al.* (2009) with permission of ACS Publications.

constrained morphology, were detected within a few hours because of NW impalement, and then returned to normal. An Annexin V binding assay, aimed at detecting cells expressing phosphatidylserine (an event found in apoptosis) was then performed following the deposition on the NW. Approximately 10% to 20% of apoptotic cells were observed. In parallel, a trypan blue dye staining, capable of traversing the membrane of dead cells only, suggests that the membrane integrity was not affected after 24 hours; however, it is regrettable that the assay was not quantified.

These observations were complemented by another study (Garipcan *et al.*, 2010) that investigated *in vitro* biocompatibility of undoped and *n*-type SiNW on L929 mouse fibroblasts. A 3-(4,5-dimethylthiazol-2-yl)-2,5-diphenyltetrazolium bromide (MTT) viability assay showed that circa 80% of the cells remained viable on both NW types. Cells had a normal morphology, although the authors noticed a more flattened morphology on doped SiNW. This change was attributed to the increased charge density susceptible to alter the charge distribution of cell surfaces. It was also noticed that a low nanowire density provided fewer interaction points to the cells, thus causing this 'flattening' effect. Unfortunately, the authors did not describe the nanowire dimensions nor their density and orientation.

More recently, Piret *et al.* (2011) cultured Chinese hamster ovary (CHO) cells on SiNW arrays exposing superhydrophilic/superhydrophobic contrasts, depending on the chemical treatment used. It was found that superhydrophilic, that is hydroxylated, areas favoured cell adhesion, with a cell viability of 98% after 48 hours, whereas superhydrophobic areas derivatized with octadecyl groups prevented cell adhesion (Fig. 4.2). These results clearly underline the importance of the chemical functionalization of the NW over cell behaviour. More importantly, the authors noticed that superhydrophilic SiNW were subjected to chemical dissolution in contact with cell culture medium while the superhydrophobic SiNW remained intact.

This phenomenon is typically observed in porous silicon-based materials (Anglin *et al.*, 2008) produced by an electrochemical etching process. Anderson *et al.* (2003) noticed that porous Si films in physiological conditions released $Si(OH)_4$ that is considered metabolically tolerant *in vivo* and promotes calcification on Si. For instance, Brammer *et al.* (2009) synthesized porous SiNW arrays (diameter ~ 10–40 nm, length ~ 1-3 μm) using electroless etching for mediating model antibiotic delivery. The authors observed the erosion of the NW after 42 days in simulated body fluid. Surprisingly, it was found that the porous SiNW had the ability to prevent the adhesion of MC3T3-E1 mouse fibroblasts inducing a cellular 'ball up' effect. Previous work has demonstrated that Si is susceptible to dissolve in biological fluids depending on the salinity, pH and enzymatic activity (Jarvis *et al.*, 2012). The electrochemical process and/or the thermal oxidation induce a strain, thus explaining the mechanical instability of porous Si. Whether this effect is partially present or not in non-porous SiNW is still a pending question.

In parallel, a few studies involved free-standing SiNW that, unlike surface-fixed wires, risk the formation of cytotoxic aggregates or accumulation in cell

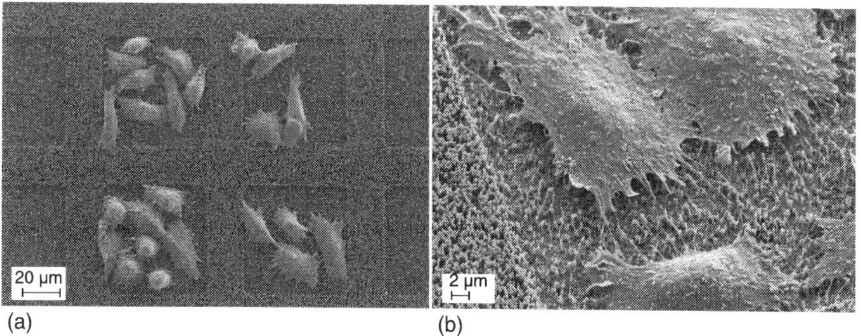

4.2 (a) Top view and (b) tilted view scanning electron microscope images of Chinese hamster ovary cells trapped within the superhydrophilic silicon nanowire patterns. Adapted from Piret *et al.* (2011) with permission of RSC Publishing.

organelles. Yang's group (Qi *et al.*, 2007) studied the toxicity of SiNW (diameter ~ 20–30 nm, length ~ 1 μm) on human hepatocellular carcinoma cells (HepG2) and normal cell lines. Cell viability started to decrease in a dose-dependent manner from 0.1 μg mL^{-1} on, with a 10% decrease at 25 μg mL^{-1}. Flow cytometry experiments did not reveal noticeable changes in the cell cycle. Treatment of cells with SiNW suspensions resulted in a decrease of the adherence of the cells and in changes of the adhesion and spreading morphologies for a compact and round shape. These changes were associated with the down-regulation of adhesion-associated genes (focal adhesion kinase and type I collagen). In contrast, α-actin and type III collagen expression profiles remained unchanged. As a reminder, the integrin family of extracellular matrix receptors regulates cell adhesion and migration. These processes in turn narrowly depend on organization of the actin cytoskeleton, which is one of the most important targets of integrin-mediated signalling. The authors hypothesized that the cell viability decrease was related to the reduced adhesion ability of the cells on NW rather than a potential influence on cell cycle. They also suggest that SiNW might affect HepG2 cell adhesion and spreading through the integrin-type I collagen pathway, rather than the actin-mediated pathway. Such a hypothesis is premature in light of the very low number of genes considered in this study and the complex pathways regulating integrin–actin interactions (for a review, see Wiesner *et al.*, 2005).

Su *et al.* further explored the potential of free-standing SiNW decorated with gold nanoparticles (AuNP-SiNW) as efficient hyperthermia agents for cancer cell destruction (Su *et al.*, 2012). They showed that human oral squamous carcinoma (KB) cell viability was reduced down to 75% when the AuNP-SiNW concentration reached 10 μg/mL. In contrast, cells retained 80% of viability when the AuNP-SiNW structures were PEGylated, thus underlining the impact of the coating on the biocompatibility of SiNW.

4.2.2 Osseointegration

Bone consists mainly of a fibrous organic matrix composed of 90% type 1 collagen (Col 1) and an inorganic matrix composed mainly of hydroxyapatite (HA). In this assembly, osteoblasts synthesize Col 1 and alkaline phosphatase (ALP) as well as other organic and inorganic components of the bone extracellular matrix, while osteoclasts are responsible for bone breakdown and resorption. These narrowly coupled processes are involved in the normal remodelling of bone. Together, Col 1 and HA form a nanotextured surface that is responsible for further promoting osteoblast function and for the mechanical properties of bone (Bilezikian *et al.*, 2008).

The goal of artificial implants is to serve as a scaffold on which tissue development can be initiated and accelerated. The lack of osseointegration, that is the bonding between the implant and the bone, is a clinical complication that leads to short lifespans of orthopaedic implants. Novel approaches to the generation of implant materials take advantage of the unique properties of nanomaterials. As a result of their large surface area, nanowires can provide additional contact points between the implant and the bone, improving the chances for osseointegration.

Ideally, SiNW-based implants for bone regeneration should promote early mineralization and support new bone formation. Consequently, molecules indicating different aspects of the process, from proliferation (alkaline phosphatase ALP), matrix formation (A2 pro-collagen type 1) to mineralization (osteocalcin, bone morphogenic protein-4) are used as markers, along general cytocompatibility assays, to determine the effectiveness of an implant (Table 4.2).

Nagesha *et al.* (2005) were the first to exploit the semiconducting properties of SiNW as a matrix to promote calcification in simulated body fluid (SBF). This fluid has inorganic ion concentrations similar to those of human blood plasma and is typically used for the evaluation of bioactivity of artificial materials *in vitro* under biomimetic conditions. Sb-doped SiNW grown on Si(100) substrates were immersed in SBF under the influence of an electrical bias (1.1 mA cm^{-2}) for 3 hours. Calcium phosphate formation was then monitored by electronic microscopy and energy dispersive X-ray (EDX) spectroscopy; small clusters of 0.64 μm were observed while, in absence of bias, only residual sodium chloride was detected. These data demonstrate that SiNW are theoretically capable of facilitating the growth of synthetic bone coatings on their surface.

More recently, Desai's group (Popat *et al.*, 2006) presented a breakthrough study that set the stage for future orthopaedic applications. The authors evaluated the response of human fetal osteoblasts (hFOB 1.19) cultured on SiNW immobilized on silica substrates. Parameters such as the length of the NW and the NW density were varied to find optimal conditions for cell adhesion and proliferation. Total protein content, alkaline phosphatase (ALP) activity and matrix production were also quantified. High density-long NW (10 NW/μm^2, length ~ 20 μm) supported the highest cell proliferation and viability over 4 days and showed better ALP activity and calcification after 4 weeks compared with the other surfaces. In addition, osteoblasts followed a natural differentiation behaviour. Although very encouraging, this study lacks comparative results with positive controls such as cells cultured in a Petri dish or cells deposited on hydroxyapatite surfaces. Additionally, the use of a relevant control such as titanium, which is widely used in medical implants, would have been appropriate to assess globally the efficiency of the investigated surfaces.

4.2.3 Haemocompatibility

A haemocompatible material must not adversely interact with any blood components. Most nano-objects feature large surface areas, and after entering into the body will probably easily adsorb blood components such as proteins and glycoproteins. As the blood circulation system is the most likely first port of entry

Table 4.2 Summary of *in vitro* osseointegration properties of silicon nanowires

SiNW formulation	Cell lines	Techniques	Observations	Aim	Author
Vertical Sb-doped SiNW Ø 160 nm L up to 100 μm	Human kidney fibroblasts (HEK293)	Calcification assay Cell proliferation observation EDX spectroscopy Optical microscopy	Normal proliferation	Calcification study under an electrical bias	Nagesha (2005)
Vertical SiNW L <1 μm and 20 μm Ø 40 nm Densities from 1 to 10 NW/μm^2	Human fetal osteoblasts (hFOB1.19)	MTT assay BCA assay ALP activity Calcium determination assay and XPS SEM	Improved properties for high density/long NW	Osteoblast adhesion and proliferation on SiNW	Popat (2006)*

Abbreviations: EDX, energy-dispersive X-ray spectroscopy; BCA, bicinchoninic acid protein assay; ALP, alkaline phosphatase; XPS, X-ray photoelectron spectroscopy.

*Statistical analysis provided.

for biomaterials once the human body is exposed during medical application, it is crucial to characterize the behaviour of nanoscale objects in contact with blood.

Indeed, an adverse interaction would typically lead to inappropriate activation or destruction of blood components. Among them, platelets (or thrombocytes) are blood cells that play an important role in blood clotting, granting protection for vascular bed integrity and sealing of blood leaks. In certain conditions, this primary function might lead to thrombosis, happening when a blood clot obstructs a vein or an artery. Experimentally, haemocompatibility tests are typically related to several fundamental aspects of blood physiology such as coagulation, platelet status, haemolysis and activation of innate immunity complement proteins (Seyfert *et al.*, 2002). If coagulation and platelet status in response to SiNW have been investigated so far, the other categories are definitely missing (Table 4.3).

Garipcan *et al.* (2010) performed a series of coagulation tests to determine the clotting properties of blood in presence of undoped and doped SiNW grown on Si wafer. The tests performed using human plasma showed that SiNW have slightly higher coagulation times than tissue culture flasks. The results were not expressed using the international normalized ratio (INR), used clinically to standardize results, and should be considered as exploratory data. In addition, SiNW showed a 20-times lower fibrinogen adsorption ($\sim 115 \pm 0.3$ mg mL^{-1} versus 325.4 ± 0.8 mg mL^{-1}). Fibrinogen is a plasma glycoprotein that plays an important role in coagulation and promotes platelet adhesion. An over-adsorption of fibrinogen can potentially accelerate platelet adhesion and result in thrombosis, explaining why a low fibrinogen binding is desirable.

Table 4.3 Summary of haemocompatibility properties of silicon nanowires

Formulation	Techniques	Viability and observations	Author
Vertical *n*-type and undoped SiNW on Si (100) wafer. No dimensions provided, no control over the diameter and orientation.	Prothrombin time test. Activated partial thromboplastin time test. Thrombin time test. Fibrinogen adsorption.	Coagulation times similar to human plasma alone. Low fibrinogen adsorption.	Garipcan (2010)
Vertical SiNW coated with PNIPAAm on Si wafer Ø 70 ± 4.3 nm L 25 ± 3.3 μm	Platelet adhesion test.	Lower platelet-adhesion with PNIPAAm than Si wafer.	Chen (2009)

Abbreviations: PNIPAAm, poly(N-isopropylacrylamide).

Chen *et al.* (2009) demonstrated that SiNW arrays coated with the thermally responsive polymer poly(*N*-isopropylacrylamide) (PNIPAAm) showed reduced platelet adhesion *in vitro*. PNIPAAm is a thermoresponsive polymer that exposes an elongated hydrophilic chain conformation below its lower critical solution temperature (LCST, 32 °C) and undergoes a transition to hydrophobic aggregates above the LCST. Such transitions can be used to release drugs entrapped in the polymer. Unfortunately, previous work had demonstrated that platelets adhere to the PNIPAAm surface above the LCST, while platelet adhesion is inhibited below the LCST. This is a problem as at body temperature, the polymer would cause accumulation of platelets and thus cause blood coagulation and thrombosis. To circumvent this issue, the authors investigated the impact of nanostructures such as SiNW on platelet adhesion. It turns out that PNIPAAm-decorated SiNW significantly reduced platelet adhesion regardless of the temperature, in contrast to substrates without SiNW and/or PNIPAAm. Finally, contact angle and adhesive force measurements revealed that the nanoscale topography retained a high content of water on the polymer surface, thus inducing reduction of platelet adhesion.

4.3 *In vivo* biocompatibility of SiNWs

As simplified models, *in vitro* testings do not exhibit the pathogenic effects that are likely to be observed *in vivo*. Phenomena such as inflammation, issues of translocation and coordinated tissue responses are absent and need to be taken into account. The expected behaviour of nanostructures (NS) *in vivo* can be summarized as follows (Fischer and Chan, 2007): (1) NS can enter the body via six principal routes: intravenous, dermal, subcutaneous, inhalation, intraperitoneal and oral; (2) afterward they can translocate and reach other target organs and may remain intact, be modified or metabolized; (3) they enter the cells of the organ and reside there for an unknown amount of time before moving toward other organs or being excreted. Once NS have translocated to the circulatory system, they can be distributed throughout the body. The liver is the major distribution site, followed by the spleen as another organ of the reticulo-endothelial system. Distribution to heart, kidney and immune-modulating organs (spleen, bone marrow) has also been reported (Oberdörster, 2010). This distribution is often followed by rapid clearance, predominantly by action of the liver and splenic macrophages. Clearance and opsonization is directly affected by size and surface characteristics and will result in variations of clearance rates and macrophage sequestration. These multiple potential routes and parameters illustrate the difficulty in measuring adverse effects of SiNW on host systems. In this regard, *in vivo* toxicity studies performed hitherto are very fragmentary (Table 4.4).

Linsmeier *et al.* (2009) investigated the brain-tissue response to nanowire implantations as a neural interface on a rat model. To do so, the authors used GaP nanowires, known to be chemically non-biocompatible material, covered with SiO_x. The resulting suspension was implanted in rat striatum (3×10^5 NW/rat) forming

Table 4.4 Summary of *in vivo* biocompatibility studies of silicon nanowires

Formulation	Animal model	Administration mode	Techniques	Author
Free-standing GaP(111)B NW covered with 20 nm SiO_x Ø 100 nm L 2 μm	Adult female Sprague-Dawley rat	Bilateral injection in the striatum (3×10^5 NW in HBSS/rat) for 12 weeks	Immunohistochemical staining Fluorescence microscopy	Eriksson Linsmeier (2009)*
Free-standing PEGylated (900 Da) SiNW Ø 5 nm L 5 μm	BALB/c mouse	Tail injection (100 μL of 0.1–1 pM solutions in PBS) for 1 hour	Multimodal non-linear optical microscopy	Jung (2009)
Free-standing SiNW Ø 20–30 nm L 2–15 μm	Male Sprague-Dawley rats	Intratracheal instillation 10 → 250 μg for 91 days	Primary alveolar macrophages counting Measurement of albumin and lactate dehydrogenase activity Cytokines and chemokines quantification Measurement of the production of reactive oxidant species Histopathology Lung collagen determination Morphometric lung tissue analysis SiNW distribution and clearance	Roberts (2012)*

*Statistical analysis provided.

part of the basal ganglia of the brain. Quantification of pivotal cells in the brain, such as microglia, astrocytes and neurons, was done by immunohistochemistry over 12 weeks. Microglia are immune cells found only in the brain and spinal cord that can detect damaged neurons and infectious agents. Astrocytes are the most abundant cells in the brain and are essential for normal neurological function. Results showed that the implantation induced a large astrocyte and microglia response at the scar location during the first week, which then declined over time. The authors hypothesize that the astrocytes were initially activated to maintain a barrier that would prevent further NW diffusion in the brain. Additional confocal microscopy observations revealed that SiO_x-coated NW underwent a partial degradation after 6 weeks, and that most of them had been phagocytized by the microglia after 12 weeks. The data underline the relative fragility of the SiO_x coating and suggest that some NW were able to cross the blood-brain barrier and leave the brain. In conclusion, if no evidence of subacute or chronic NW toxicity was observed, this study suggests short- and long-term adverse effects requiring further investigation.

Jung *et al.* (2009) explored another portal of entry by performing intravital imaging of PEG-modified SiNW (diameter ~ 5 nm, length ~ 5 μm) on a mouse model using four-wave mixing (FWM) and third-harmonic generation (THG) emissions. These signals permitted real-time monitoring of the SiNW in the peripheral blood of a mouse after tail vein injection. The authors observed shorter circulation time than for various PEGylated carbon nanotubes (0.5 of an hour versus 1.5–15 hours). Such difference may originate from the low molecular weight of the PEG used and/or intrinsic material properties. The biodistribution was then studied at 1 hour post injection. It turns out that the SiNW circulated through liver and spleen and were absent from the kidney. It is likely that the SiNW were captured by the reticuloendothelial system and sent for clearance to the liver.

Roberts *et al.* (2012) assessed the potential toxicity following pulmonary exposure to SiNW (diameter ~ 20–30 nm, length ~ 2–15 μm) in a rat model. A dose-response time course study was conducted using rats that were intratracheally instilled with doses of SiNW ranging from 10 to 250 μg per rat (~0.04–1 mg/kg of body weight). Parameters of lung toxicity and disease, including lung injury, inflammation and fibrotic response, were evaluated from 1 to 91 days after exposure. Treatment resulted in dose-dependent increase in lung injury and temporary inflammation. As an expression of fibrosis (which is the formation of excess connective tissue in a reparative or reactive process), lung collagen was increased after 91 days at the maximum dose. Additionally, 70% of the SiNW were cleared in 28 days by alveolar macrophages (Fig. 4.3). Particle clearance mediated by alveolar macrophages is the most prevalent clearance mechanism observed in the alveolar region. As the efficiency of this process is size-dependent, we can hypothesize that the smallest (or non-aggregated) fraction of the SiNW interacted instead with epithelial cells or translocated to the interstitium (Oberdörster *et al.*, 2005).

Interestingly, in this same study, the authors compared their observations with an extensively studied nanomaterial, namely carbon nanotubes (CNT), and

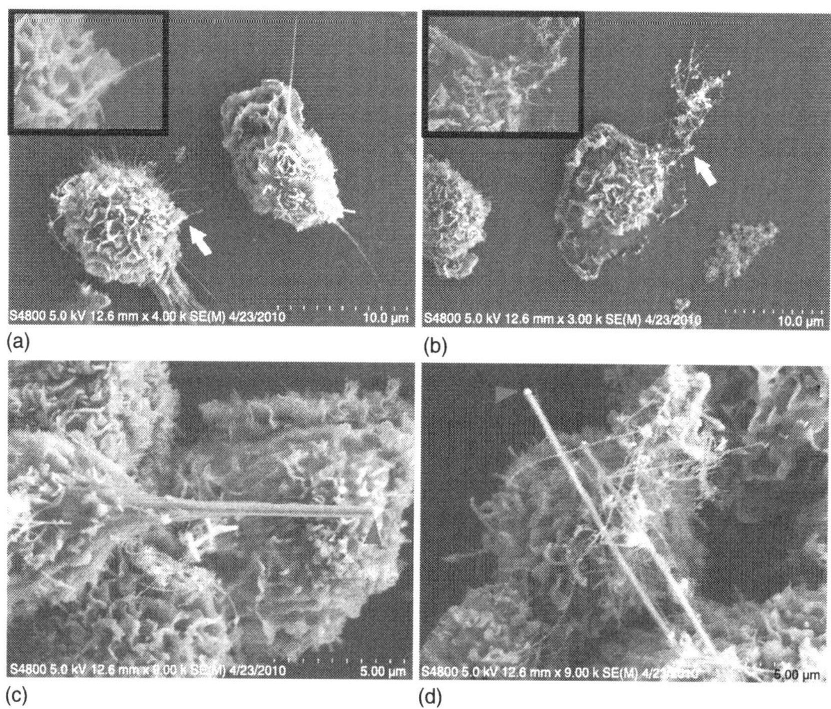

4.3 Scanning electron microscope images of lung tissue from rats intratracheally instilled with 100 µg free-standing silicon nanowire 1 (a), 7 (b) and 28 (c, d) days after exposure. Grey arrows in (c, d) highlight the nanowire enclosing by primary alveolar macrophages. Reprinted from Roberts *et al.* (2012) with permission of Hindawi Publishing.

noticed that inflammation resolved at earlier time points with SiNW (28 days versus 56 days). SiNW were also cleared faster from the lungs than CNT with similar dimensions. This trend was confirmed by measuring *ex vivo* the generation of reactive oxidant species (ROS) via chemiluminescence. It turns out that the phagocytes that appeared in the lungs following SiNW exposure produced ROS, inducing an early inflammation that regressed over time. It is noteworthy that the SiNW composition, that is the content of gold and SiO_2, apparently plays no critical role in the collagen increase.

4.4 Methodology issues

4.4.1 Improper material characterization

According to a myriad of studies and review articles, the biological impact and biokinetics of nanowires for *in vitro* and *in vivo* toxicity studies are dependent on the nanowire dimension and morphology, size distribution, nanowire composition, surface area and chemistry, and the local biological environment (e.g. blood or lymph). These parameters taken together can modify biomolecule binding, cellular uptake, translocation from portal of entry to the target site and the possibility of causing tissue inflammation (Nel *et al.*, 2006). These crucial factors need consequently a thorough characterization before undertaking any toxicity assessment. This point is even more blatant in the case of SiNW where the NW can exhibit a wide variety of shapes, variable degrees of porosity and different types of doping.

This pre-requisite has been highlighted by Murdock *et al.* (2008), who underlined the importance of developing an adequate set of protocols for the physico-chemical characterization of nanomaterials. Indeed, these properties are typically estimated using electron microscopy and X-ray photoelectron spectroscopy (XPS) in 'just received' conditions, that is as a powder or suspended in pure solvents, thus limiting the significance of the resulting experiments. For instance, recent toxicological studies concluded that nanowire surface area is a better dosemetric than mass or concentration for comparing the cytotoxicity of NW of different sizes and different types (Oberdörster, 2010). Additionally, a more consistent NW characterization and a classification scheme to support it would enable cross comparisons between the results obtained from different research groups.

As suggested by Cho *et al.* (2011), the state of aggregation is also a parameter typically underestimated for nanostructure biocompatibility studies, especially for *in vivo* tests where the SiNW are mostly used in a free-standing form. In typical *in vitro* experiments, cells are cultured at the bottom of a Petri dish and incubated with the NW. The NW are then assumed to be well-dispersed, resulting in a fixed concentration all over the culture medium. However, NW may aggregate and sediment because of early interactions with cell culture components, meaning that the material concentration may be locally higher and reflective of the bulk material properties. This in turn might affect the cellular uptake of the NW, which may give rise to erroneous and misleading data.

Concomitantly, determining the actual NW size in biological fluids *in vivo* is essential for the interpretation of toxicological results as size is a key factor that influences translocation across cell barriers and inhalation through the respiratory tract. When any biomaterial comes in contact with biological fluids, proteins and glycoproteins are, within seconds, the first biomolecules adsorbed onto the biomaterial's surface and may also be displaced by other proteins in a matter of a few minutes. Subsequent cellular events, including the biodispersion of NW across barriers and into target tissues and cells, are most likely mediated by this bio-corona rather than the material surface itself.

Recent studies have identified adsorption of different serum/plasma components on nanoparticles, their affinity and exchange rates so that their biological identity through the formation of the corona could be defined (Monopoli *et al.*, 2011). So far, this type of study has not been conducted for SiNW and is crucial to determine the aggregation state of the NW at the deposition site.

4.4.2 *Modus operandi* issues

Given the aforementioned issues, several elements directly related to the *modus operandi* are worth pointing out in order to facilitate the interpretation of SiNW biocompatibility data. First, before assessing the cytotoxic effects of SiNW on a given cell type, standard growth curve data should be collected to determine baseline growth properties of selected cells. When comparing cells with each other using this method, they can be classified according to their growth rates, which may help to explain the results of cytotoxicity experiments. The same principle could be applied to morphological evaluation of exposed cells by calculating the average length to width proportions relative to any control.

Of great importance is also the need for standard reference materials as a discriminatory tool for the validation of standardized *in vitro* and *in vivo* tests in evaluating the biocompatibility of SiNW. Similar to the work of Roberts *et al.* (2012), experimental results should ideally be compared with a nanomaterial of known toxicological properties and comparable dimensions, namely CNT. Even better would be carbon black and asbestos nanotubes that are often used as toxicity standards in the literature (e.g. Murr *et al.*, 2005).

Another example of inaccuracy is provided by the free-standing SiNW doses and administration modes, particularly if these doses are administered as a bolus (meaning in the form of a single large dose). As noticed for nanoparticles, the dose rate is often extremely high and it is delivered in a very short time (within a second) in contrast to realistic exposure, which may take several hours, days and even weeks to deliver the same dose (Oberdörster, 2010). It is common practice to dose primary cells *in vitro* without any discussion of realistic *in vivo* exposures. As observed in Table 4.1, a 100 μg/mL culture medium dose is very high and unlikely to resemble realistic *in vivo* exposures. What would be a realistic starting dose? Is the study dedicated to determine if SiNW can be used as a vector and, if so, what is the envisaged portal of entry? For instance, specific biomedical applications for drug delivery and diagnostic purposes will require intravenous, subcutaneous or intramuscular administration. In contrast, the study of unintentional exposure during manufacture of SiNW will require inhalation, ingestion and dermal administration modes. These questions encompass a much larger topic: what is the intended application? Indeed, the intended application will ultimately define the appropriate toxicological model to be used and the tests to be performed, especially as SiNW can be benign in certain cell types and induce negative effects in others.

4.5 Future trends

The research direction of interfacing SiNW and living cells is currently one of the most exciting topics and is rapidly developing. Many new discoveries are expected when such tools will be used to tackle real biological problems. Along the process, these discoveries will be accompanied by toxicity studies that will match the intended applications. Indeed, a careful review of the past studies highlights gaps between the reported experimental design and the envisaged applications. Cytotoxicity assays measure only finite effects on cells during the first 24 or 48 hours after exposure to SiNW. However, many biological reactions *in vivo* are not simply cytotoxic and are propagated beyond 48 hours. Examples are improper cells programmed apoptosis and necrosis caused by DNA damage, inflammatory and immune reactions (Oberdörster, 2010). For example, most *in vivo* tests are high, single-dose, acute studies and are of limited relevance to NW exposure, which should examine the impact of repeated low doses over a long period of time. There is consequently an urgent need for chronic toxicity data, especially if SiNW present occupational risks. One can predict that future trends in the use of SiNW will be paralleled with more oriented biocompatibility assessments. Below are some examples of what can be expected.

4.5.1 Lack of data about the bio-corona

As mentioned previously, the surface properties of SiNW, including surface composition, wettability, surface oxidation, morphology and portal of entry, may influence protein adsorption and subsequently the cellular responses. A number of recent studies have demonstrated that, despite the size of the nanostructures, they do not freely go into all biological systems but are instead governed by the functional molecules added to their surfaces (Lynch and Dawson, 2008). The same SiNW administered to the lung or intravenously will interact with different biological media and will be decorated with different secondary coatings, mostly proteins and glycoproteins. The adsorption behaviour of proteins is based on complex dynamic equilibria that directly impact the SiNW stability. For instance, Sabuncu *et al.* (2012) showed that serum components increase the dispersion quality of gold nanoparticles via steric effects. These biomolecules possess a wide range of affinities for SiNW surface, resulting in variable adsorption kinetics. Consequently, the composition of the bio-corona will be determined by the on and off rates of each molecule for the NW (Lynch and Dawson, 2008). For instance, proteins with high concentration and high association constants are expected to adsorb quickly on SiNW, but may then dissociate immediately to be substituted by proteins of lower concentration and lower exchange rate.

In response to these mechanisms, researchers claim that it is possible to engineer nanostructures to direct the intracellular or *in vivo* biodistribution but strategies for avoiding secondary behaviours of SiNW are lacking. The use of primary coatings, such as poly(ethylene glycol), provides evidence that SiNW can be made anti-fouling to avoid undesired serum protein coating and/or opsonization. Elaborating on the role played by specific sub-sets of serum proteins in a given biological environment once in contact with SiNW is thus of great importance to design biocompatible materials.

4.5.2 Genotoxicity profiling

Previous studies have shown that SiNW arrays have the ability to gain direct access to the cell interior and reach into the cytosol (Shalek *et al.*, 2010; Kim *et al.*, 2007). For many therapeutic applications, direct access into the cell interior is highly advantageous. Examples include targeted delivery of proteins, DNA and RNA, or electrical conduits for measuring and stimulating ion channel activity. On top of that, recent achievements were made in the development of SiNW for *in situ* single cell study. For instance, Lieber's group (Patolsky *et al.*, 2006) has integrated SiNW transistor arrays for the detection, stimulation and inhibition of neuronal signal propagation. Following this trend, Li *et al.* (2009) presented a SiNW-based system for quantifying mechanical behaviour of cell lines, revealing force information of cancer cells. Park's group (Robinson *et al.*, 2012) went further by intracellularly recording and stimulating neuronal activity on vertical SiNW arrays.

In this context, SiNW are logically expected in the close future to penetrate into the cell nucleus to sense and gain control over biochemical signalling networks and transcription machinery. However, if the nucleus is to be reached, genotoxicity issues might arise especially if the NW slowly degrade. To date, most *in vitro* investigations of SiNW biocompatibility have been confined to cellular responses while genotoxicity has definitely been forgotten. The possibility that SiNW may damage DNA and show asbestos-like behaviour must be envisaged as SiNW share common properties with asbestos, for example small fibre diameter and length. It must be mentioned that asbestos-like behaviour would be the worst case scenario as asbestos is known to produce chromosomal aberrations and cause mesothelioma, a variety of lung cancer (Kane and Hurt, 2008).

Consequently, genotoxicity assessment of SiNW should be a priority as genotoxic effects may eventually lead to abnormal and reduced cell growth, even if the cells initially appear viable. Preliminary *in vitro* assays (for a review, see Hillegass *et al.*, 2010) such as the chromosomal aberration assay and Ames bacterial reverse mutation assay will be a requirement if the SiNW are to be used in 'real world' biomedical experiments.

4.5.3 Potential production of reactive oxygen species

The ability of nanostructures to induce the toxic generation of ROS in cellular environments is a feature that has been observed for a long time (Arora *et al.*, 2012). The formation of radicals such as O_2^-, HO and hydrogen peroxide H_2O_2, which are generated in cells under normal conditions, may destabilize cellular ability to produce and detoxify the ROS. However, the exact mechanisms are not well understood. If particles such as Q-dots spontaneously induce ROS formation because of their electronic configuration, it seems that cells can also generate ROS in contact with nanostructures lacking this property.

As a result, ROS are both cytotoxic mediators and crucial signalling molecules. As a result of their high chemical reactivity, they can damage DNA, proteins and carbohydrates and cause cell death. Their impact on cellular physiology is thus of great importance. To the best of our knowledge, only Roberts *et al.* (2012) investigated the generation of ROS in response to SiNW exposure. Further studies in this unexplored area of research will be required in future. In the aforementioned study, the authors interestingly opened the debate to determine if the production of ROS was attributable to the presence of gold catalyst nanoparticles on the SiNW. Control experiments performed in acellular systems in the absence of gold particles showed that the production of free radicals was not significantly elevated. There is a need for *in vivo* studies considering that a complex interplay between cells, SiNW and ROS may affect the biocompatibility of SiNW. This point is directly related to a paradigm of toxicology: the absence of apparent cytoxicity does not mean that cellular functions are not affected.

4.6 Conclusion

SiNW appear to be a promising tool for nanomedicine, with a multitude of putative applications in drug delivery, diagnostics and biosensors. A considerable amount of work has been done so far to elaborate on the biocompatibility of SiNW, including some very complete toxicological studies; however, most of the data are still too fragmentary to be conclusive about SiNW biocompatibility. The present chapter touches a few of the, too many, complex aspects of determining the biocompatibility of SiNW. In this regard, the available findings are most useful in illustrating difficulties in evaluating toxicity of the heterogeneous SiNW family.

Indeed, several key parameters have considerable impact on the behaviour of SiNW in the biological environment and have not yet been fully investigated. First, a complete understanding of the size, shape, composition and aggregation-dependent interactions of SiNW with biological systems is lacking. Also absent are studies where more than one of these properties are varied in conjunction with one another to determine associative properties. Second, several toxicological factors need to be taken into consideration when considering whether a nanomaterial potentially raises safety issues, notably persistence/bio-accumulation in the body, anti-microbial activity and level of reactivity. One can consequently predict that future trends will ideally lead to a better comprehension of the interaction of SiNW with biomolecules and to systematic studies that investigate a range of chronic and acute exposure doses of SiNW with different physico-chemical variables.

To conclude, two major challenges will need to be addressed before SiNW can be considered as suitable for biological applications: (1) from a fundamental point-of-view, what are the mechanisms underlying the interactions between SiNW and a specific biological environment? (2) Which oriented toxicological studies should be conducted to address specific biological problematics? Only if these challenges are to be addressed will the SiNW reach the status of biomedical products.

4.7 References

Anderson, S. H. C., Elliott, H., Wallis, D. J., Canham, L. T. and Powell, J. J. (2003) 'Dissolution of different forms of partially porous silicon wafers under simulated physiological conditions', *Phys Status Solidi A*, 197, 331–5.

Anglin, E. J., Cheng, L., Freeman, W. R. and Sailor, M. J. (2008) 'Porous silicon in drug delivery devices and materials', *Adv Drug Deliv Rev*, 60, 1266–77.

Arora, S., Rajwade, J. M. and Paknikar, K. M. (2012) 'Nanotoxicology and in vitro studies: The need of the hour', *Toxicol Appl Pharmacol*, 258, 151–65.

Bilezikian, J. P., Raisz, L. G. and Martin, T. J. (2008) *Principles of bone biology (third edition)*, San Diego, Academic Press.

Brammer, K. S., Choi, C., Oh, S., Cobb, C. J., Connelly, L. S., *et al.* (2009) 'Antibiofouling, sustained antibiotic release by Si nanowire templates', *Nano Lett*, 9, 3570–4.

Chen, L., Liu, M., Bai, H., Chen, P., Xia, F., *et al.* (2009) 'Antiplatelet and thermally responsive poly(N-isopropylacrylamide) surface with nanoscale topography', *J Am Chem Soc*, 131, 10467–72.

Cho, E. C., Zhang, Q. and Xia, Y. (2011) 'The effect of sedimentation and diffusion on cellular uptake of gold nanoparticles', *Nature Nanotechnol*, 6, 385–91.

Eriksson Linsmeier, C., Prinz, C. N., Pettersson, L. M. E., Caroff, P., Samuelson, L., *et al.* (2009) 'Nanowire biocompatibility in the brain – Looking for a needle in a 3D stack', *Nano Lett*, 9, 4184–90.

Fischer, H. C. and Chan, W. C. W. (2007) 'Nanotoxicity: The growing need for in vivo study', *Curr Opin Biotechnol*, 18, 565–71.

Gad, S. E. (2005) 'Biocompatibility', in *Encyclopedia of toxicology (second edition)*, New York, Elsevier, 280–5

Garipcan, B., Odabas, S., Demirel, G., Burger, J., Nonnenmann, S. S., *et al.* (2010) '*In vitro* biocompatibility of n-type and undoped silicon nanowires', *Adv Eng Mater*, 13, B3–B9.

Han, S. W., Nakamura, C., Obataya, I., Nakamura, N. and Miyake, J. (2005) 'A molecular delivery system by using AFM and nanoneedle', *Biosens Bioelectron*, 20, 2120–5.

Hillegass, J. M., Shukla, A., Lathrop, S. A., Macpherson, M. B., Fukagawa, N. K. and Mossman, B. T. (2010) 'Assessing nanotoxicity in cells in vitro', *Wiley Interdiscip Rev Nanomed Nanobiotechnol*, 2, 219–31.

Jarvis, K. L., Barnes, T. J. and Prestidge, C. A. (2012) 'Surface chemistry of porous silicon and implications for drug encapsulation and delivery applications', *Adv Colloid Interface Sci*, 175, 25–38.

Jung, Y., Tong, L., Tanaudommongkon, A., Cheng, J.-X. and Yang, C. (2009) '*In vitro* and *in vivo* nonlinear optical imaging of silicon nanowires', *Nano Lett*, 9, 2440–4.

Kane, A. B. and Hurt, R. H. (2008) 'Nanotoxicology: The asbestos analogy revisited', *Nature Nanotechnol*, 3, 378–9.

Kim, W., Ng, J. K., Kunitake, M. E., Conklin, B. R. and Yang, P. (2007) 'Interfacing silicon nanowires with mammalian cells', *J Am Chem Soc*, 129, 7228–9.

Li, Z., Song, J., Mantini, G., Lu, M.-Y., Fang, H., *et al.* (2009) 'Quantifying the traction force of a single cell by aligned silicon nanowire array', *Nano Lett*, 9, 3575–80.

Lynch, I. and Dawson, K. A. (2008) 'Protein-nanoparticle interactions', *Nano Today*, 3, 40–7.

Monopoli, M. P., Walczyk, D., Campbell, A., Elia, G., Lynch, I., *et al.* (2011) 'Physical-chemical aspects of protein corona: Relevance to in vitro and in vivo biological impacts of nanoparticles', *J Am Chem Soc*, 133, 2525–34.

Murdock, R. C., Braydich-Stolle, L., Schrand, A. M., Schlager, J. J. and Hussain, S. M. (2008) 'Characterization of nanomaterial dispersion in solution prior to *in vitro* exposure using dynamic light scattering technique', *Toxicol Sci*, 101, 239–53.

Murr, L., Garza, K., Soto, K., Carrasco, A., Powell, T., *et al.* (2005) 'Cytotoxicity assessment of some carbon nanotubes and related carbon nanoparticle aggregates and the implications for anthropogenic carbon nanotube aggregates in the environment', *Int J Environ Res Publ Health*, 2, 31–42.

Nagesha, D. K., Whitehead, M. A. and Coffer, J. L. (2005) 'Biorelevant calcification and non-cytotoxic behavior in silicon nanowires', *Adv Mater*, 17, 921–4.

Nel, A., Xia, T., Mädler, L. and Li, N. (2006) 'Toxic potential of materials at the nanolevel', *Science*, 311, 622–7.

Oberdörster, G. (2010) 'Safety assessment for nanotechnology and nanomedicine: concepts of nanotoxicology', *J Intern Med*, 267, 89–105.

Oberdörster, G., Oberdörster, E. and Oberdörster, J. (2005) 'Nanotoxicology: An emerging discipline evolving from studies of ultrafine particles', *Environ Health Perspect*, 113, 823–39.

Patolsky, F., Timko, B. P., Yu, G., Fang, Y., Greytak, A. B., *et al.* (2006) 'Detection, stimulation, and inhibition of neuronal signals with high-density nanowire transistor arrays', *Science*, 313, 1100–4.

Piret, G., Galopin, E., Coffinier, Y., Boukherroub, R., Legrand, D. and Slomianny, C. (2011) 'Culture of mammalian cells on patterned superhydrophilic/superhydrophobic silicon nanowire arrays', *Soft Matter*, 7, 8642–9.

Popat, K. C., Daniels, R. H., Dubrow, R. S., Hardev, V. and Desai, T. A. (2006) 'Nanostructured surfaces for bone biotemplating applications', *J Orthopaed Res*, 24, 619–27.

Qi, S., Yi, C., Chen, W., Fong, C.-C., Lee, S.-T. and Yang, M. (2007) 'Effects of silicon nanowires on HepG2 cell adhesion and spreading', *ChemBioChem*, 8, 1115–18.

Qi, S., Yi, C., Ji, S., Fong, C.-C. and Yang, M. (2009) 'Cell adhesion and spreading behavior on vertically aligned silicon nanowire arrays', *ACS Appl Mater Interfaces*, 1, 30–4.

Roberts, J. R., Mercer, R. R., Chapman, R. S., Cohen, G. M., Bangsaruntip, S., *et al.* (2012) 'Pulmonary toxicity, distribution, and clearance of intratracheally instilled silicon nanowires in rats', *J Nanomater*, 2012, 17.

Robinson, J. T., Jorgolli, M., Shalek, A. K., Yoon, M.-H., Gertner, R. S. and Park, H. (2012) 'Vertical nanowire electrode arrays as a scalable platform for intracellular interfacing to neuronal circuits', *Nature Nanotechnol*, 7, 180–4.

Sabuncu, A. C., Grubbs, J., Qian, S., Abdel-Fattah, T. M., Stacey, M. W. and Beskok, A. (2012) 'Probing nanoparticle interactions in cell culture media', *Colloids Surf B*, 95, 96–102.

Seyfert, U. T., Biehl, V. and Schenk, J. (2002) 'In vitro hemocompatibility testing of biomaterials according to the ISO 10993–4', *Biomol Eng*, 19, 91–6.

Shalek, A. K., Robinson, J. T., Karp, E. S., Lee, J. S., Ahn, D.-R., *et al.* (2010) 'Vertical silicon nanowires as a universal platform for delivering biomolecules into living cells', *Proc Nat Acad Sci USA*, 107, 1870–5.

Su, Y., Wei, X., Peng, F., Zhong, Y., Lu, Y., *et al.* (2012) 'Gold nanoparticles-decorated silicon nanowires as highly efficient near-infrared hyperthermia agents for cancer cells destruction', *Nano Lett*, 12, 1845–50.

Wiesner, S., Legate, K. R. and Fässler, R. (2005) 'Integrin-actin interactions', *Cell Mol Life Sci*, 62, 1081–99.

Part II

Silicon nanowires for tissue engineering and delivery applications

5

Functional semiconducting silicon nanowires for cellular binding and internalization

W. ZHANG and C. YANG, Purdue University, USA

DOI: 10.1533/9780857097712.2.89

Abstract: This chapter discusses the *in vivo* imaging and *in vitro* cellular interaction of functionalized silicon nanowires (SiNW) studied using the non-linear optical (NLO) signals of SiNW. The chapter first discusses the motivation for studying the cellular interactions of SiNW in a quest for rational design in nanomedicine and introduces the methodology of NLO imaging and surface functionalization used in the studies. The main focus of the chapter is to apply functional SiNW to *in vivo* imaging and *in vitro* studies of cellular binding and internalization.

Key words: silicon nanowires, non-linear optical imaging, cellular interaction, surface functionalization.

5.1 Motivation: developing a nano-bio model system for rational design in nanomedicine

Nanomaterials with dimensions comparable with cell compartments, which are in the sub-micron or nano size domain, have unparalleled advantages in biomedical applications. The simple size compatibility allows us to use nanomaterials as small probes to decipher cellular machinery with minimal interference (Taton, 2002). A wide range of nanomaterials including liposomes (Immordino *et al.*, 2006), polymeric micelles (Duncan *et al.*, 2006), quantum dots (Alivisatos, 1996), carbon nanotubes (Kam *et al.*, 2005), gold nanostructures (e.g. nanospheres (Huang *et al.*, 2007; Qian *et al.*, 2008), nanoshells (Hirsch *et al.*, 2006), nanocages (Skrabalak *et al.*, 2008) and nanorods (Murphy *et al.*, 2008)), and nanowires (Kim *et al.*, 2007; Tian *et al.*, 2010; Shalek *et al.*, 2010) have shown exciting potential as imaging and sensing probes, drug and gene delivery carriers, and therapeutic agents. Although much progress has been made in this nano-bio direction, the translation of nanomedicine based on these nanostructures to a clinical setting has been hampered by the limited fundamental knowledge of the interactions between nanomaterials and biological systems. Understanding the interactions of nanomaterials with biological systems through systematic analysis is crucial to control and predict endocytosis as well as potential toxicity, and therefore enable rational design in nanomedicine.

To establish this fundamental understanding, much effort has been spent using various organic and inorganic nanoparticles (NP). Organic NP, including liposomes, polymers and micelles have been widely used for nano-bio interaction studies. For instance, labeled liposomes were used to study the effect of ligand density on the binding efficiency. Using liposomes that contain folate-conjugated polyethyleneglycolphosphatidylethanolamine (folate-PEG-PE) (Lee and Low, 1994; Goren *et al.*, 2000; Paulos *et al.*, 2004), it was found that 0.5% folate produced the most efficient binding of the liposomes to KB (a subline of the tumor cell line HeLa) carcinoma cells that overexpress the folate receptor. Using labeled spherical polymer particles (Zauner *et al.*, 2001; Prabha *et al.*, 2002; Rejman *et al.*, 2004), the size effect on cellular uptake efficiency of nanoparticles was reported. Labeled polymer nanoparticles have also been used to reveal the significant roles of particle shape in endocytosis and phagocytosis (Geng *et al.*, 2007; Gratton *et al.*, 2008; Champion and Mitragotri, 2006). Discher and co-workers showed that filamentous polymer micelles persisted in the blood circulation up to 1 week after intravenous injection, about 10 times longer than the spherical counterparts (Geng *et al.*, 2007). Using cationic poly(ethylene glycol)-based particles fabricated by top-down lithography, DeSimone and co-workers showed that rod-like particles have a higher cell internalization rate than the more symmetric counterparts (Gratton *et al.*, 2008). Using polystyrene particles as a model, Champion and Mitragotri, (2006) revealed an unexpected role of shape in phagocytosis.

Recently, inorganic nanostructures, such as silica NP, gold NP, quantum dots (QD), carbon nanotubes (CNT) and nanowires (NW), have also gained more attention in studies of cellular interactions because of their unique material- and size-dependent physiochemical properties, which could be complementary to lipid- or polymer-based nanoparticles. Fluorescently labeled silica NP with variable aspect ratio, size and surface functionalization were used to study NP translocation and cellular uptake and behavior (Huang *et al.*, 2010; Rancan *et al.*, 2012). Gold nanoparticles, including low-aspect-ratio gold nanorods and nanoshells, with strong luminescence and versatile surface chemistry have enabled researchers to investigate the effects of shape and surfactant coating on cellular uptake and cytotoxicity (Chithrani *et al.*, 2006; Hauck and Chan, 2007; Leonov *et al.*, 2008; Tong *et al.*, 2007). Semiconductor QD with unique fluorescence properties reveal a remarkable size and surface functionalization effect on cellular adsorption and endocytosis (Osaki *et al.*, 2004; Tan *et al.*, 2010; Park *et al.*, 2011). CNT with intrinsic near-infrared fluorescence (Cherukuri *et al.*, 2004) and spontaneous Raman scattering (Liu *et al.*, 2008) have been tracked in live cells and live animals for studies of cellular interactions and CNT biodistribution in organisms. Very recently, NW were also used to explore cellular response. Magnetic nanowires were utilized to exploit their interaction with living cells in which cells were labeled with fluorescent probes (Safi *et al.*, 2011). ZnO nanowires with intrinsic fluorescence have also been applied to the molecularly targeted imaging of cancer cells (Hong *et al.*, 2011).

Despite these advances, research along this direction faces several challenges (Sanhai *et al.*, 2008). First, current studies are challenged by lack of a strong intrinsic signal from the nanostructure. Although fluorescent agents are routinely used as labels for visualization of nanoscale drug carriers such as liposomes, polymer particles and copolymer micelles, they often suffer from the photobleaching problem. More importantly, interpretation of fluorescence data can be further complicated because of dissociation of probes from the objects to be studied. Recent research by Cheng and co-workers revealed an unexpected release of lipophilic dyes from copolymer micelles (Chen *et al.*, 2008b, a) and poly(lactic-co-glycolic acid) nanoparticles (Xu *et al.*, 2009) to lipid-rich structures during cellular uptake or blood circulation. To avoid labeling, near infrared (NIR) fluorescence (Cherukuri *et al.*, 2006) and spontaneous Raman scattering (Liu *et al.*, 2008) have been used to study CNT in live cells and live animals. Intrinsic multiphoton luminescence (Farrer *et al.*, 2005; Wang *et al.*, 2005; Park *et al.*, 2008) has been used to track gold nanorods and nanoshells in live cells (Tong *et al.*, 2007) and implanted tumors (Park *et al.*, 2008). Nevertheless, although the two-photon luminescence from gold nanostructures is as bright as the luminescence from quantum dots (Wang *et al.*, 2005), more than 95% of the excitation energy is actually converted into heat, causing effective photo-toxicity to cells (Huang *et al.*, 2006; Tong *et al.*, 2007). Therefore, an intense and intrinsic optical signal with low damage potential is desired to follow a nanostructure in a biological system.

Second, it is difficult to simultaneously control both the aspect ratio and surface chemistry of a nanostructure. For hydrogel particles fabricated by the soft lithography method, current studies have been restricted to non-targeting particles (Gratton *et al.*, 2008; Glangchai *et al.*, 2008; Champion and Mitragotri, 2006). For single-walled CNT, although various surface modification schemes have been developed, it is hard to vary the diameter and/or control the length of the tube. Consequently, current studies have been focused on nanotubes with diameter limited to a few nanometers and length shorter than 200 nm (Liu *et al.*, 2008; Kang *et al.*, 2008). For gold nanorods, despite the versatile surface chemistry (Tong *et al.*, 2009), it is difficult to prepare nanorods longer than 100 nm by the commonly used seeded growth method (Sau and Murphy, 2004; Nikoobakht and El-Sayed, 2003).

Based on the above discussion, one would expect that an ideal model system for interrogating the cell-nanostructure interactions should meet three criteria: 1 an intensive and intrinsic signal that allows real-time visualization of single nanostructures with 3D submicron spatial resolution; 2 a surface capable of being modified in a mild condition to produce a controlled density of ligands or charges; 3 precise control of size as well as shape, that is aspect ratio for a rod shape, through fabrication or synthesis.

Here, we introduce a newly developed excellent nano-bio model system based on functionalized silicon NW (SiNW) for study of the cellular response to 1D nanostructures. Our studies suggest that SiNW demonstrate three unique features meeting the criteria discussed above, including unparalleled dimension-control properties, intensive intrinsic NLO signals for imaging and flexible surface chemistry. The precise control of dimensions and aspect ratios of SiNW can be achieved through a metal nanocluster catalyzed chemical vapor deposition (CVD) method (Cui *et al.*, 2001). In this chemical approach, monodisperse gold NP are used as the catalyst to control the diameter of SiNW. Growth pressure and growth time are optimized to produce SiNW with desirable lengths ranging from a few hundred nanometers to tens of micrometers (Yang *et al.*, 2005). We also discovered strong and stable third-order non-linear optical (NLO) signals (including four-wave mixing (FWM) and third-harmonic generation (THG)) from SiNW, which, associated with deep penetration enabled by near-infrared or infrared pump beams and high 3D spatial resolution, can be used for both *in vivo* applications and label-free and non-invasive detection of the NW in biological experiments in real time. Additionally, SiNW with a native oxide layer on the surface can be easily functionalized based on well-studied silica surface chemistry (Iler, 1979). Collectively, we demonstrate that functionalized single crystalline SiNW can be employed to interrogate how a 1D nanostructure interacts with cells *in vivo* and *in vitro*.

5.2 Methods: non-linear optical characterization and surface functionalization of silicon nanowires (SiNWs)

5.2.1 Non-linear optical imaging of SiNWs

As a result of the large third-order susceptibility of crystalline silicon, SiNW can emit strong intrinsic THG and FWM signals, which can be used to image SiNW in different biological environments (Jung *et al.*, 2009).

NLO signals of SiNW were acquired on a multiphoton multimodal imaging platform (Plate I(a), see colour section between pp. 94 and 95). A femtosecond (fs) laser (Mai Tai, Spectra-Physics, Fremont, CA, USA) generated 130-fs pulse at a repetition rate of 80 MHz. An optical parametric oscillator (OPO) (Spectra-Physics, Fremont, CA, USA) pumped by 80% of the Mai Tai output at 790 nm generates a signal beam at 1290 nm and an idler beam at 2036 nm. The frequency of the idler beam was doubled to 1018 nm by a Periodically poled lithium niobate (PPLN) crystal. The 1018 nm beam was used as the Stokes beam. The other 20% of the Mai Tai beam was used as the pump beam. These two beams were collinearly combined with the Stokes beam passing through a delay line. The combined beams were sent into a FV1000 laser-scanning microscope (Olympus America Inc., PA, USA) and focused into a sample using a 60× water objective lens with a numerical aperture of 1.2.

The THG and FWM emission spectra that were recorded from individual SiNW with a diameter of 40 nm displayed peaks at 428 nm (Plate I(b)) and 645 nm (Plate I(c)), respectively. These peak positions were in agreement with the emission wavelengths of THG, 430 nm, produced by 1290 nm excitation, and FWM, 645 nm, generated by a collinearly combined pump field (790 nm) and Stokes field (1018 nm), confirming that the contrast in the images (Plate I(b, c)) arose from THG and FWM, respectively. Significantly, the THG and FWM signals from SiNW were 10 times higher than 60 nm silver nanoparticles (Plate II) and more than 220 times higher than 1 μm melamine beads at the same laser power, respectively. The intensive THG and FWM emission from SiNW could be attributed to the large third-order susceptibility of crystal silicon, which is c.a. 1–2 orders of magnitude higher than that of other materials such as crystal CdS, TiO_2 and Au (Boyd, 2003). Additionally, an important advantage of the NLO imaging method is the inherent 3D spatial resolution provided by NLO microscopy. Plate I(d, e) shows the lateral and axial intensity profiles of a representative SiNW, respectively. The full widths at half maxima of Gaussian fittings of measured data for the lateral and axial intensity profiles are 0.30 μm and 1.43 μm, respectively. This 3D resolution allows detection of individual SiNW on a complex and rough scaffold, such as a 3D collagen array.

In addition to strong intensity and spatial resolution, the THG and FWM signals from SiNW are quite stable. Silver nanoparticles (NP), one of the strongest THG emitters (Tai *et al.*, 2007; Cho *et al.*, 2009), were selected for comparison. As shown on Plate II, during continuous scanning of SiNW and silver NP for 70 seconds with 8.6 mW of the 1290 nm laser, the THG signals from silver NP quenched quickly and few signals were observed after 60-second ultrafast pulses (Plate II(b)). The rapid decrease of the THG intensity is possibly caused by melting of the NP by the ultrafast pulses, while the THG signals from SiNW remained consistent over time (Plate II(a)). For quantitative analysis of the intensity levels, the THG intensities of seven SiNW and seven silver NP versus the scanning time are plotted in Plate II(d, e), respectively. Under the same condition of excitation, the THG intensity at the beginning of scanning ranged from 600 to 2300 au for the SiNW and from 200 to 450 au for the silver NP. The intensity difference between individual NW is caused by the orientation variation. For silver NP, the intensity distribution can be the result of various aggregates. This comparison shows that the SiNW produced a THG signal approximately ten times stronger than the silver NP. The intensive and stable NLO signal allows us to monitor the intracellular trafficking of SiNW with a single particle tracking method.

5.2.2 Functionalization of SiNWs

Functionalization not only renders SiNW to be more biocompatible, but also immobilizes biomolecules to the SiNW surface to enable specific targeting and thus cellular regulation. In our studies, folate and amino functionalization were identified for cellular interaction. Successful folate conjugation on SiNW surface achieved the specific targeting of certain types of cancer cells, as malignant cells have primarily been found to express large numbers of folate receptors on their membrane surface (Park, 2002). Amino-functionalized SiNW were used to study the non-specific charge–charge interaction between cells and SiNW (Zhang *et al.*, 2012).

As-synthesized SiNW were free-standing on the substrate with a random orientation, as shown in the SEM image of Fig. 5.1. This free-standing configuration enables conformally effective surface functionalization of NW. Amino-modified SiNW (SiNW-NH_2) and folate-modified SiNW (SiNW-folate) were obtained through exploiting the surface chemistry of silica, as illustrated in Fig. 5.1. NW samples were first treated with oxygen plasma to produce a clean and oxidized silica-like surface for further modification. The plasma-treated SiNW on the substrate were then immersed into 1% (v/v) 3-aminopropylmethoxysilane (APTMS) in pure ethanol for about 24 hours. Then the substrate was rinsed with ethanol and transferred into an 80 °C oven for 2 hours to stabilize the immobilized groups. The amino-modified samples (1) (denoted as 'SiNW-NH_2') were washed with methanol and dried with nitrogen gas (Pham *et al.*, 2002). The resulting products (1) were treated with triethylamine (TEA) and N-[β-maleimidopropyloxy] succinimide ester (BMPS, 3 mg/mL in anhydrous DMSO) for 30 minutes. Excess acetic anhydride was used to block the unreacted amino groups. After rinsing and drying, 1 mM folate-cysteine in MES buffer (pH 6.5) was added to the substrate and allowed to react for about 3 hours (Niculescu-Duvaz *et al.*, 2008). Excess folate-cysteine was washed away with phosphate-buffered saline (PBS). The obtained folate-modified product (2) was denoted as 'SiNW-folate'.

To estimate the coverage of functional groups on SiNW surface, we started with the reported coverage of 1.7 molecules/nm^2 for amino groups based on the total covalently bonded APTES coverage on silica (Vrancken *et al.*, 1995). After considering the yield of reaction between amino groups and succinimidyl ester of BMPS (83%) (Zhang *et al.*, 2009), and the yield of addition reaction between thiol group of folate-cysteine and maleimide group of BMPS (85%) (Kirpotin *et al.*, 1997), the maximum coverage of folate was estimated to be 1.2 molecules/nm^2 calculated by multiplying the coverage of amino groups by the two reaction yields. Using a value of 2.6 nm for the length of folate-cys-BMPS-APTMS (obtained from Chem3D using MM2 minimization), a maximum surface area of approximately 21 nm^2 (Reddy *et al.*, 2002), and the area of a circle with radius of 2.6 nm, we obtained the minimum coverage of folate as approximately 0.05 molecules/nm^2.

In order to confirm surface functionalization, zeta potential measurements were conducted to monitor the changes for each sample after modification using a Malvern Zeta Sizer Nano-ZS90 (Malvern Instruments). SiNW were removed from the substrates into ethanol through sonication, and then washed with pure ethanol and water by the centrifugation. Finally, NW were suspended in ultrapure water to form a well-dispersed suspension for zeta potential measurements. The results are shown in Table 5.1. Untreated SiNW with the negative oxide layer on its surface had a zeta potential of -31.8 ± 16.7 mV. After modification, the zeta potentials of SiNW-NH_2 and SiNW-folate changed to 24.2 ± 16.7 mV and -20.9 ± 9.72 mV, respectively. As the pK_a of alkyl ammonium and carboxylic

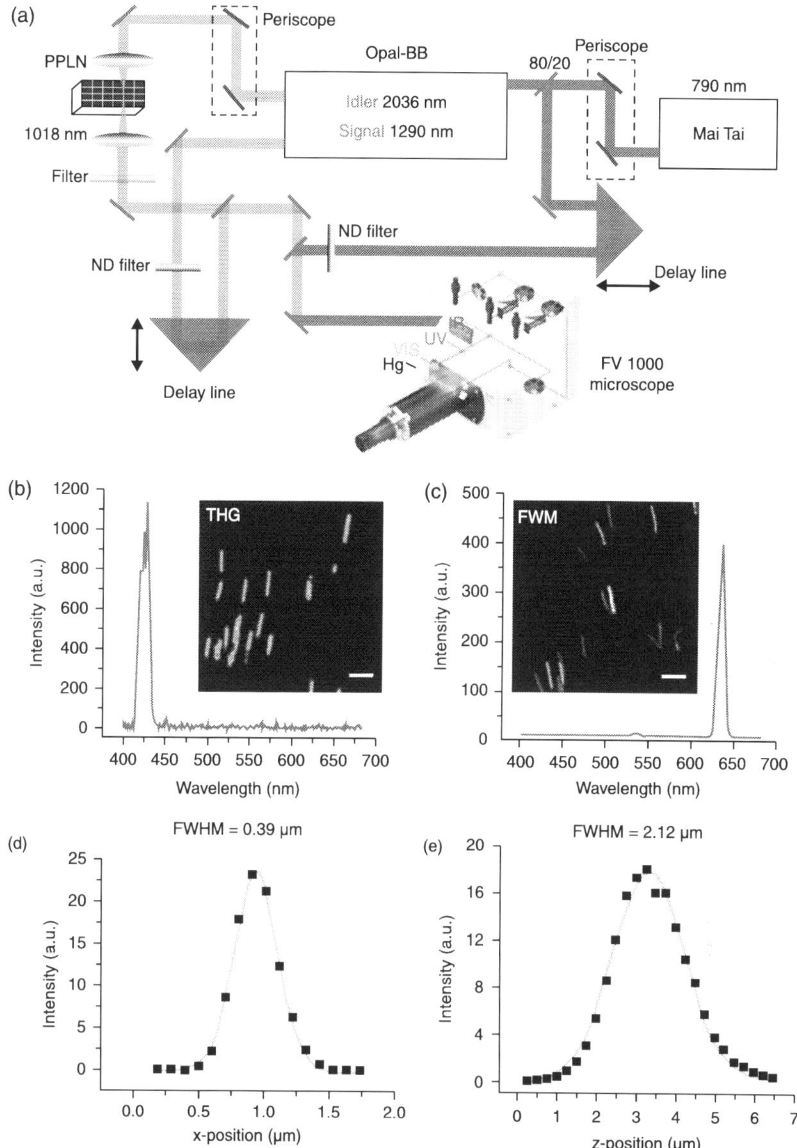

Plate I (Chapter 5) (a) Schematic diagram of a multimodal NLO microscopy system with an fs laser source. (b) THG image and spectrum of SiNWs. (c) FWM image and spectrum of SiNWs. Scale bars, 2 μm. (d), (e) The lateral and axial intensity profiles of FWM signals of a representative SiNW, respectively. ■ represent measured data. Red curves are Gaussian fittings of measured data. The full widths at half maxima (FWHM) for (d) and (e) are 0.30 μm and 1.43 μm, respectively. Modified and reproduced with permission from Jung *et al.* (2009). Copyright (2009) American Chemical Society.

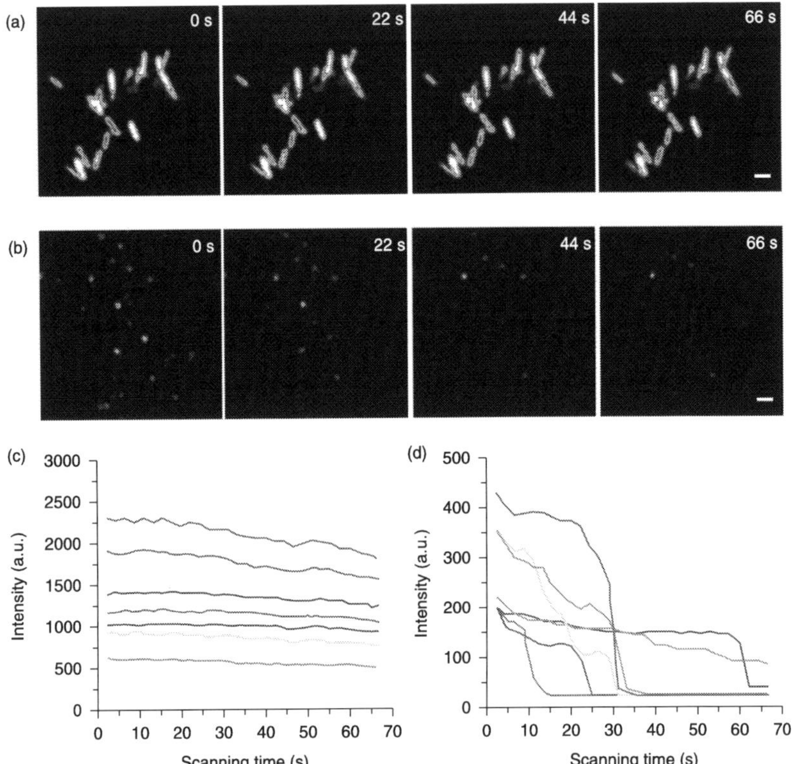

Plate II (Chapter 5) THG intensity and photostability of SiNWs and silver NPs. (a) THG images of SiNWs recorded at different scanning times. (b) THG images of silver NPs at different scanning times. The scanning time is indicated in each image. Scale bars, 2 μm. (c) THG intensity of seven representative SiNWs as a function of the scanning time. (d) THG intensity of seven representative silver NPs as a function of the scanning time. Reprinted with permission from Jung *et al.* (2009). Copyright (2009) American Chemical Society.

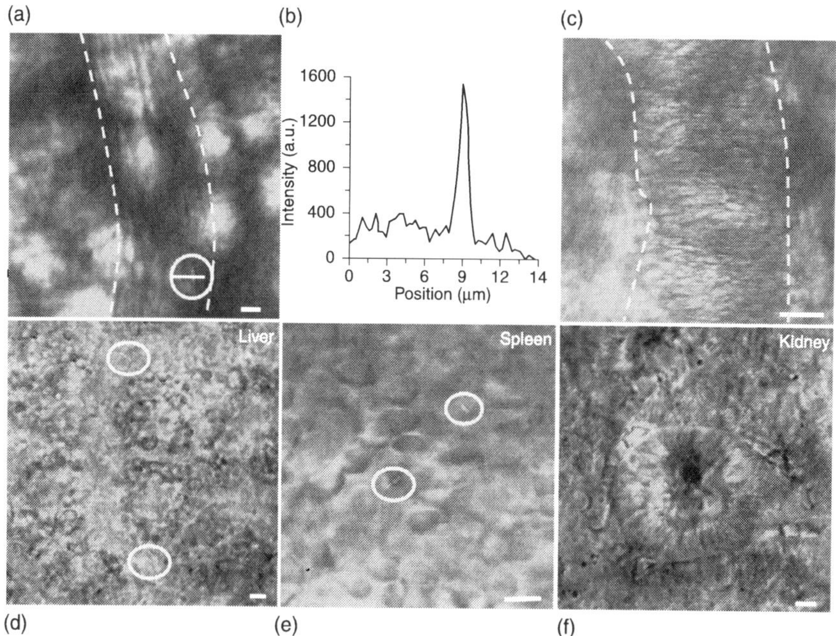

Plate III (Chapter 5) *In vivo* FWM images of SiNWs.
(a) FWM image (red) of the peripheral blood of a living mouse taken at 20 min post injection of SiNW-PEG_{900} PBS solution. Yellow dashed lines mark the blood vessel. The white solid line indicates the scan line for the intensity profile shown in (b). (b) FWM intensity profile from the line scan along the flowing SiNW. (c) FWM image of the peripheral blood of a living mouse taken post injection of PBS. (d-f) FWM images of SiNWs (red) deposited in liver (d), spleen (e), and kidney (f) explanted at 1 h post injection. All FWM images are superimposed with transmission images (cyan) taken simultaneously. The SiNWs were highlighted by yellow cycles. Scale bars, 5 μm. Reprinted with permission from Jung *et al.* (2009). Copyright (2009) American Chemical Society.

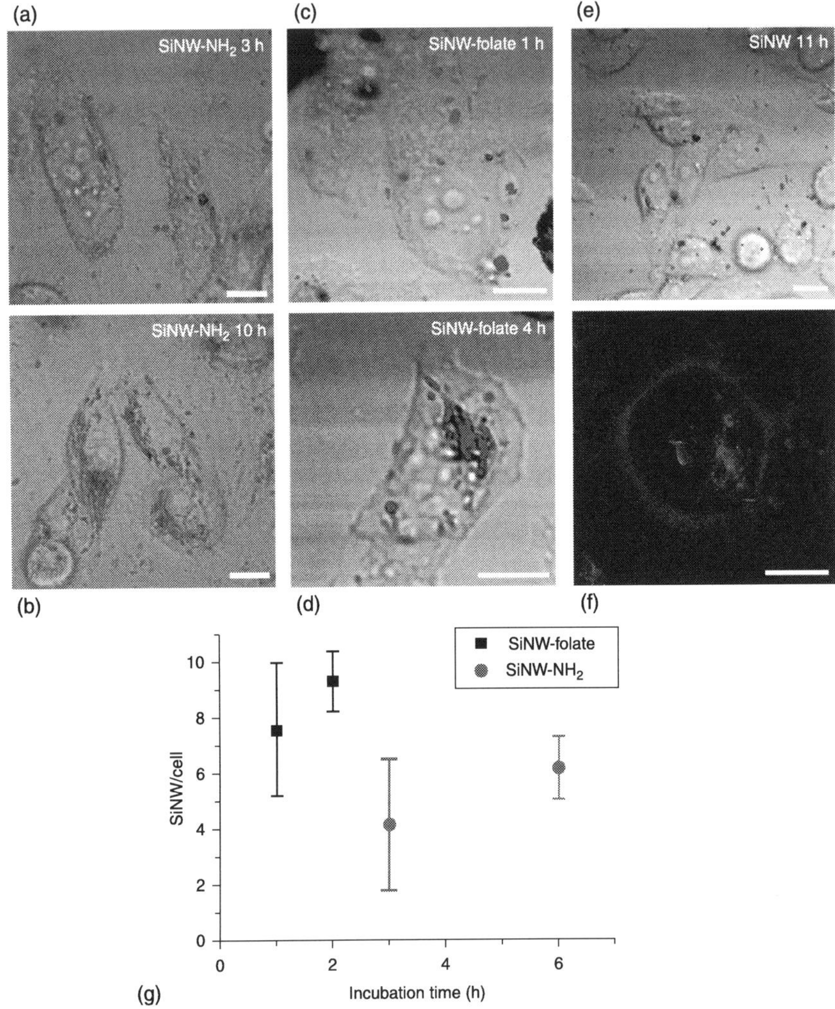

Plate IV (Chapter 5) Binding and internalization of SiNWs (red) in CHO-β cells. Overlay of FWM and transmission images of (a and b) SiNW-NH_2 after incubation for 3 and 10 h, respectively; (c and d) SiNW-folate after incubation for 1 and 4 h, respectively; and (e) unmodified SiNWs after incubation for 11 h. (f) Overlay of FWM and fluorescence image of a CHO-β cell after internalization of SiNWs. Green, fluorescence from folate-FITC labeled cell membrane. (g) Average number of bound NWs per cell before the onset of internalization as a function of incubation time. Scale bars in a–f, 10 μm. Reprinted with permission from Zhang *et al.* (2012). Copyright (2012) American Chemical Society.

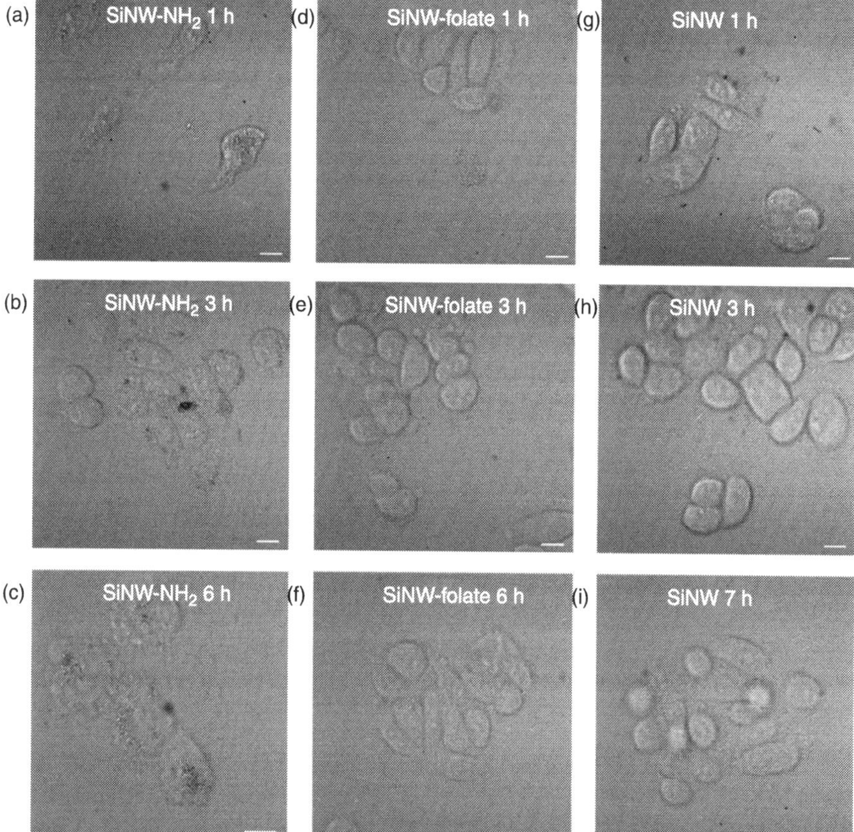

Plate V (Chapter 5) Surface binding and internalization of SiNWs (red) in CHO cells. Overlay of FWM (red) and transmission (gray) images of (a–c) SiNW-NH_2, (d–f) SiNW-folate after incubation for 1, 3 and 6 h, and (g–i) unmodified SiNWs after incubation for 1, 3 and 7 h, respectively. Scale bars, 10 μm. Reprinted with permission from Zhang *et al.* (2012). Copyright (2012) American Chemical Society.

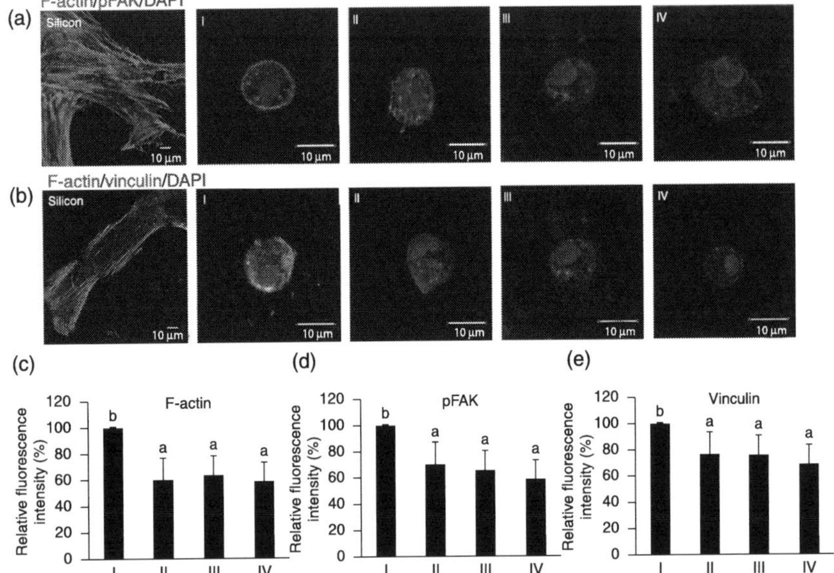

Plate VI (Chapter 7) Distribution and quantification of F-actin, pFAK and vinculin from hMSCs grown on the various groups of SiNWs. (a) F-actin (green), tyrosine-397 phosphorylated focal adhesion kinase (pFAK, red) and DAPI (nuclei, blue). (b) F-actin (green), vinculin (red) and DAPI (blue) were co-stained on the hMSCs grown on 2-D flat Si and on Group I, II, III and IV SiNWs after 72 h culture. Relative quantitative protein expression of (c) F-actin, (d) pFAK and (e) vinculin acquired from the whole cells; the average fluorescence intensity was calculated by immunofluorescence staining and compared to Group I SiNW. Data represent mean ± S.D. $p < 0.05$, $n = 3$. (Reprinted from *Biomaterials*, 33, Kuo *et al.*, 'Regulation of the fate of human mesenchymal stem cells by mechanical and stereo-topographical cues provided by silicon nanowires', 5013–5022, 2012, with permission from Elsevier.)

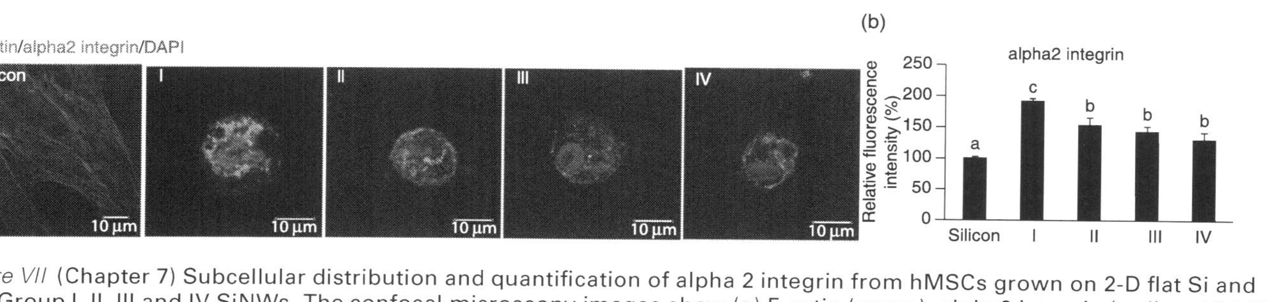

Plate VII (Chapter 7) Subcellular distribution and quantification of alpha 2 integrin from hMSCs grown on 2-D flat Si and on Group I, II, III and IV SiNWs. The confocal microscopy images show (a) F-actin (green), alpha2 integrin (red) and DAPI were co-stained in hMSCs grown on 2-D flat Si and SiNWs after 72 h culture. (b) Average fluorescence intensity was utilized to measure protein expression of alpha2 integrin. Data represent mean ± S.D. $p < 0.05$, $n = 3$. (Reprinted from *Biomaterials*, 33, Kuo *et al.*, 'Regulation of the fate of human mesenchymal stem cells by mechanical and stereo-topographical cues provided by silicon nanowires', 5013–5022, 2012, with permission from Elsevier.)

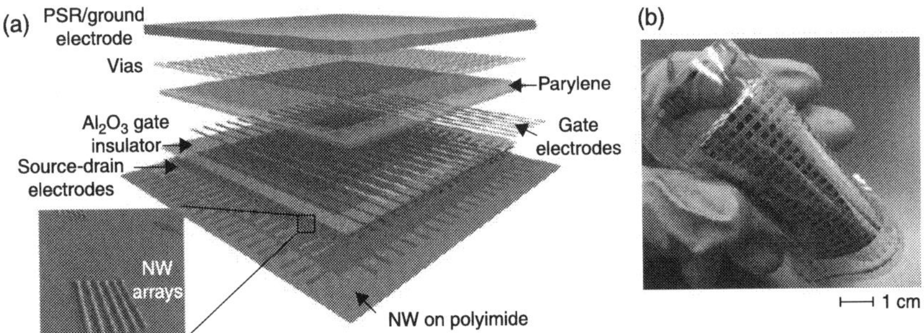

Plate VIII (Chapter 11) Nanowire-based macroscale flexible devices. (a) Schematic of the passive and active layers of NW e-skin. (b) Optical photograph of a fully fabricated e-skin device under bending. Reproduced with permission from Takei *et al.* (2010).

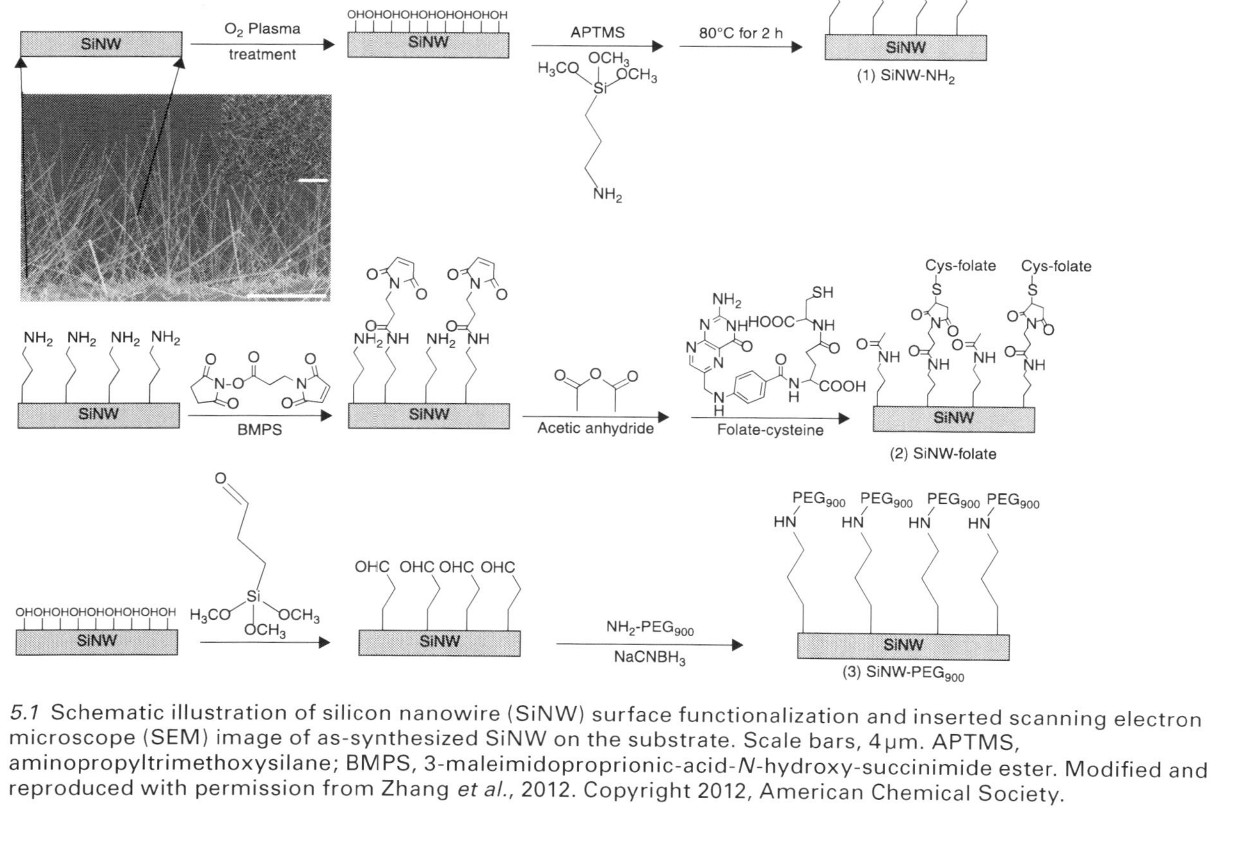

5.1 Schematic illustration of silicon nanowire (SiNW) surface functionalization and inserted scanning electron microscope (SEM) image of as-synthesized SiNW on the substrate. Scale bars, 4 μm. APTMS, aminopropyltrimethoxysilane; BMPS, 3-maleimidoproprionic-acid-*N*-hydroxy-succinimide ester. Modified and reproduced with permission from Zhang *et al.*, 2012. Copyright 2012, American Chemical Society.

Table 5.1 Zeta potential of silicon nanowires (SiNW)

NW samples	Unmodified SiNW	SiNW-NH_2	SiNW-folate
Zeta potential	-31.8 ± 16.7 mV	24.2 ± 16.5 mV	-20.9 ± 9.7 mV

groups are about 10.6 and 4.0, respectively, under neutral conditions, the amino groups of SiNW-NH_2 were protonated, and thus the SiNW-NH_2 surface was positively charged. The folate ligand features both a positive ammonium group (protonated) and two negative carboxylic groups (deprotonated), resulting in an overall negative charge. These zeta potential results indicated successful surface functionalization of the SiNW.

We have also demonstrated polyethylene glycol (PEG) functionalization on SiNW to prolong the blood circulation time for SiNW for *in vivo* study (Jung *et al.*, 2009). As-synthesized SiNW on substrate were first modified with 1% (v/v) 3-(trimethoxysilyl) propyl aldehyde in ethanol for 0.5 hours, followed by reaction with 0.1% PEG (MW 900 Da) in the presence of sodium cyanoborohydride for 24 hours. PEG_{900} was chosen because of its high biliary clearance in previous *in vivo* studies (Webster *et al.*, 2007; Friman *et al.*, 1995; Roma *et al.*, 1991). The PEGylated NW were removed from the substrate by sonication into PBS. The resultant product (3) was denoted as 'SiNW-PEG_{900}'.

5.3 Applications: *in vivo* imaging and *in vitro* cellular interaction of functional SiNWs

5.3.1 Intravital imaging of SiNWs circulating in blood vessels

The intensive intrinsic FWM emission of SiNW with 3D spatial resolution and high photostability opens up exciting opportunities for using SiNW as a novel *in vivo* imaging agent (Jung *et al.*, 2009). Here, we first demonstrated the potential of SiNW in intravital imaging by real time imaging of SiNW circulating in the blood vessels inside a mouse earlobe.

Suspensions (100 μL) of SiNW-PEG_{900} in PBS were injected into an anesthetized BALB/c mouse through the tail vein. PEGylation has been found to be effective to promote blood circulation for other nanosystems including liposomes, Au nanorods and CNT (Lasic and Needham, 1995; Niidome *et al.*, 2006; Liu *et al.*, 2007). A laser beam with power of 23 mW for the pump beam and 2 mW for the Stokes beam was focused on the ear lobe using a 40× water-immersion objective. Transmission illumination was used to visualize the blood vessel and surrounding tissues, while the epi-detected FWM was applied simultaneously to monitor the circulating SiNW with a scanning rate of 2 μs/pixel and 256 × 256 pixels/frame (see Plate III(a), for one frame). The FWM signal showing an elongated shape with c.a. 5 μm in the elongated direction agrees well with the length of SiNW synthesized, and the intensity profile of FWM across the flowing SiNW shows a peak intensity of 1500 a. u., five times larger than the background (~300 a. u.) from the blood (Plate III(b)). No FWM signal was detected in the control mouse injected with 100 μL of pure PBS (Plate III(c)).

Thirty minutes post injection, the FWM signals from the SiNW were no longer detected in the blood stream. We then studied the distribution of SiNW in the organs explanted using the FWM signals at 1 hour post injection. To do this, the mouse was euthanized, and organs including liver, spleen and kidney were explanted and fixed in 4% formalin solution to preserve the tissue architecture, and then cut into small pieces by blade for imaging. From the FWM images superimposed with transmission images taken simultaneously, we could clearly find SiNW in both liver and spleen, but not in the kidney (Plate III(d–f)). These results indicated that the injected SiNW were captured by the reticuloendothelial system such as the macrophage in the liver and spleen, but not filtered through the kidneys. After studying the bio-distribution of SiNW in these organs, we further investigated cellular response to SiNW.

5.3.2 *In vitro* cellular response to SiNWs

Various functional surfaces combined with strong intrinsic NLO signals of SiNW allow us to monitor different cellular responses to SiNW (Zhang *et al.*, 2012).

For cellular interaction studies, immortalized Chinese hamster ovary (CHO) cells stably transfected with folate receptor (FR) β (CHO-β) and normal CHO cells were used in our experiments. Typically, a 1 mL suspension of cells (10^5 cells/mL) was seeded onto a glass-bottomed petri dish and cultured at 37 °C in a humidified atmosphere containing 5% CO_2. Then, a 1 mL suspension of SiNW (100 g/mL) in cell culture medium was added to the petri dish and the dish was transferred into an incubator at 37 °C with periodic monitoring.

SiNW-NH_2 with a positively charged surface under cell culture conditions (pH 7.4) was used to study non-specific charge–charge interactions between NW and cells. Binding of folate to FR could be utilized to investigate the specific interaction between SiNW-folate and CHO-β cell with FR on its membrane.

Binding and uptake of SiNW-NH_2 and SiNW-folate and unmodified SiNW with the same length by CHO-β cells were monitored using FWM combined with optical transmission imaging. As shown on Plate IV(a, b), SiNW-NH_2 clearly accumulated on the cell membrane after incubation for 3 hours, and further uptake of SiNW-NH_2 by cells was observed at 10 hours. For SiNW-folate, fast binding was finished within the initial 1 hour of incubation because of the rapid conjugation of folate with FR (Plate IV(c)), which was in agreement with the previous report for folate derivatives (Paulos *et al.*, 2004). Almost all the SiNW-folate was internalized by cells and accumulated locally inside the cells, likely in the perinuclear area where microtubule organization center localizes (Shukla *et al.*,

2005), when incubated for 4 hours (Plate IV(d)). FWM imaging superimposed with fluorescence imaging confirmed internalization of NW using folate-FITC labeled CHO-β cells after internalization of SiNW (Plate IV(f)). Quantitative analysis of binding was done by counting the bound NW to get the average numbers of both functionalized NW per cell before the onset of internalization. As shown on Plate IV(g), the bound number for each functionalized NW increased over incubation time, and the binding rate of SiNW-folate was much higher than that of SiNW-NH_2. In contrast, no obvious binding and internalization of unmodified SiNW were observed even after 11 hours of incubation (Plate IV(e)).

Differences of binding rates for different NW could be attributed to the effects of surface properties on NW-cell interactions. Nanomaterials with positively charged surfaces have been proven to be more favorable in cellular interactions than those with neutral or negative surfaces because of attractive electrostatic interactions with negatively charged cell surfaces (Ryman-Rasmussen *et al.*, 2007; Verma and Stellacci, 2010; Cho *et al.*, 2009). Our results of no binding and thus no internalization of unmodified SiNW and moderate binding and internalization rates of SiNW-NH_2 are consistent with these observations. However, from the perspective of charge–charge interactions, the repulsive electrostatic force would prevent SiNW-folate with an overall negatively charged folate from binding to the cell surface. The observed fastest binding and internalization rates of SiNW-folate were contradictory to this prediction; we thus claimed that strong ligand (folate)-receptor (FR) interaction with a strength of about 1 nN (Girish *et al.*, 2009) could overcome the repulsive force. The strength of the attractive force between an amino group and the cell could be approximately estimated to be 4×10^{-11} N using a simple single point charge model assuming the diameter of the cell to be about 10 μm when considering that a cell membrane carries negative charge of approximately 8.3×10^{-9} C/cm^2 (Elul, 1967). In addition to a greater strength, the irreversibility of folate-FR binding interaction would prevent the release of the bound NW from cell membrane. Thus, our results indicate that strong specific ligand-receptor binding is able to overcome the non-specific charge–charge interaction, resulting in fast binding and efficient uptake of the SiNW-folate into CHO-β cells.

To further confirm our explanation, a control experiment was conducted using normal CHO cells without β FR. Binding and internalization of these three kinds of NW were again examined using our optical system (Plate V). After incubation for 1 hour, no binding was observed for all three samples (Plate V(a, d, g)). Similar to previous results using CHO-β cells, SiNW-NH_2 clearly accumulated on the cellular membrane at 3 hours, and was further internalized by CHO cells with a slightly shorter incubation time, occurring at 6 hours (Plate V(b, c)). Inversely, SiNW-folate remains suspended in the cell culture medium instead of binding to cell surface, even with incubation times up to 6 hours (Plate V(e, f)), which was similar to unmodified SiNW (Plate V(h, i)). The result of this control experiment clearly confirmed that the strong specific interaction between folate and FR facilitated the absorption and uptake of SiNW-folate by CHO-β cells.

5.4 Conclusions and future trends

In summary, we have demonstrated for the first time intrinsic FWM and THG signals from dimension-controllable SiNW. The NLO signal observed from the SiNW is highly intensive and photostable. Taking the advantages of depth resolution offered by NLO imaging, the signal enabled intravital imaging of NW circulating in the peripheral blood of a mouse and mapping of NW accumulated in organs. In addition, combined with versatile surface chemistry of SiNW, the strong intrinsic NLO signal of modified SiNW with different functional groups (NH_2, folate) could be used to monitor cellular binding and internalization of SiNW. Our pilot studies not only demonstrate that SiNW could be a valid *in vivo* and *in vitro* imaging agent, but also provide better understanding of cellular interactions with 1D nanostructures.

In addition to these achievements, this newly developed nano-bio model system based on functional SiNW allows us to study many factors (e.g. ligand density, surface functionalization, concentration of SiNW) expected to affect the uptake efficiency. It also enables one to delineate the detailed endocytotic pathways for 1D nanostructures, which are quite complex and less well known. Furthermore, the intracellular trafficking of functionalized SiNW (F-SiNW) can be monitored using real-time FWM imaging and analyzed by a single particle tracking method. Additionally, binding of SiNW on the cell membrane could act as a mechanical perturbation model to study the changes of cell shape and the mechanosensory response during mitosis and cytokinesis, which are the essential processes during cell division. More importantly, our work offers new opportunities for the study of nano-bio interactions with intrinsic signals and controllable structures. With techniques and protocols established through this system, we can expect to work on additional new nano-bio systems to address potential limitations as well as to probe new nano-bio interactions. Finally, further research on pharmacokinetics, biodistribution in tumors and major organs, and cytotoxicity of functional nanomaterials will provide insights on the diagnosis and rational design of nano-sized drug carriers for targeted chemotherapy.

5.5 References

Alivisatos, A. P. (1996) 'Semiconductor clusters, nanocrystals, and quantum dots', *Science*, 271, 933–7.

Boyd, R. W. (2003) *Nonlinear Optics*, Boston, Academic Press.

Champion, J. A. and Mitragotri, S. (2006) 'Role of target geometry in phagocytosis', *Proc. Natl. Acad. Sci. USA*, 103, 4930–4.

Chen, H., Kim, S., He, W., Wang, H., Low, P. S., *et al.* (2008a) 'Fast release of lipophilic agents from circulating PEG-PDLLA micelles revealed by in vivo Förster resonance energy transfer imaging', *Langmuir*, 24, 5213–17.

Chen, H., Kim, S., Li, L., Wang, S., Park, K. and Cheng, J. X. (2008b) 'Release of hydrophobic molecules from polymer micelles into vell membranes revealed by Förster resonance energy transfer imaging', *Proc. Natl. Acad. Sci. USA*, 105, 6596–601.

Cherukuri, P., Bachilo, S. M., Litovsky, S. H. and Weisman, R. B. (2004) 'Near-infrared fluorescence microscopy of single-walled carbon nanotubes in phagocytic cells', *J. Am. Chem. Soc.*, 126, 15638–9.

Cherukuri, P., Gannon, C., Leeuw, T., Schmidt, H., Smalley, R., *et al.* (2006) 'Mammalian pharmacokinetics of carbon nanotubes using intrinsic near-infrared fluorescence', *Proc. Natl. Acad. Sci. USA*, 103, 18882–6.

Chithrani, B. D., Ghazani, A. A. and Chan, W. C. W. (2006) 'Determining the size and shape dependence of gold nanoparticle uptake into mammalian cells', *Nano Lett.*, 6, 662–8.

Cho, E. C., Xie, J., Wurm, P. A. and Xia, Y. (2009) 'Understanding the role of surface charges in cellular adsorption versus internalization by selectively removing gold nanoparticles on the cell surface with a I2/KI etchant', *Nano Lett.*, 9, 1080–4.

Cui, Y., Lauhon, L. J., Gudiksen, M. S., Wang, J. and Lieber, C. M. (2001) 'Diameter-controlled synthesis of single-crystal silicon nanowires', *Appl. Phys. Lett.*, 78, 2214–16.

Duncan, R., Ringsdorf, H. and Satchi-Fainaro, R. (2006) 'Polymer therapeutics – polymers as drugs, conjugates and gene delivery systems: past, present and future opportunities', *Adv. Polym. Sci.*, 192, 1–8.

Elul, R. (1967) 'Fixed charge in the cell membrane'. *J. Physiol.*, 189, 351–65.

Farrer, R. A., Butterfield, F. L., Chen, V. W. and Fourkas, J. T. (2005) 'Highly efficient multiphoton-absorption-induced luminescence from gold nanoparticles', *Nano Lett.*, 5, 1139–42.

Friman, S., Thune, A., Nilsson, B. and Svanvik, J. (1995) 'Medication with ursodeoxycholic acid enhances the biliary clearance of polyethylene glycol 900, but not Mannitol', *Digestion*, 56, 382–8.

Geng, Y., Dalhaimer, P., Cai, S., Tsai, R., Tewari, M., *et al.* (2007) 'Shape effects of filaments versus spherical particles in flow and drug delivery', *Nat. Nano*, 2, 249–55.

Girish, C. M., Binulal, N. S., Anitha, V. C., Nair, S., Mony, U. and Prasanth, R. (2009) 'Atomic force microscopic study of folate receptors in live cells with functionalized tips', *Appl. Phys. Lett.*, 95, 223703.

Glangchai, L. C., Caldorera-Moore, M., Shi, L. and Roy, K. (2008) 'Nanoimprint lithography based fabrication of shape-specific, enzymatically-triggered smart nanoparticles', *J. Control. Release*, 125, 263–72.

Goren, D., Horowitz, A. T., Tzemach, D., Tarshish, M., Zalipsky, S. and Gabizon, A. (2000) 'Nuclear delivery of doxorubicin via folate-targeted liposomes with bypass of multidrug-resistance efflux pump', *Clin Cancer Res*, 6, 1949–57.

Gratton, S. E. A., Ropp, P. A., Pohlhaus, P. D., Luft, J. C., Madden, V. J., *et al.* (2008) 'The effect of particle design on cellular internalization pathways', *Proc. Natl. Acad. Sci. USA*, 105, 11613–18.

Hauck, T. S. and Chan, W. C. W. (2007) 'Gold nanoshells in cancer imaging and therapy: towards clinical application', *Nanomedicine*, 2, 735–8.

Hirsch, L. R., Gobin, A. M., Lowery, A. R., Tam, F., Drezek, R. A., *et al.* (2006) 'Metal nanoshells', *Annals Biomed. Eng.*, 34, 15–22.

Hong, H., Shi, J., Yang, Y., Zhang, Y., Engle, J. W., *et al.* (2011) 'Cancer-targeted optical imaging with fluorescent zinc oxide nanowires', *Nano Lett.*, 11, 3744–50.

Huang, X., El-Sayed, I. H., Qian, W. and El-Sayed, M. A. (2006) 'Cancer cell imaging and photothermal therapy in the near-infrared region by using gold nanorods', *J. Am. Chem. Soc.*, 128, 2115–20.

Huang, X., El-Sayed, I. H., Qian, W. and El-Sayed, M. A. (2007) 'Cancer cells assemble and align gold nanorods conjugated to antibodies to produce highly enhanced, sharp, and polarized surface Raman spectra: a potential cancer diagnostic marker', *Nano Lett.*, 7, 1591–7.

Huang, X., Teng, X., Chen, D., Tang, F. and He, J. (2010) 'The effect of the shape of mesoporous silica nanoparticles on cellular uptake and cell function', *Biomaterials*, 31, 438–48.

Iler, R. K. (1979) *The Chemistry of Silica: Solubility, Polymerization, Colloid and Surface Properties and Biochemistry of Silica*, New York, Wiley.

Immordino, M. L., Dosio, F. and Cattel, L. (2006) 'Stealth liposomes: review of the basic science, rationale, and clinical applications, existing and potential', *Int. J. Nanomedicine*, 1, 297–315.

Jung, Y., Tong, L., Tanaudommongkon, A., Cheng, J.-X. and Yang, C. (2009) 'In vitro and in vivo nonlinear optical imaging of silicon nanowires', *Nano Lett.*, 9, 2440–4.

Kam, N. W. S., O'Connell, M., Wisdom, J. A. and Dai, H. (2005) 'Carbon nanotubes as multifunctional biological transporters and near-infrared agents for selective cancer cell destruction', *Proc. Natl. Acad. Sci. USA*, 102, 11600–5.

Kang, B., Yu, D., Chang, S., Chen, D., Dai, Y. and Ding, Y. (2008) 'Intracellular uptake, trafficking and subcellular distribution of folate conjugated single walled carbon nanotubes within living cells', *Nanotechnology*, 19, 375103.

Kim, W., Ng, J. K., Kunitake, M. E., Conklin, B. R. and Yang, P. (2007) 'Interfacing silicon nanowires with mammalian cells', *J. Am. Chem. Soc.*, 129, 7228–9.

Kirpotin, D., Park, J. W., Hong, K., Zalipsky, S., Li, W.-L., *et al.* (1997) 'Sterically stabilized anti-HER2 immunoliposomes: design and targeting to human breast cancer cells in vitro', *Biochemistry*, 36, 66–75.

Lasic, D. D. and Needham, D. (1995) 'The 'stealth' liposome: a prototypical biomaterial', *Chem. Rev.*, 95, 2601–28.

Lee, R. J. and Low, P. S. (1994) 'Delivery of liposomes into cultured KB cells via folate receptor-mediated endocytosis', *J. Biol. Chem.*, 269, 3198–204.

Leonov, A. P., Zheng, J., Clogston, J. D., Stern, S. T., Patri, A. K. and Wei, A. (2008) 'Detoxification of gold nanorods by treatment with polystyrenesulfonate', *ACS Nano*, 2, 2481–8.

Liu, Z., Cai, W., He, L., Nakayama, N., Chen, K., *et al.* (2007) 'In vivo biodistribution and highly efficient tumour targeting of carbon nanotubes in mice', *Nat. Nano*, 2, 47–52.

Liu, Z., Davis, C., Cai, W., He, L., Chen, X. and Dai, H. (2008) 'Circulation and long-term fate of functionalized, biocompatible single-walled carbon nanotubes in mice probed by Raman spectroscopy', *Proc. Natl. Acad. Sci. USA*, 105, 1410–15.

Murphy, C. J., Gole, A. M., Hunyadi, S. E., Stone, J. W., Sisco, P. N., *et al.* (2008) 'Chemical sensing and imaging with metallic nanorods', *Chem. Commun.*, 5, 544–57.

Niculescu-Duvaz, D., Getaz, J. and Springer, C. J. (2008) 'Long functionalized poly(ethylene glycol)s of defined molecular weight: synthesis and application in solid-phase synthesis of conjugates', *Bioconjugate Chem.*, 19, 973–81.

Niidome, T., Yamagata, M., Okamoto, Y., Akiyama, Y., Takahashi, H., *et al.* (2006) 'PEG-modified gold nanorods with a stealth character for in vivo applications', *J. Control. Release*, 114, 343–7.

Nikoobakht, B. and El-Sayed, M. A. (2003) 'Preparation and growth mechanism of gold nanorods (NRs) using seed-mediated growth method', *Chem. Mater.*, 15, 1957–62.

Osaki, F., Kanamori, T., Sando, S., Sera, T. and Aoyama, Y. (2004) 'A quantum dot conjugated sugar ball and its cellular uptake. On the size effects of endocytosis in the subviral region', *J. Am. Chem. Soc.*, 126, 6520–1.

Park, J., Estrada, A., Sharp, K., Sang, K., Schwartz, J. A., *et al.* (2008) 'Two-photon-induced photoluminescence imaging of tumors using near-infrared excited gold nanoshells', *Opt. Express*, 16, 1590–9.

Park, J., Nam, J., Won, N., Jin, H., Jung, S., *et al.* (2011) 'Compact and stable quantum dots with positive, negative, or zwitterionic surface: specific cell interactions and non-specific adsorptions by the surface charges', *Adv. Funct. Mater.*, 21, 1558–66.

Park, Y. (2002) 'Tumor-directed targeting of liposomes', *Biosience Reports*, 22, 267–81.

Paulos, C. M., Reddy, J. A., Leamon, C. P., Turk, M. J. and Low, P. S. (2004) 'Ligand binding and kinetics of folate receptor recycling in vivo: impact on receptor-mediated drug delivery', *Mol. Pharmacol.*, 66, 1406–14.

Pham, T., Jackson, J. B., Halas, N. J. and Lee, T. R. (2002) 'Preparation and characterization of gold nanoshells coated with self-assembled monolayers', *Langmuir*, 18, 4915–20.

Prabha, S., Zhou, W.-Z., Panyam, J. and Labhasetwar, V. (2002) 'Size-dependency of nanoparticle-mediated gene transfection: studies with fractionated nanoparticles', *Int. J. Pharm.*, 244, 105–15.

Qian, X., Peng, X.-H., Ansari, D. O., Yin-Goen, Q., Chen, G. Z., *et al.* (2008) 'In vivo tumor targeting and spectroscopic detection with surface-enhanced Raman nanoparticle tags', *Nat. Biotech.*, 26, 83–90.

Rancan, F., Gao, Q., Graf, C., Troppens, S., Hadam, S., *et al.* (2012) 'Skin penetration and cellular uptake of amorphous silica nanoparticles with variable size, surface functionalization, and colloidal stability', *ACS Nano*, 6, 6829–42.

Reddy, J. A., Abburi, C., Hofland, H., Howard, S. J., Vlahov, I., *et al.* (2002) 'Folate-targeted, cationic liposome-mediated gene transfer into disseminated peritoneal tumors', *Gene Ther.*, 9, 1542–50.

Rejman, J., Oberle, V., Zuhorn, I. S. and Hoekstra, D. (2004) 'Size-dependent internalization of particles via the pathways of clathrin- and caveolae-mediated endocytosis', *Biochem. J.*, 377, 159–69.

Roma, M. G., Marinelli, R. A. and Rodr Guez Garay, E. A. (1991) 'Biliary excretion of polyethylene glycol molecular weight 900: evidence for a bile salt-stimulated vesicular transport mechanism', *Biochem. Pharmacol.*, 42, 1775–81.

Ryman-Rasmussen, J. P., Riviere, J. E. and Monteiro-Riviere, N. A. (2007) 'Variables influencing interactions of untargeted quantum dot nanoparticles with skin cells and identification of biochemical modulators', *Nano Lett.*, 7, 1344–8.

Safi, M., Yan, M., Guedeau-Boudeville, M.-A., Conjeaud, H. L. N., Garnier-Thibaud, V., *et al.* (2011) 'Interactions between magnetic nanowires and living cells: uptake, toxicity, and degradation', *ACS Nano*, 5, 5354–64.

Sanhai, W. R., Sakamoto, J. H., Canady, R. and Ferrari, M. (2008) 'Seven challenges for nanomedicine', *Nat. Nano*, 3, 242–4.

Sau, T. K. and Murphy, C. J. (2004) 'Seeded high yield synthesis of short Au nanorods in aqueous solution', *Langmuir*, 20, 6414–20.

Shalek, A. K., Robinson, J. T., Karp, E. S., Lee, J. S., Ahn, D.-R., *et al.* (2010) 'Vertical silicon nanowires as a universal platform for delivering biomolecules into living cells', *Proc. Natl. Acad. Sci. USA*, 107, 1870–5.

Shukla, R., Bansal, V., Chaudhary, M., Basu, A., Bhonde, R. R. and Sastry, M. (2005) 'Biocompatibility of gold nanoparticles and their endocytotic fate inside the cellular compartment: a microscopic overview', *Langmuir*, 21, 10644–54.

Skrabalak, S. E., Chen, J., Sun, Y., Lu, X., Au, L., *et al.* (2008) 'Gold nanocages: synthesis, properties, and applications', *Acc. Chem. Res.*, 41, 1587–95.

Tai, S. P., Wu, Y., Shieh, D. B., Chen, L. J., Lin, K. J., *et al.* (2007) 'Molecular imaging of cancer cells using plasmon-resonant-enhanced third-harmonic-generation in silver nanoparticles', *Adv. Mater.*, 19, 4520–3.

Tan, S. J., Jana, N. R., Gao, S., Patra, P. K. and Ying, J. Y. (2010) 'Surface-ligand-dependent cellular interaction, subcellular localization, and cytotoxicity of polymer-coated quantum dots', *Chem.Mater.*, 22, 2239–47.

Taton, T. A. (2002) 'Nanostructures as tailored biological probes', *Trends Biotechnol.*, 20, 277–9.

Tian, B., Cohen-Karni, T., Qing, Q., Duan, X., Xie, P. and Lieber, C. M. (2010) 'Three-dimensional, flexible nanoscale field-effect transistors as localized bioprobes'. *Science*, 329, 830–4.

Tong, L., Wei, Q., Wei, A. and Cheng, J. X. (2009) 'Gold nanorods as contrast agents for biological imaging: optical properties, surface conjugation and photothermal effects', *Photochem. Photobiol.*, 85, 21–32.

Tong, L., Zhao, Y., Huff, T. B., Hansen, M. N., Wei, A. and Cheng, J. X. (2007) 'Gold nanorods mediate tumor cell death by compromising membrane integrity', *Adv. Mater.*, 19, 3136–41.

Verma, A. and Stellacci, F. (2010) 'Effect of surface properties on nanoparticle–cell interactions', *Small*, 6, 12–21.

Vrancken, K. C., Possemiers, K., Van Der Voort, P. and Vansant, E. F. (1995) 'Surface modification of silica gels with aminoorganosilanes', *Colloids Surf. A*, 98, 235–241.

Wang, H., Huff, T. B., Zweifel, D. A., He, W., Low, P. S., *et al.* (2005) 'In vitro and in vivo two-photon luminescence imaging of single gold nanorods', *Proc. Natl. Acad. Sci. USA*, 102, 15752–6.

Webster, R., Didier, E., Harris, P., Siegel, N., Stadler, J., *et al.* (2007) 'PEGylated proteins: evaluation of their safety in the absence of definitive metabolism studies', *Drug Metab. Dispos.*, 35, 9–16.

Xu, P., Gullotti, E., Tong, L., Highley, C. B., Errabelli, D. R., *et al.* (2009) 'Intracellular drug delivery by poly(lactic-co-glycolic acid) nanoparticles, revisited', *Mol. Pharm.*, 6, 190–201.

Yang, C., Zhong, Z. and Lieber, C. M. (2005) 'Encoding electronic properties by synthesis of axial modulation-doped silicon nanowires', *Science*, 310, 1304–7.

Zauner, W., Farrow, N. A. and Haines, A. M. R. (2001) 'In vitro uptake of polystyrene microspheres: effect of particle size, cell line and cell density', *J. Control. Release*, 71, 39–51.

Zhang, W., Cui, J., Tao, C.-A., Wu, Y., Li, Z., *et al.* (2009) 'A strategy for producing pure single-layer graphene sheets based on a confined self-assembly approach', *Angew. Chem., Int. Ed.*, 48, 5864–8.

Zhang, W., Tong, L. and Yang, C. (2012) 'Cellular binding and internalization of functionalized silicon nanowires', *Nano Lett.*, 12, 1002–6.

6 Functional semiconducting silicon nanowires and their composites as orthopedic tissue scaffolds

J. L. COFFER, Texas Christian University, USA

DOI: 10.1533/9780857097712.2.104

Abstract: This chapter describes the unique opportunities afforded to silicon nanowires (SiNW) in the field of tissue engineering, with a focus on orthopedics. We begin with a description of the necessary surface modification processes for the nanowire, either from an electrochemical etching or surface functionalization approach. This is followed by a discussion of options in composite formulation of SiNW with biocompatible polymer scaffolds to improve pragmatic processibility, and finally studies of relevant cell line attachment, proliferation and differentiation *in vitro* for such scaffolds. Finally, we conclude with a brief mention of future directions for use of SiNW in this specific field of therapeutic bionanotechnology.

Key words: silicon, nanowire, tissue engineering, bone repair, stem cell, differentiation.

6.1 Introduction

Facilitating self-healing is an enthralling concept, engaging scientists and physicians who are employing a broad number of approaches from both fundamental and clinical perspectives to address this goal (Lanza *et al.*, 2007). The challenges are well defined: fabrication of three-dimensional scaffolds with proper architecture, porosity, surface chemistry and, in some cases, mechanical strength, to permit vasculature, neuronal in-growth, cell attachment, proliferation and differentiation – all on the path to tissue or organ regeneration. At the same time, scaffold degradation kinetics must accommodate the above processes in an ideally seamless manner. Consequently, a great deal of effort has been expended in scaffold design and evaluation, much of it centered on organic macromolecule-based systems (Atala *et al.*, 2012).

One long-term challenge in the development of suitable tissue engineering scaffolds is the design of so-called smart materials, structures that adapt to and can be modified by an external stimulus – chemical, optical, electrical, mechanical. One logical option for such a platform is to use a standard electronic device material, such as elemental silicon, in such a configuration. Yet, as pointed out in Chapter 1, bulk crystalline Si is traditionally viewed as bio-inert, with an accompanying *in vivo* reponse of fibrotic capsule formation.

However, silicon in nanowire form is different. The physical dimensions of a silicon nanowire (SiNW), with an impressive associated surface area and diverse surface chemistry, provide multiple advantages. Great synthetic diversity in terms of nanowire width and length is possible with Si, ranging from widths as small as ~10 nm to as large as micrometers; lengths ranging typically from hundreds of nanometers to tens of micrometers are possible (Schmidt *et al.*, 2009). A very useful spectrum of surface modification reactions, both for inorganic as well as organic species, has also been developed (see Chapter 3). For scaffolds relying on electronic device function, both *n*- (Wang *et al.*, 2005) and *p*- (Cui *et al.*, 2000) type doping into the nanowire has been established, with associated diodes and transistors also constructed. Finally, it must be emphasized that the knowledge base associated with Si-based complementary metal-oxide-semiconductor (CMOS) devices is extensive and available for practical use in the long term.

Bone repair is a particular example of tissue engineering for which Si is well suited. Precedent lies in two different materials: Ca-rich silica glass phases (Hench *et al.*, 1971) as well as mesoporous Si films (Coffer *et al.*, 2005). For the former material, the presence of surface silanols (Si-OH species) has been shown to induce calcium phosphate, and ultimately hydroxapatite, formation. For the case of porous Si, demonstration of its ability to resorb *in vivo* (Bowditch *et al.*, 1999) into silicic acid and also stimulate calcification *in vitro* (Canham, 1995) provide clear indications of its ability to engage in osteoinductive processes.

In order for SiNW to play an active role in bone repair, several key properties must be addressed. First, the proper surface chemistry must be in place to achieve osteoinduction (active induction of bone growth (or mineralization) at the synthetic surface) in an acellular fashion, then *in vitro*, and ultimately, *in vivo*. Second, the nanowire and the accompanying matrix must be structurally modified such that controlled degradation of the matrix will occur (e.g. through porosification or a similar modification process). Finally, it is also necessary to have useful quantities of such skeletal regeneration biomaterials easily processible, readily achieved using formulations with known medically accepted polymers.

In this chapter, efforts to achieve all three goals are described. As it will be shown, in some cases, the ability to create degradable Si structures can be attempted through etching processes that concomitantly alter the surface chemistry of the nanowire.

6.2 Nanowire surface etching processes to induce biomineralization

The majority of the SiNW used in fundamental investigations of this type have employed a vapor-liquid-solid type route, as described by Zheng and Xu in Chapter 2. The surface termination in this type of nanowire is that of native oxide (SiO_2). Let us begin with an analysis of as-prepared SiNW, in the order of 100 nm in diameter, with regard to their ability to undergo calcification in simulated plasma. This type of accellular assay is a convenient screen for a threshold evaluation of the suitability of a given synthetic material for orthopedic applications. Simulated body fluid (SBF) comprises basic cation/anion combinations present in serum, minus the plasma proteins (Kokubo *et al.*, 1990). If successful, exposure to a given scaffold to a SBF solution at physiological temperature at selected time intervals will yield detectable amounts of calcified deposits that typically compositionally analyze as some phase of calcium phosphate, in some cases hydroxyapatite (HA, the mineral phase associated with bone).

Initial evaluation of these as-prepared SiNW to SBF at 37 °C over several weeks' exposure (>1 month) reveals no evidence for calcified deposit formation; in selected cases, only random formation of occasional NaCl crystals can be observed (Nagesha *et al.*, 2005). However, it has been shown that anodic etching of crystalline silicon (under the proper conditions) will produce mesoporous features, thereby transforming the material from a bio-inert to a bioactive form (Canham, 1995); associated with this transformation to porous character is an ability to resorb physiologically, induce calcium phosphate formation through the release of silicic acid ($Si(OH)_4$) (Anderson *et al.*, 2003) and exhibit a negligible inflammatory response *in vivo*. Hence experiments were undertaken to anodize as-formed SiNW under low current conditions and electrolyte concentrations that would ideally produce mesoporous features on the nanowire.

Typical results are shown in Fig. 6.1. A combination of nanowire electropolishing and coarsening is obtained, with the diameter of etched SiNW structures reduced from an original value of nearly a micron to structures ranging from 50 to 150 nm

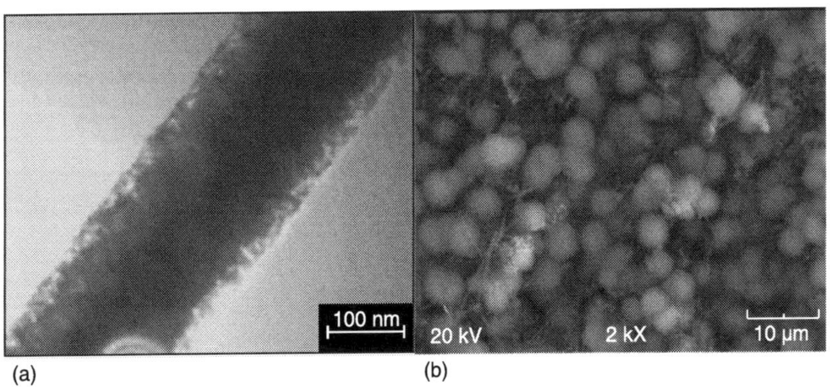

6.1 (a) Transmission electron microscope image of a silicon nanowire (SiNW) etched in 1:1 HF/EtOH at 0.45 mA/cm^2 for 5 minutes. A roughened/porous surface layer is evident. (b) Effect of simulated body fluid (SBF) exposure of these etched SiNW for 30 days.

in width. Interestingly, these etched nanowires are found to release monomeric silicic acid ($Si(OH)_4$) at the micromolar level; subsequent exposure of these etched SiNW to a solution of SBF at 37 °C results in ready formation of calcium phosphate spherulites across the scaffold (Fig. 6.1(b)).

Long-term assessment of etched SiNW degradation under *in vitro* or *in vivo* conditions has not yet been performed, but the ease of calcification provides optimism for such behavior. It should also be pointed out that it is possible to form mesoporous Si nanowires by electroless etching of Si in the presence of HF and silver (Ag^-) ions (Hochbaum *et al.*, 2009; Qu *et al.*, 2009). The resorptive character of these porous nanowires in biological media is not yet known either, however.

6.3 Nanowire surface functionalization strategies to induce biomineralization

6.3.1 Electrochemically assisted surface functionalization

We next turn our attention to chemical modification of the SiNW surface for the purpose of inducing biorelevant calcification. A two-fold strategy is employed. The first option again involves the use of simple electrochemistry, this time in the form of applying a low cathodic current (1.1 mA/cm^2) in the presence of Ca^{2+} ions for a brief period (60 minutes); such an effect results in the effective 'seeding' of a critical density of calcium species onto the nanowire surface (Nagesha *et al.*, 2005). Subsequent exposure of this cationic SiNW to SBF, again at physiological temperature, results in uniform formation of calcium phosphate, hundreds of nanometers thick, along the cylindrical surface (Fig. 6.2). However, extended

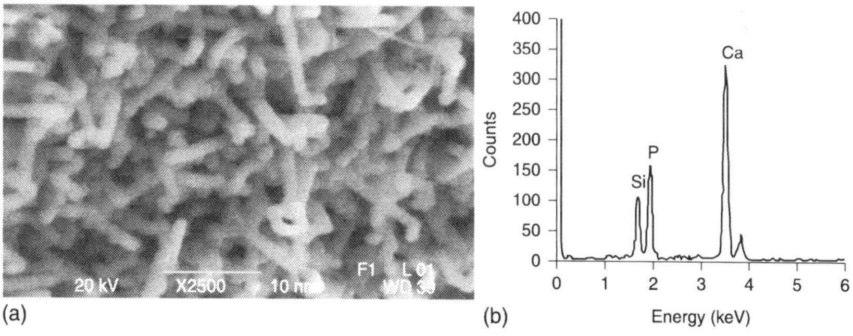

6.2 Scanning electron microscope image of calcium phosphate-coated silicon nanowires (SiNW), prepared by exposure of SiNW to 1 mM Ca^{2+} at a cathodic bias of 1 mA/cm^2 for 60 minutes, followed by simulated body fluid (SBF) exposure for 1 week; (b) corresponding energy dispersive X-ray spectrum indicating the strong presence of calcium and phosphorus in the sample.

application of bias in SBF to these scaffolds results in non-uniform deposition of clusters of calcium phosphate at scattered regions across the nanowire network (data not shown). Therefore, the kinetics of initial nucleation of calcium centers is critical to uniform formation of calcified phases on the Si structure.

In terms of possible mechanisms, it is logical to propose that calcification is induced in a manner similar to that of calcium-rich silica phases, whereby surface silanol groups mediate calcium phosphate nucleation; the released Ca^{2+} presumably accelerates growth via an increased ion activity product for a hydroxyapatite phase in the SBF. For these relatively large (>100 nm) nanowires, no detectable wire corrosion is observed by electron microscopy. Nevertheless, these results are encouraging from the perspective of bone bonding and providing necessary osteoconduction in an implant.

6.3.2 Covalent surface functionalization of silicon nanowires (SiNWs) for osteocompatibility

The topic of SiNW surface modification is of course, in both scope and significance, broad and deep enough to warrant its own chapter, which appears earlier in this volume (Chapter 3). We restrict our comments here to functionalization processes relevant to biomineralization/osteocompatibility.

As one example of this approach, we analyzed the impact of the presence of the bisphosphonate drug alendronate on *in vitro* calcification of the SiNW in SBF

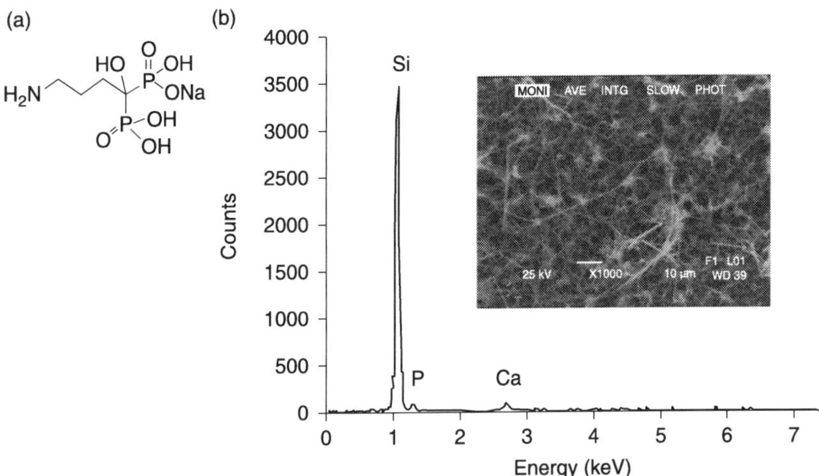

6.3 (a) Structure of alendronate. (b) Scanning electron microscope image of Ca^{2+}- seeded silicon nanowire sample exposed to 2.5 mM alendronate prior to extended immersion in simulated body fluid (SBF). Negligible calcification can be detected in this type of sample.

and the specific impact of this functionality on SiNW cytocompatibility (Jiang *et al.*, 2009). Nitrogen-containing bisphosphonates such as alendronate (Fig. 6.3(a)) are a class of anti-osteoporosis drugs whose mode of action serves to inhibit farnesyl synthetase activity; osteoclast resorption onto the bone surface is inhibited as a consequence. Alendronate has a high affinity for calcium phosphate (through the P-O functionality). Surface modification of calcium phosphate-coated SiNW with alendronate is readily achieved by a facile immersion of CaP/SiNW in aqueous alendronate solution of milimolar concentration.

In one example of an analysis of impact of alendronate on SiNW calcification in SBF, SiNW were biased in regular SBF electrolyte (1 hour) to produce the requisite Ca^{2-} seed layer, then immersed in 2.5 mM aqueous alendronate solution at room temperature for 24 hours, then soaked in regular SBF solution for an extended period (for extended periods up to 4 weeks). Interestingly, it was found that in these experiments involving the presence of alendronate resulted in no detectable calcification of SiNWs at the time scales typically measured (1–4 weeks) (Jiang *et al.*, 2009) (Fig. 6.3(b)). This inhibition of SiNW calcification in the presence of alendronate is likely a consequence of the strong affinity of alendronate for any exposed calcium centers, which leads to a significant suppression of necessary nucleation sites for precipitation of calcium phosphate from SBF solution under zero bias.

There is a corresponding impact of phosphonate identity on the SiNW surface associated with cytocompatibility. Several types of surfaces were analyzed with respect to mesechymal stem cells obtained from mouse stroma. In addition to as-prepared SiNW as a control, calcium phosphate (CaP)-coated SiNW, and alendronate modified CaP/SiNW, a new bisphosphonate containing a glucose moiety was prepared by Montchamp and co-workers, and this bisphosphonate adsorbed onto the surface of CaP/SiNW was examined *in vitro* as well. Stromal cell proliferation was monitored for up to 1 week (Jiang *et al.*, 2009). The relative cytocompatibility of these nanowires was found to follow the order: glucose-bisphosphonate-CaP/SiNW ~ CaP/SiNW ~ SiNW >> alendronate CaP/SiNW. Thus in terms of tolerance of surface functionality, the bound alendronate (with the exposed primary amine) produces the relatively strongest cytotoxic response. Such results, along with the observed inhibition of calcification in the alendronate-modified CaP/SiNW assays, are consistent with the *in vivo* results of Bodde and co-workers, who found that the presence of alendronate in synthetic bone cement did not increase bone formation in femoral defects present in a rabbit model (Bodde *et al.*, 2008). This cytotoxic response is readily reversed when the NH_2 species of the alendronate derivative is subsequently replaced by a more cytocompatible glucose-bisphosphonate/CaP/SiNW species. For the other nanowire surfaces, there is no significant difference observed between the behavior of the calcium phosphate-coated SiNW and that of the as-prepared silicon oxide-terminated SiNW, apparently a consequence of cellular response to oxygen-rich surfaces.

Overall, such results confirm our expectation that it is possible to sensitively 'tune' the *in vitro* performance of these scaffolds with surface chemistry, specifically with regard to FDA-approved therapeutic agents such as bisphosphonates (e.g. alendronate).

6.4 Construction of silicon nanowire (SiNW)-polymer scaffolds: mimicking trabecular bone

Recall that in a legitimate tissue engineering scaffold, a suitable cell line must proliferate, attach and subsequently differentiate to achieve the intended goal. Extensive prior work has established the sensitivity of such cellular processes to scaffold architecture and composition, so the ability to sculpt the nanowire in a three-dimensional context in a facile manner is crucial for pragmatic use in the long term. As a starting point, we have investigated fabricating SiNW in composite form with FDA-approved polymers such as polycaprolactone (PCL). To date, three different approaches have been pursued: 1 nanowire transfer onto highly porous polymer surfaces; 2 uniform nanowire transfer onto porous polymer surfaces with horizontally oriented NW; 3 vertical Si nanowire arrays on patterned polymer substrates. These are described in more detail below.

6.4.1 SiNW transfer onto highly porous polymer surfaces

Although fabrication of vertical, carefully positioned nanowire arrays is optimal for numerous applications, non-woven interconnected nanowire networks, that is 'carpets' are far easier to fabricate and can prove useful for this type of intended material. The use of large Au catalyst domains (micron-sized), along with locally high Si reactant concentrations, can provide SiNW/microwire films thick enough such that it is possible to manipulate free-standing structures of millimeter dimensions with a pair of tweezers. These Si 'carpets' can then be physically transferred to a porous polymer scaffold, such as one constructed from PCL. The base porous polymer scaffold is prepared using well-established techniques such as salt leaching (Thomson *et al.*, 2000; Whitehead *et al.*, 2008). In this method, a water-soluble porogen is suspended in the proper mass ratio, with the polymer dissolved in a polar organic solvent such as chloroform, followed by evaporation; removal of the porogen takes place by aqueous extraction, thereby leaving behind a flexible construct with porosities as high as 90%. The nanowire 'carpet' is then heated to a temperature slightly below that of the melting point of the polymer, then placed in physical contact with the porous polymer; the flexible non-woven nanowires thus adapt to the morphology of the porous polymer scaffold. This is exemplified by our experiments with 90% porous PCL sponges prepared by the salt leaching process and embedded with Si microwires (Fig. 6.4(a)). The

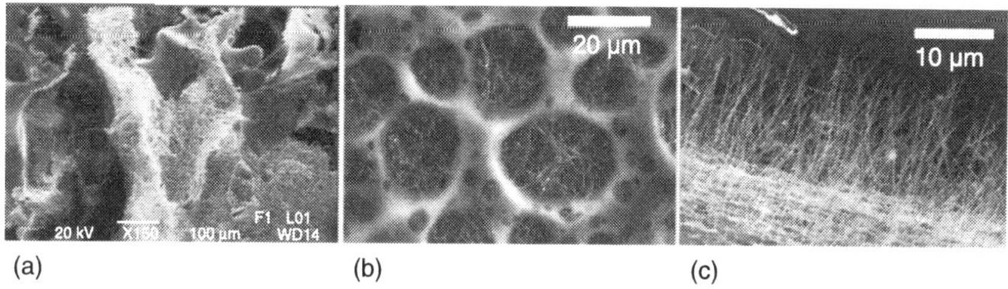

6.4 Different silicon nanowires (SiNW)/polycaprolactone (PCL) composite scaffolds: (a) SiNW 'carpet'/porous PCL structures; (b) horizontal SiNW/porous PCL films; (c) vertical SiNW arrays on solid PCL substrates.

morphological similarity of such structures to the microstructure of trabaceular bone is intriguing.

6.4.2 Uniform nanowire transfer onto porous polymer surfaces with horizontally oriented nanowires

An alternative approach to producing horizontally aligned NW in a porous polymer surface has been developed, with this approach ideally providing a more uniform distribution of NW in the porous polymer network (Jiang *et al.*, 2013). It entails placement of a PCL film on top of SiNW array and heating at 110 °C for 1 hour; after cooling, the sample is rinsed with chloroform drop-wise until the SiNW are exposed, thereby generating a fairly uniform morphology with pore features approaching the dimension of individual cells (Fig. 6.4(b)). The goal is thus to allow gravity and the pore location to guide cellular interaction with the nanowire construct present at the bottom of the polymer pit.

6.4.3 Vertical SiNW arrays on patterned polymer substrates

In order to probe the role of nanowire orientation on *in vitro* response, it is also important to produce vertical nanowire arrays in a given polymer susbtrate. This is possible in principle by imprinting solid structures of low melting point polymers such as PCL with a given type of nanowire (Jiang *et al.*, 2013). For example, a film of PCL ($\sim 1 \times 1$ cm^2) and a SiNW array are initially heated independently at 110 °C; once the contacting surface of PCL pellet starts to melt, the heating is stopped and the semitransparent PCL film is placed on top of SiNW. After cooling, the bulk PCL film can be physically removed, and a vertical array of SiNW embedded in the polymer surface remains behind (Fig. 6.4(c)).

6.5 The role of SiNW orientation in cellular attachment, proliferation and differentiation in the nanocomposite

6.5.1 Cell attachment assays with mesenchymal stem cells (MSCs)

Another functional requirement of a successful bone tissue engineering scaffold requires the ready attachment of mesenchymal stem cells (MSC) and their subsequent differentiation along an osteogenic pathway. Thus, an initial assessment of cell adhesion and proliferation on these SiNW/polymeric composites is crucial, as variations in nanoscale topography of a given substrate can impact focal adhesion assembly – a key step in the attachment of cells on biomaterials (Bershadsky *et al.*, 2006). In the results described here, the attachment of MSC on the SiNW/PCL scaffolds was examined by scanning electron microscopy (SEM) as a function of culturing time. PCL- and SiNW-only control samples of similar surface topographies were used to properly evaluate effects of SiNW/PCL composite surface topography on MSC attachment.

Figure 6.5 shows the representative morphology of MSC after being cultured for 7 days on horizontal and vertically aligned SiNW in PCL as well as SiNW-only films. To probe the possible role of surface oxide thickness on cell attachment, SiNW controls with two different oxide thicknesses were examined (2–3 nm (native oxide) and 10 nm). When cultured on SiNW alone (synthesized by a VLS method), MSC showed a rounded shape and restricted spreading (Fig. 6.5(a), (b)), independent of oxide thickness. In contrast, MSC on SiNW/PCL composites with the NW in a horizontal configuration showed elongated shapes with multiple filopodia stretching out and attaching to the composite underneath (Fig. 6.5(c)). For MSC grown on SiNW/PCL composites with the NW in a vertical array (Fig. 6.5(d)), the penetration of SiNW through cells is clearly visible. The shape of the cell was distorted and apparently affected by the presence of the surrounding SiNW. It should also be pointed out that in the case of horizontal SiNW/PCL composites, long-range alignment of cells is readily observed on extended periods of exposure (3 weeks or more).

The attachment of cells on these SiNW/PCL composites likely takes place through formation of specialized supramolecular protein complexes called focal adhesions (Bershadsky *et al.*, 2006). These adhesions not only serve as the mechanical linkages to the surface of biomaterial, but also as a biochemical signaling hub to concentrate and direct numerous signaling proteins at sites of integrin binding and clustering. Variations in nanoscale topography of the substrate can modulate adhesion assembly (Arnold *et al.*, 2004; Cavalcanti-Adam *et al.*, 2007; Spatz and Geiger, 2007), resulting in the regulation of cell behavior by altering signaling pathways (Bershadsky *et al.*, 2006). In our experiments, SiNW with no PCL content greatly restricted the cell spreading, presumably because of

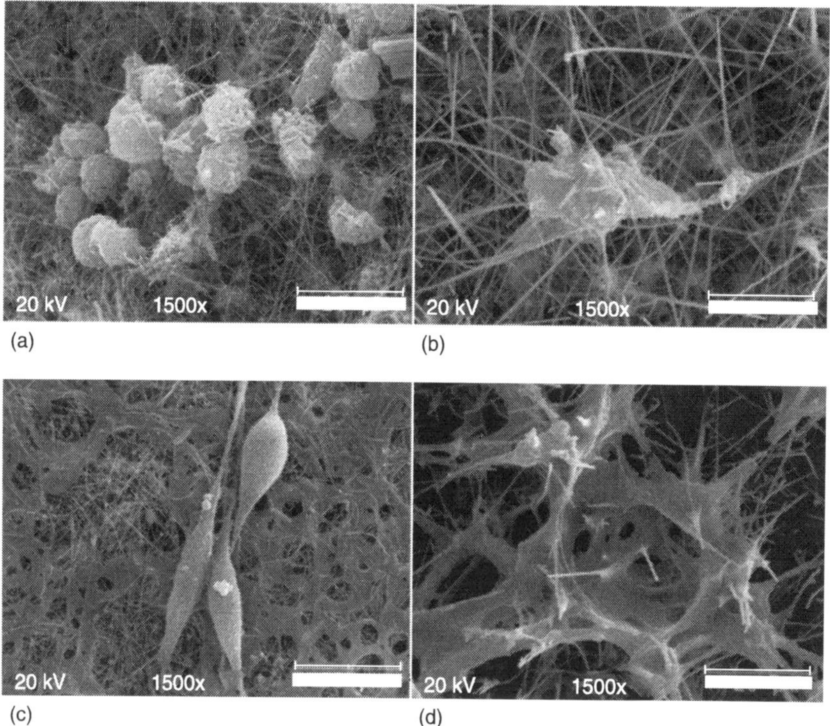

6.5 Scanning electron microscope images of mesenchymal stem cells (MSC) cultured on (a) silicon nanowire (SiNW) possessing a relatively thick oxide surface layer (prepared by VLS method); (b) SiNW with a thinner surface oxide layer; (c) SiNW/polycaprolactone (PCL) composite with a horizontal NW configuration; (d) SiNW/PCL composite with a vertical array of SiNW. Scale bar 20 μm.

a lack of focal adhesion and filopodia formation thereby limiting MSC attachment and migration. The role of the PCL phase coupled with SiNW provides a proper platform for cells to attach and migrate.

While the horizontally oriented SiNW in a PCL matrix can successfully support cell adhesion and spreading, vertically aligned SiNW on flat PCL substrates can also support cell growth without inducing significant cell death by penetration. Such results are consistent with those reported previously by the Yang group who showed that the engulfment of vertically oriented SiNWs (on Si surfaces) by cells did not induce cell death; rather DNA immobilized on SiNW could be delivered into cells with the nanowire acting as a vector (Kim *et al.*, 2007). Other studies have shown that vertical SiNW can be used to guide neuronal progenitor growth, knock down transcript levels, inhibit apoptosis and introduce targeted proteins to specific organelles by impaling and delivering biomolecules into living cells (Shalek *et al.*, 2010).

6.5.2 Viability assays of MSCs on SiNW/polycaprolactone (PCL) composites

Simple colorimetric assays are available to test the viability of MSC on SiNW/PCL composites, such as the MTT assay (Mosmann, 1983). As shown in Fig. 6.6, cells grown on SiNW/PCL composites exhibited increased viabilities as the culturing time increases. The viability of MSC on SiNW/PCL composites is higher than those grown on SiNW surfaces, with composites fabricated by the printing method exhibiting a higher viability than those grown on composites fabricated by the embedding method.

The observed lowest viability of cells cultured on SiNW surfaces is consistent with cell attachment experiments in which stromal cells generally showed rounded shapes as a sign of their unhealthy status. It is unclear why MSC grown on vertical SiNW exhibit higher viability than those grown on horizontal SiNW/PCL surfaces. As reported previously, the transcript levels of five common housekeeping genes of HeLa S3 cells and human fibroblasts cultured on vertical SiNW are similar to those in cells cultured on multiwell plates, indicating that perturbations of cell growth induced by vertical SiNW might be negligible (Shalek *et al.*, 2010). Although the observed deviations might be attributed to a cell-dependent manner or high sensitivity of cells to changes of external surface features, details of cell physiology response to nanowire penetration are a topic of future investigation.

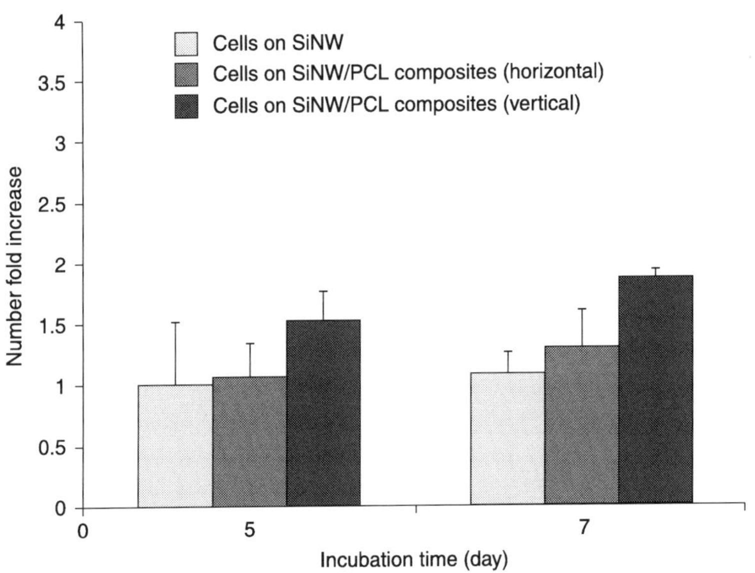

6.6 Viability of mesenchymal stem cells (MSC) cultured on silicon nanowire (SiNW) (MTT assays). The cells are affected by the surface features of different substrates. SiNW as a control substrate were synthesized by a standard VLS method.

6.5.3 Differentiation of MSCs on SiNW/PCL composites

Osteogenic differentiation of MSC cultured on SiNW/PCL composites can be readily assessed by an analysis of alkaline phosphatase (ALP) activity spectrophotometrically (Bessey *et al.*, 1946). As shown in Fig. 6.7, the ALP activity of MSC cultured on SiNW/PCL composites increased steadily with increasing culturing time, and there is a small, but statistically insignificant greater level of ALP expression for stromal cells interacting with SiNW/PCL composites prepared by the printing method. Based on chemical similarities between platforms, similar differentiation behavior is anticipated. In the future, additional refinement of differentiation behavior in these nanowire composites can possibly be achieved by additional chemical modifications such as coupling of synthetic biomimetic peptides (Egusa *et al.*, 2009) or bone morphogenetic protein (Sampath *et al.*, 1992; Bostrom and Camacho, 1998; Ripamonti and Duneas, 1998) onto composite surfaces. As a consequence, more detailed assays of relevant protein marker expression (such as RT PCR, Western blot, etc.) will be required.

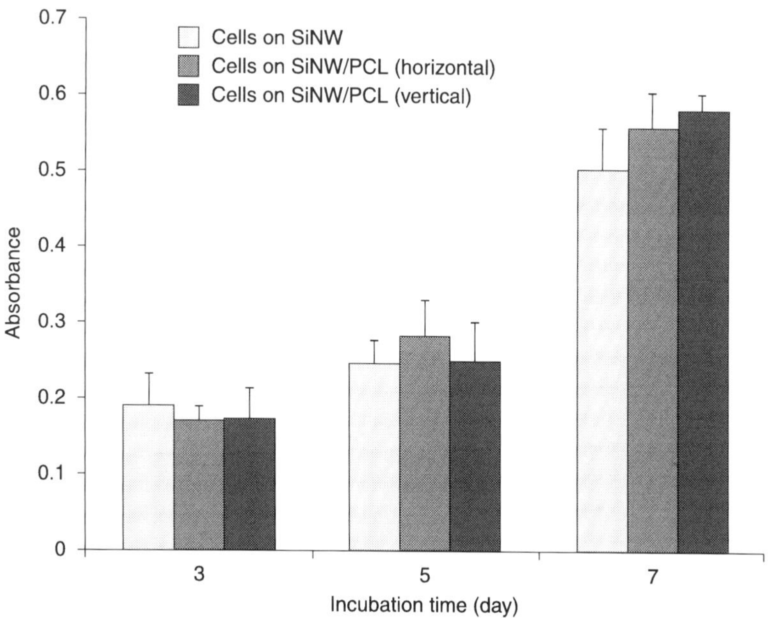

6.7 Alkaline phosphatase (ALP) activity of mesenchymal stem cells (MSC) cultured on different substrates. Absorbance is an indicator of cellular differentiation. Silicon nanowires (SiNW) as a control substrate were synthesized by a standard VLS method.

6.6 Conclusions and future trends

The fundamental properties of SiNW relevant to tissue engineering have been clearly identified, along with straightforward surface modification routes and methods for the fabrication of SiNW/PCL composites with either horizontally or vertically aligned SiNW arrays. Future opportunities will likely entail more detailed studies on the impact of chemical modification on the cell-specific *in vitro* response, as well as fabrication of 3-D composite scaffolds with appropriate porosities and mechanical characteristics.

6.7 Acknowledgement

The author gratefully acknowledges the Robert A. Welch Foundation (Grant P-1212) for their generous support of this research.

6.8 References

Anderson, S., Elliott, H., Wallis, D., Canham, L.T., and Powell, J.J. (2003), Dissolution of different forms of silicon of partially porous silicon wafers under simulated physiological conditions. *Phys. Stat. Solidi* (a), **197**, 331–5.

Arnold, M., Cavalcanti-Adam, E.A., Glass, R., Blummel, J., Eck, W., *et al.* (2004), Activation of integrin function by nanopatterned adhesive interfaces. *Chemphyschem.*, **5**, 383–8.

Atala, A., Kasper F.K., and Mikos A.G. (2012), Engineering complex tissues. *Sci. Transl. Med.* 14, 160rv12. doi: 10.1126/scitranslmed.3004890.

Bershadsky, A., Kozlov, M., and Geiger, B. (2006), Adhesion-mediated mechanosensitivity: a time to experiment, and a time to theorize. *Curr. Opinion Cell Biol.*, **8**, 472–81.

Bessey, O.A., Lowry, O.H., and Brock, M.J. (1946), A method for the rapid determination of alkaline phosphates with five cubic millimeters of serum. *J. Biol. Chem.*, **164**, 321–9.

Bodde, E. W. H., Kowalski, R. S. Z., Spauwen, P. H. M., and Jansen, J. A. (2008), No increased bone formation around alendronate or omeprazole loaded bioactive bone cements in a femoral defect. *Tissue Eng. A*, **14**, 29–39.

Bostrom, M.P., and Camacho, N.P. (1998), Potential role of bone morphogenetic proteins in fracture healing. *Clinical Orthopaedics and Related Research*, **355**, S274–82.

Bowditch, A., Waters, K., Gale, H., Rice, P., Scott, E., *et al.* (1999), In vivo assessment of tissue compatibility and calcification of bulk and porous silicon. *Mater. Res. Soc. Symp. Proc.*, **536**, 149–56.

Canham, L. T. (1995), Bioactive silicon structure fabrication through nanoetching techniques. *Adv. Mater.*, **7**, 1033–7.

Cavalcanti-Adam, E.A., Volberg, T., Micoulet, A., Kessler, H., Geiger, B., and Spatz, J.P. (2007), Cell spreading and focal adhesion dynamics are regulated by spacing of integrin ligands. *Biophys. J.*, **92**, 2964–74.

Coffer, J.L., Whitehead, M.A., Nagesha, D.K., Mukherjee, P., Akkaraju, G., *et al.* (2005), Porous silicon-based scaffolds for tissue engineering and other biomedical applications. *Phys. Stat. Sol. (a)*, **202**, 1451–5.

Cui, Y., Duan, X., Hu, J., and Lieber, C.M. (2000), Doping and electrical transport in silicon nanowires. *J. Phys. Chem. B*, **104**, 5213–16.

Egusa, H., Kaneda, Y., Akashi, Y., Hamada, Y., Matsumoto, T., *et al.* (2009), Enhanced bone regeneration via multimodal actions of synthetic peptide SVVYGLR on osteoprogenitors and osteoclasts. *Biomaterials*, **30**, 4676–86.

Hench, L.L., Splinter, R.J., Allen, W.C., and Greenlee, T.K. (1971), Bonding mechanisms at the interface of ceramic prosthetic materials. *J. Biomed. Res. Symp.*, **2**, 117–41.

Hochbaum, A. I., Gargas, D., Hwang, Y. J., and Yang, P. D. (2009), Single crystalline mesoporous silicon nanowires. *Nano Lett.*, **9**, 3550–4.

Jiang, K., Fan, D., Belabassi, Y., Akkaraju, G., Montchamp, J.L., and Coffer, J.L. (2009), Medicinal surface modification of silicon nanowires: impact on calcification and stromal cell proliferation. *ACS Appl. Mater. Interfaces*, **1**, 266–9.

Jiang, K., Akkaraju, G., and Coffer, J.L. (2013), Silicon nanowire/polycaprolactone composites and their impact on stromal cell function. *J. Mater. Res.*, **28**, 185–92.

Kim, W., Ng, J. K., Kunitake, M. E., Conklin, B. R., and Yang, P. (2007), Interfacing silicon nanowires with mammalian cells. *J. Am. Chem. Soc.*, **129**, 7228–9.

Kokubo,T., Kushitrani, H., Sakka, S., Kitsigi, T., and Yamamuro, T. (1990), Solutions able to reproduce in-vivo surface structure changes in bioactive glass-ceramic A-W. *J. Biomed. Mater. Res.*, **24**, 721–34.

Lanza, R.P., Langer, R., and Vacanti, J. (2007) *Principles of Tissue Engineering*, Academic Press, NY, 3rd edition.

Mosmann, T. (1983), Rapid colorimetric assay for cellular growth and survival: application to proliferation and cytotoxicity assays. *J. Immunol. Meth*, **65**, 55–63.

Nagesha, D. K., Whitehead, M.A., and Coffer, J.L. (2005), Biorelevant calcification and non-cytotoxic behavior in silicon nanowires. *Adv. Mater.*, **17**, 921–4.

Qu, Y., Liao, L., Li, Y., Zhang, H., Huang, Y., and Duan, X. (2009), Electrically conductive and optically active porous silicon nanowires. *Nano Lett.*, **9**, 4539–43.

Ripamonti, U., and Duneas, N. (1998), Tissue morphogenesis and regeneration by bone morphogenetic proteins. *Plastic and Reconstructive Surgery*, **101**, 227–39.

Sampath, T.K., Maliakal, J.C., Hauschka, P.V., Jones, W.K., Sasak, H., *et al.* (1992), Recombinant human osteogenic protein-1 (hOP-1) induces new bone formation in vivo with a specific activity comparable with natural bovine osteogenic protein and stimulates osteoblast proliferation and differentiation in vitro. *J. Biol. Chem.*, **267**, 20352–62.

Schmidt, V., Joerg V. Wittemann, J.V., Senz, S., and Gösele, U. (2009). Silicon nanowires: a review on aspects of their growth and their electrical properties. *Adv. Mater.*, **21**, 2681–702.

Shalek, A.K., Robinson, J.T., Karp, E.S., Lee, J.S., Ahn, D.R., *et al.* (2010), Vertical silicon nanowires as a universal platform for delivering biomolecules into living cells. *Proc. Nat. Acad. Sci. USA*, **107**, 1870–5.

Spatz, J.P., and Geiger, B. (2007), Molecular engineering of cellular environments: cell adhesion to nano-digital surfaces. *Meth. Cell Biology*, **83**, 89–111.

Thomson, R.C., Shung, A.K., Yaszemski, M.J., and Mikos, A.G. (2000), 'Polymer scaffold processing', in *Principles of Tissue Engineering*, Lanza, R., Langer, R., and Vacanti, J. (eds). Academic Press, NY, 2nd edition, pp. 251–62.

Wang, Y., Lew, K.K., Ho, T.T., Pan, L., Novak, S.W., *et al.* (2005), Use of phosphine as an n-type dopant source for vapor–liquid–solid growth of silicon nanowires. *Nano Lett.*, **5**, 2139–43.

Whitehead, M.A., Fan, D., Mukherjee, P., Akkaraju, G.R., Canham, L.T., and Coffer, J.L. (2008), High-porosity poly(epsilon-caprolactone)/mesoporous silicon scaffolds: calcium phosphate deposition and biological response to bone precursor cells. *Tissue Eng. A.*, **14**, 195–206.

7
Mediated differentiation of stem cells by engineered semiconducting silicon nanowires

T.-J. YEN and H.-I LIN, National Tsing Hua University, Taiwan R. O. C.

DOI: 10.1533/9780857097712.2.118

Abstract: This chapter begins with a general introduction regarding approaches to the use of biophysical and biochemical stimuli in regulating the fates of human mesenchymal stem cell (hMSC) differentiation (such as neuron cells, muscle cells, bone cells, etc.). Among these approaches, mechanical stimulation provides an innovative strategy to specific inductions of hMSC. We then describe the application of the specific biophysical stimulation of spring constant obtained from silicon nanowires (SiNW) to induce the fate regulation effect on hMSC. This is achieved in the absence of induction medium through alpha 2 beta 1 integrin heterodimers, vinculin and focal adhesion kinase.

Key words: human mesenchymal stem cell (hMSC), electroless metal deposition (EMD), silicon nanowires (SiNW), spring constant, mechanical stimulation.

7.1 Introduction

For the past two decades, an ongoing need for the storage of umbilical cord blood has existed because this cord blood contains lots of precious unadulterated stem cells (SC) that could help repair certain adverse physiological problems, such as hematopoietic, genetic disorders and organ failure. SC, mostly contributed from bone marrow, adipose tissue and umbilical cord blood in mammals, are the most critical biological cells that are responsible for recovery, proliferation and differentiation of all multicellular organisms. According to their different functions, SC can be classified into three dominant types: 1 totipotent SC, 2 pluripotent SC, and 3 unipotent SC. Embryonic SC (ESC), a kind of totipotent SC, represent a full-functioned cell differentiation into all the specialized cells and further individual growth. Yet, so far researchers involved with manipulation of ESC physiology are struggling with isolation of the inner cell mass of embryos (blastocysts) (Ankey, 2008), and are restricted by the government policies and ethical issues (Ankey, 2008; McLaren, 2000; Fuchs and Segre, 2000; Dai *et al.*, 2011).

Mesenchymal stem cells (MSC), a kind of pluripotent SC derived from various adult tissues, enables multilineage differentiation potential and thereafter function as a good material for studies in the field of stem cell therapy (Dai *et al.*, 2011;

Vija *et al.*, 2009; Ozawa *et al.*, 2008; Shah, 2012). As for unipotent SC, they can only differentiate into their own cell type. Fortunately, unipotent SC, such as skin stem cells and muscle stem cells, still possess a self-renewal property that is absent from normal non-stem cell types such that injured tissue can recover.

In comparing the differentiation potential among the three major types of SC, totipotent SC possesses the highest differentiation potential beyond the pluripotent and the unipotent SC, suggesting that they are the best candidate for the study of tissue engineering and clinical therapy. Nevertheless, because of the aforementioned limitations of totipotent SC, pluripotent SC, which are capable of differentiating into various cell lineages, provides another choice for tissue engineering and clinical therapy.

In fact, pluripotent MSC can differentiate into various cell lineages, such as osteocytes, adipocytes, chondrocytes, etc. (Friedenstein *et al.*, 1987; Yang *et al.*, 2008; Ghaedi *et al.*, 2011; Mohsin *et al.*, 2011; Danišovič *et al.*, 2012; Wu and Tao, 2012; Charbord, 2010; Kuo *et al.*, 2012), as shown in Fig. 7.1. The mechanisms of triggering these differentiation pathways of MSC are diverse and

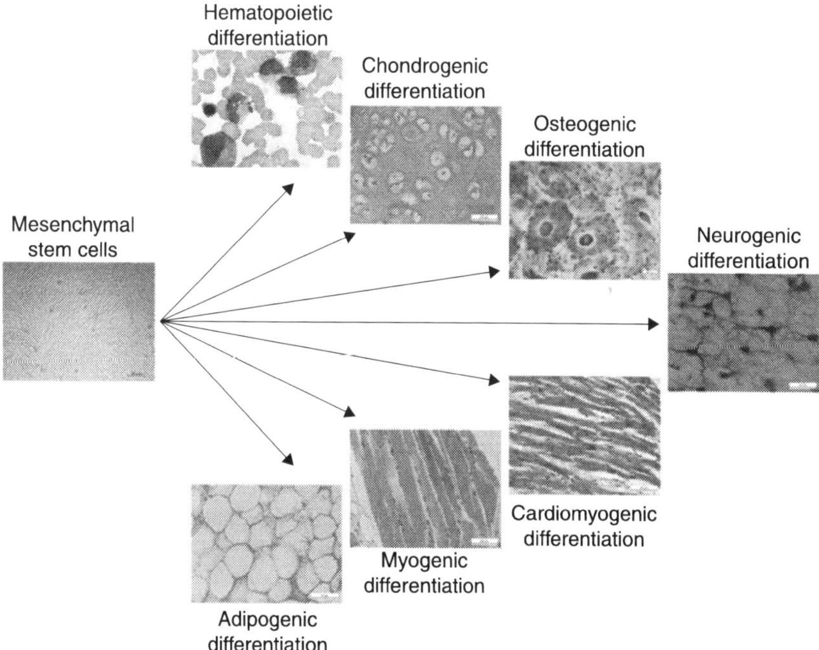

7.1 Plasticity of mesenchymal stem cells (MSC). MSC undergo multilineage differentiation under proper conditions, both *in vivo* and *in vitro* (reprinted from *Tissue and Cell*, 44, Danišovič *et al.*, 'Growth factors and chondrogenic differentiation of mesenchymal stem cells, 69–73, 2012, with permission from Elsevier).

can be classified into two typical types: biophysical stimulation (Liu *et al.*, 2010; Li *et al.*, 2011; Her *et al.*, 2012; Wang *et al.*, 2012; Even-Ram *et al.*, 2006), and biochemical stimulation (Kim *et al.*, 2008; Mathews *et al.*, 2012; Cheng *et al.*, 2012; Indrawattana *et al.*, 2004; Jiang *et al.*, 2012). For biophysical stimulation, the elasticity (stiffness) of the substrates to culture MSC atop is usually regarded as a key biophysical stimulation approach, especially given the elasticity of polymeric materials. For the polymeric materials, the elasticity can be modified easily through the degree of cross-linking and amount of additives. A clear relationship between the substrate elasticity and MSC fate regulation (Even-Ram *et al.*, 2006) is presented in Fig. 7.2. By culturing MSC on polymeric materials with various elasticities (1, 10 and 100 KPa), this specific extracellular mechanical stimulation can regulate MSC to neuronal cell, muscle cell and bone cell lineages, respectively. Following this interesting tendency, substrate elasticity promises itself as a sufficient biophysical stimulation in regulating the fates of MSC. Additionally, biochemical stimulation depends on the use of growth factors (Kim *et al.*, 2008; Cheng *et al.*, 2012; Indrawattana *et al.*, 2004; Jiang *et al.*, 2012; Kim *et al.*, 2012), protein mediation (Mathews *et al.*, 2012) and drug release methods (Shi *et al.*, 2010; Kim *et al.*, 2012) to force specific differentiations of MSC. The real mechanism(s) of biochemical stimulation, however, remains controversial.

In order to understand the mechanisms of biophysical and biochemical stimulation, researchers have employed nanomaterials to conduct their stimuli in regulating the fates of MSC. Nanomaterials with various structures, including nanowires/nanofibers/nanopillars (Kuo *et al.*, 2012; Jiang *et al.*, 2012; Brammer *et al.*, 2011), nanopores (Her *et al.*, 2012; Kim *et al.*, 2012) and nanotubes (Zhao *et al.*, 2012; Tay *et al.*, 2010; Rodrigues *et al.*, 2012; Hu *et al.*, 2012; Zhao *et al.*,

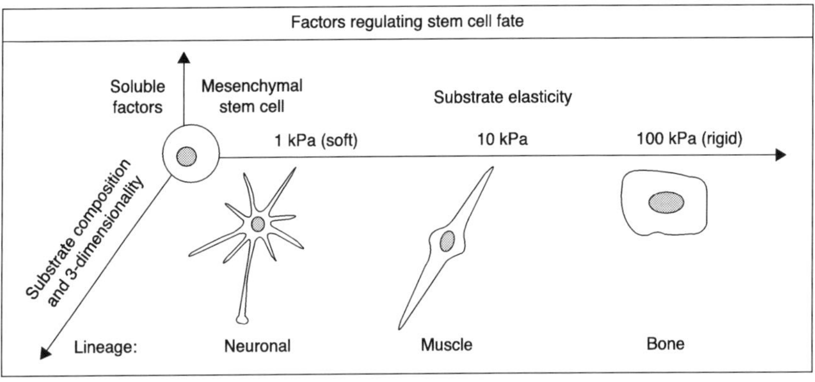

7.2 Controlling stem cell fate (reprinted from *Cell*, 126, Even-Ram *et al.*, 'Matrix control of stem cell fate', 645–7, 2006, with permission from Elsevier).

2013) give rise to extraordinary mechanical properties and/or some chemical additives to allow MSC to receive the extracellular stimulating signals and thus settle on their final differentiation pathways. Among these nanomaterials, silicon nanowires (SiNW) have been attracting great interest because of their useful properties, such as abundance in the Earth's crust, a wide range of conductance, controllable super-hydrophilic/hydrophobic properties, and a comparable size with biomolecules, such that recently SiNW have been employed in sensitive biological/chemical sensors and microfluidic systems. For example, the construction of typical biosensors with basic components of bioreceptors, electrical interfaces, signal amplifiers, signal processors and displays, and the relative scale of biological cells, viruses, proteins, nucleic acids and nanowires are presented in Figs. 7.3 and 7.4 (Chen *et al.*, 2011). As a consequence it is clear that the nanowires possess a comparable diameter to biological substances (viruses and proteins) that can not only input the stimulation but also receive chemical responses from biological cells. More recently, researchers further reported that by controlling the dimension (diameter, length and topography) of SiNW, one can enhance cellular viability and physiological functions of the cultured cells

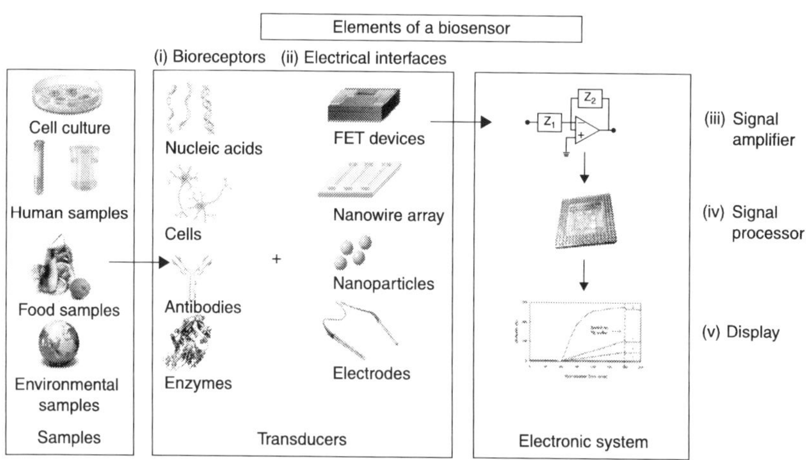

7.3 The construction of typical biosensors with elements and selected components. The procedures are described as follows: (i) receptors specifically bind the analyte; (ii) an interface architecture where a specific biological event takes place and gives rise to a signal recorded by (iii) the transducer element; (iv) computer software to convert the signal into a meaningful physical parameter; finally, the resulting quantity is displayed through (v) an interface to the human operator (reprinted from *Nano Today*, 6, Chen *et al.*, 'Silicon nanowire field-effect transistor (FET)-based biosensors for biomedical diagnosis and cellular recording investigation', 131–54, 2011, with permission from Elsevier).

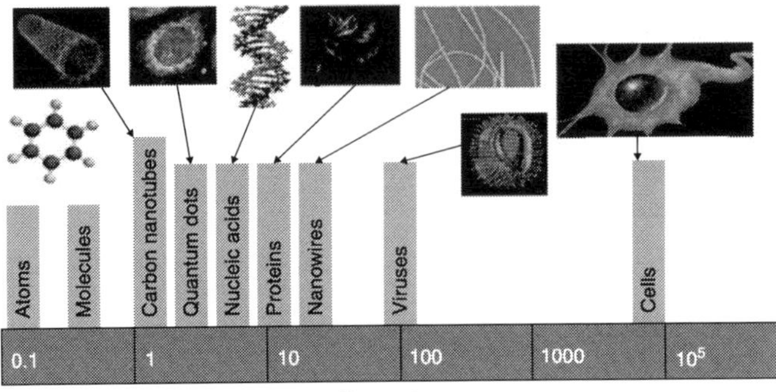

7.4 The sizes of nanomaterials (NW and NT) in comparison with some biological entities, such as bacteria, viruses, proteins and DNA (reprinted from *Nano Today*, 6, Chen *et al.*, 'Silicon nanowire field-effect transistor-based biosensors for biomedical diagnosis and cellular recording investigation', 131–54, 2011, with permission from Elsevier).

(Kuo *et al.*, 2012; Pui *et al.*, 2011; Chen, K.I. *et al.*, 2011; Chen, C.C. *et al.*, 2011). The aforementioned characteristics of SiNW prove them to be an ideal platform for *in vitro* studies.

Accordingly, the proper dimensions of SiNW can be obtained from many methods, for instance, vapor-liquid-solid process (VLS) (Wang *et al.*, 2008; Wu *et al.*, 2012; Rogacs *et al.*, 2010; Pan *et al.*, 2005; Rosaz *et al.*, 2011; Akhtar *et al.*, 2008), solid-liquid-solid process (SLS) (Wang *et al.*, 2008; Wu *et al.*, 2012; Yan *et al.*, 2000; Yu *et al.*, 2001), oxide-assisted growth method (OAG) (Wang *et al.*, 2008; Wu *et al.*, 2012; Niu *et al.*, 2004; Yao *et al.*, 2005) and electroless metal deposition (EMD) (Kuo *et al.*, 2012; Wu *et al.*, 2012; Chen *et al.*, 2008) (see Chapter 2 of this book for additional details). According to the above methods, SiNW are fabricated in diverse morphologies shown in Fig. 7.5 (Kuo *et al.*, 2012; Pan *et al.*, 2005; Yu *et al.*, 2001; Niu *et al.*, 2004). Among these four major methods, it is laborious to fabricate well-aligned and uniform SiNW by means of VLS, SLS and OAG methods because of the drawbacks of high reaction temperature processes, random orientations, distributed diameters and impurities, which consequently hinder their applications. On the other hand, the EMD method benefits from a near room-temperature process that is free from template requirements, possesses a low cost and substrate orientation-sensitivity, and can supply very uniform and orientated SiNW, thereby encouraging a consistent mechanical property (elasticity). In this chapter, we will examine the process of fabricating one-dimensional SiNW by the EMD method, with the controlled dimension of SiNW to engineer their elasticity (i.e. spring constant, K_x), and use of the technique to regulate osteogenic differentiation of MSC as well as their corresponding differentiation pathway.

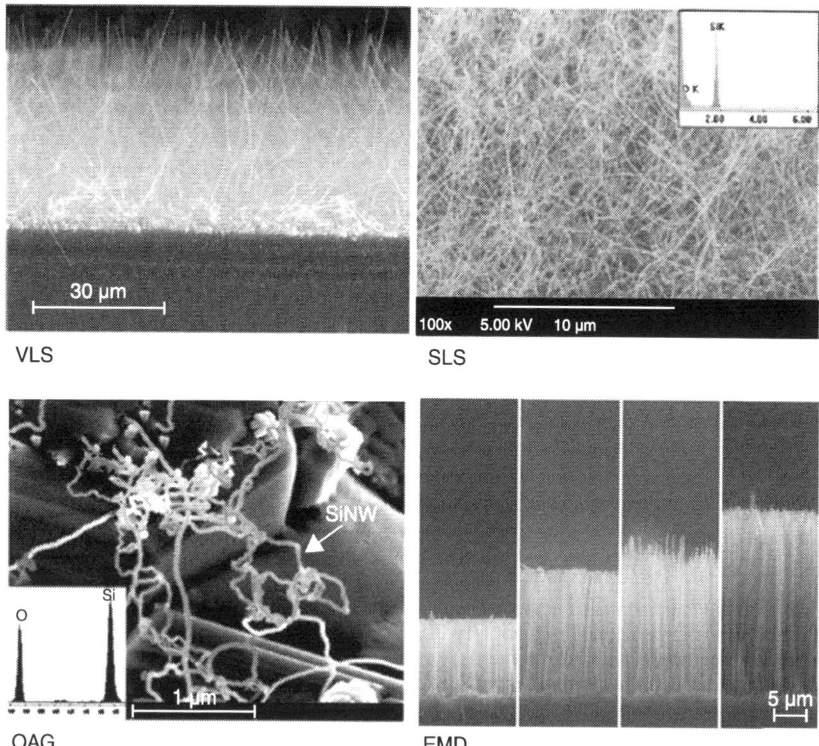

7.5 Silicon nanowire morphologies for various growth methods. (Reprinted from *Biomaterials*, 33, Kuo *et al.*, 'Regulation of the fate of human mesenchymal stem cells by mechanical and stereo-topographical cues provided by silicon nanowires', 5013–22, 2012, with permission from Elsevier; *Journal of Crystal Growth*, 277; Pan *et al.*, 'Effect of diborane on the microstructure of boron-doped silicon nanowires, 428–36, 2005, with permission from Elsevier; *Physica E*, 9, Yu *et al.*, 'Controlled growth of oriented amorphous silicon nanowires via a solid-liquid-solid (SLS) mechanism', 2001 with permission from Elsevier; *Physica E*, 24, Niu *et al.*, 'Sulfide-assisted growth of silicon nano-wires by thermal evaporation of sulfur powders', 2004 with permission from Elsevier.)

7.2 Methods for fabricating silicon nanowires (SiNWs)

7.2.1 Electroless metal deposition (EMD) method

SiNW were fabricated on (100) *n*-type Si wafer pieces (1 cm × 1 cm) by an electroless plating method as described previously (Kuo *et al.*, 2012; Chen *et al.*, 2008) (see Chapter 2 of this volume as well). The procedure produces SiNW with different lengths and diameters. The as-prepared SiNW and Si wafer piece controls were then sterilized in a steam autoclave at 121 °C for 15 minutes and washed twice with PBS for 10 minutes before being used for cell culture. The detailed parameters are tabulated in Table 7.1.

7.2.2 Biological cell culture process

Isolation of human bone marrow-derived mesenchymal stem cells (MSCs)

All biological cellular experiments and the obtained human MSC (hMSC) were conducted according to regulations stipulated by the Taipei Veterans General Hospital (TVGH) in Taiwan and operated in a qualified cell culture room at this location. Isolation of bone marrow-derived hMSC was achieved according to a previously published method (Lee K.D. *et al.*, 2004; Lee O.K. *et al.*, 2004). Briefly, human bone marrow was aspirated from the iliac crest of healthy donors during fracture fixation surgery. Institutional Review Board approval was obtained and informed consent was given by the donors before the bone marrow samples were collected. Mononuclear cells were harvested using a commercially available kit according to the manufacturer's instructions. Non-adherent cells were then washed away. hMSC were obtained by limited dilution and maintained in a commercially available expansion medium in the presence of 100 units/mL penicillin, 1000 units/mL streptomycin and 2 mmol/L L-glutamine. The surface immune-phenotype and multilineage differentiation potential of hMSC were confirmed before use in further experiments.

Cellular viability

The viability of the adherent cells on the SiNW chips was assessed by staining with calcein acetoxymethyl ester (Calcein AM) and ethidium homodimer-1 (EthD-1). The adherent live cells (green, stained with Calcein AM) and adherent dead cells (red, stained with EthD-1) were quantified from the images using the ImageJ program. hMSC were seeded on the SiNW chips for 72 hours and then incubated with 2 μM of Calcein AM and 4 μM of EthD-1 for 30 minutes (Lee *et al.*, 2008). Subsequently, epifluorescence images were collected by inverted

Table 7.1 Electroless metal deposition parameters for silicon nanowire fabrication

Group		I	II	III	IV
sEMD parameters	Period (min)	5	10	15	20
	Electrolyte		0.03 M $AgNO_3$ + 4.6 M HF		
	Temperature (°C)		50		

fluorescence microscopy. The percentage of adherent live hMSC on each SiNW was normalized by the number of total adherent population. The amount of adherent living cells on the SiNW chips were quantified by a commercially available kit. hMSC cells (5×10^4 cells/mL) were seeded on each SiNW chip and cultured for 72 hours without induction medium. SiNW alone were used as substrate controls while assay reagent was used as the blank control. After incubation, the medium was removed and the cells were washed twice with PBS, then 100 μL of reagent was added to each well containing 500 μL of medium and the cells incubated at 37 °C for 4 hours. After incubation, 100 μL of reaction medium was added to each well of a 96-well plate in order to detect fluorescence. Fluorescent intensity was measured at 560 nm excitation and 590 nm emission by a spectrophotometer. The total quantity of adhering living cells was compared against the Group I SiNW and the relative amount of each SiNW was expressed as an adherent living cells value.

Gene expression and immunofluorescence staining

Expression levels of various genes by hMSC on the various SiNW and the 2-D flat Si were assessed by quantitative real-time polymerase chain reaction (RT-PCR). hMSC cells seeded on the various SiNW and a 2-D flat Si in 24-well plates, were incubated for 72 hours and respectively extracted their total RNA using TRIzol reagent. These RNA samples were then used for reverse transcription and subsequent PCR amplification. Quantitative real-time PCR was performed by LightCycler 480 Real-Time System. Intron spanning primers specific for each gene were designed using the Universal ProbeLibrary Assay Design Center and were detected using corresponding probes from the Universal ProbeLibrary (Roche) shown in Table 7.2. The average threshold cycle (Ct) for each gene was normalized by that of glyceraldehyde 3-phosphate dehydrogenase (GAPDH).

After 24 hours of hMSC seeding, non-adherent cells were removed by PBS washing and the remaining cells attached to the SiNW cultured for 3 days. At the end of this period, the cells on the SiNW were gently washed with PBS and then fixed in 4% paraformaldehyde for 20 minutes, permeabilized with 0.5% Triton X-100 in PBS for 3 minutes and finally blocked with 10% goat serum in PBS for 60 minutes. The fixed cells were immuno-stained for F-actin, phosphorylated focal adhesion kinase (pFAK) and vinculin. An inverted confocal fluorescence microscope was used to visualize the distribution of F-actin, vinculin and pFAK after immunofluorescence staining. The amounts of F-actin, pFAK, vinculin and alpha 2 integrin were consequently determined by Olympus FV1000 image software.

Cell fixation process

Cells cultured on the various substrates were washed twice with PBS and soaked in PBS (pH 7.4) with 2% glutaraldehyde buffer for 120 minutes in order to fix the

Table 7.2 Primer sequences and probes from the Universal Probe Library used in semi-qRT-PCR analysis

Gene name	Oligonucleotide sequence	Probe number
Beta 1 integrin	5′-TGACTTCCAGATTCCAGCAA-3′	30
	5′-CCACAGTTGTTACGGCACTC-3′	
Alpha2 integrin	5′-TCGTGCACAGTTTGAAGATC-3′	7
	5′-TGGAACACTTCCTGTTGTTACC-3′	
Alpha5 integrin	5′-CCCATTGAATTTGACAGCAA-3′	55
	5′-TGCAAGGACTTGTACTCCACA-3′	
FAK	5′-GTCTGCCTTCGCTTCACG-3′	45
	5′- GAATTTGTAACTGGAAGATGCAAG-3′	
COL1A1	5′-ATGTTCAGCTTTGTGGACCTC-3′	15
	5′-CTGTACGCAGGTGATTGGTG-3′	
Runx2	5′-CTACCACCCCGCTGTCTTC-3′	29
	5′-CAGAGGTGGCAGTGTCATCA-3′	
GAPDH	5′-GCTCTCTGCTCCTCCTGTTC-3′	60
	5′-ACGACCAAATCCGTTGACTC-3′	

FAK, focal adhesion kinase; COL1A1, type I collagen alpha 1.
Source: Kuo *et al.*, 2012.

cells. The fixation solution was removed and the cells washed thoroughly with PBS. The samples were post-fixed in 1% osmium tetroxide in PBS for 60 minutes, which was followed by serial dehydration in medical grade ethanol using concentrations ranging from 30% to 100%. Subsequently, sample drying was performed in 100% ethanol using a critical point dryer.

7.2.3 Material characterization

The dimensions of as-prepared SiNW were first imaged by SEM (JSM-6390, JEOL, Japan) with the length and diameter of each SiNW measured more than 300 times. Moreover, the cell-fixed and dehydrated samples were mounted on aluminum stubs using colloidal silver, then sputter-coated with platinum and the cellular morphologies observed by SEM. A field-emission transmission electron microscope (TEM, JEM-3000F) was utilized to characterize the SiNW microstructures.

7.3 Regulated differentiation for human mesenchymal stem cells (hMSCs)

With regard to biophysical and biochemical stimulation, natural polymers with similar chemical composition to tissues can be consequently employed as good cell culture matrices for studying cellular behavior. The most important natural polymers are collagen (Her *et al.*, 2012; Bosnakovski *et al.*, 2006; Noth *et al.*, 2007), hyaluronic acid (Her *et al.*, 2012; Collins and Birkinshaw, 2013) and chitosan (Kim *et al.*, 2012; Dash *et al.*, 2011; Sinha *et al.*, 2004). However, these natural polymers typically present drawbacks with regard to relative low mechanical strength, immunogenicity, complexity, variability in mass production and the assurance of pathogen removal (Koide, 2005). On the other hand, synthetic polymers such as polyethylene (PE), polyethylene glycol (PEG), polylactide-co-glycolide (PLGA) and poly methyl methacrylate (PMMA), allow researchers to tailor their physical, chemical, mechanical and degradative properties to overcome the drawbacks that natural polymers encounter.

For example, Liu *et al.* (2010) modified the PEG hydrogel with collagen mimetic peptide (CMP) in the sequence of (GPO)4GFOGER(GPO)4GCG, whose physical properties are tabulated in Table 7.3. This hybrid hydrogel, subsequently named PEG/collagen mimetic peptide afterwards, acts as a permeable scaffold that can encapsulate hMSC for studying their proliferation and differentiation into neocartilage and/or chondrocytes. According to a series of parameters (CMP concentration, precursor concentration, pH value, gelation time, gel yield and swelling degree), the modulus of this hybrid hydrogel is related to the gel yield and precursor concentration and ranges from 3.73 KPa to 5.56 KPa. This varied modulus affects the chondrogenesis of hMSC, in which a softer matrix exhibits stronger chondrogenic differentiation and a stiffer matrix induces the opposite results instead. Therefore, this soft PEG/CMP hydrogel shows promise as a biomimetic scaffold that provides a favorable environment for the chondrogenic differentiation of hMSC and is useful for the repair of cartilage defects.

In addition to the mechanical stimulation of polymeric materials, electrical stimulation is another promising approach to regulate the fate of hMSC (Hess *et al.*, 2012; Genovese *et al.*, 2009). Employment of both biophysical (electrical field stimulation) and biochemical stimulation (collagen-based substrates) were demonstrated in order to compare their effects on the regulation of hMSC (Hess *et al.*, 2012). Here biochemical stimulation depended on the artificial

Table 7.3 Physical properties of PEG/collagen mimetic peptide hybrid hydrogel

Gel #	CMP concent. (mM)	Precursor concent. (w/v)%	pH	Gelation time (min)	Gel yield (%)	Swelling degree	Ge (Pa)
1(PEG-CMP-10)	2.0	10	7.4	30	89.2	22.1	3730.0
2(PEG-CMP-10)	2.0	10	8.0	15	91.7	19.7	4070.6
3(PEG-10)	0	10	8.0	10	95.3	20.6	4231.8
4(PEG-CMP-15)	2.0	15	8.0	12	92.4	13.8	5214.7
5(PEG-15)	0	15	8.0	8	93.2	12.5	5561.5

Source: Liu *et al.*, 2010.

extracellular matrices (aECM), which were composed of collagen I (coll) and glycosaminoglycans (GAG) like chondroitin sulfate (CS), or a high-sulfated hyaluronan derivative (sHya). All three of these different materials were coated on three-dimensional poly(caprolactone) (PCL) scaffolds. The researchers utilized four different substrates (i.e. non-coated, collagen I-coated (coll), coll/chondroitin sulfate-coated (coll/CS) and coll/high-sulfated hyaluronan derivatives-coated (coll/sHya) scaffolds) and two culturing mediums (i.e. expansion medium (exm) and osteogenic differentiation media (osm)) to determine the difference of osteogenicity in hMSC, and four bone-related proteins during MSC differentiation to indicate the level of osteogenicity of hMSC (i.e. (a) runt-related transcription factor 2 (RUNX-2), (b) alkaline phosphatase (ALP), (c) osteopontin (OPN) and (d) osteocalcin (OC)), as shown in Fig. 7.6. The expression of RUNX-2, a key transcription factor for osteogenesis, showed no significant differences under different environments (Fig. 7.6(a)). ALP expression, however, strongly relied on sHya composition: in the case of aECM with sHya, the ALP expression became threefold higher than those of non-coated, coll and coll/CS coated scaffolds (Fig. 7.6(b)). Next, OPN is also a major protein, whose expression indicates the maturation of osteoblasts (Aubin, 2001). OPN expression in expansion medium and osteogenic differentiation medium exhibited extremely opposite results (Fig. 7.6(c)). Under expansion medium, the obtained OPN showed very low expression by days 14 and 28. However, the expression level of OPN on coll/sHya-coated scaffold increased up to 20-fold already by day 14 with the presence of osteogenic differentiation medium. By day 28, these coatings all met a similar expression level of OPN under osteogenic differentiation medium. Finally, OC is one typical gene marker to represent mature osteoblasts. The level of OC expression is responsble to indicate the degree of down-regulation of hMSCs. OC expression of hMSCs exhibited higher level in osteogenic differentiation medium compared with that of hMSCs cultivated in expansion medium. In Fig. 7.6(d), we may observe a small increase of OC with the presence of osteogenic differentiation medium, presented from coll/sHya-coated scaffold compared with the others. A quick summary here indicates that both the type of aECM (coll/sHya coated scaffold) and the presence of medium (osteogenic differentiation medium) can strongly promote the osteogenic differentiation of hMSC through biochemical stimulation.

After revealing the effects of aECM on hMSC regulation, another approach to regulate the fate of hMSC is electric field introduction (Hess *et al.*, 2012). The results concerning four important gene expressions with and without electric field are provided in Fig. 7.7. First, for RUNX-2 expression, both electric field and sHya were not able to measure its expression level. However, osteogenic differentiation medium did regulate a five-fold greater promotion of expression of RUNX-2 than that by expansion medium as shown in Fig. 7.7(a). Second, with respect to ALP expression (Fig. 7.7(b)), electric field introduction indeed enhanced ALP levels and showed the highest level on coll/sHya coated scaffold. Third, OPN presented no regulation under non-stimulation conditions by day 14, even on collagen- and coll/

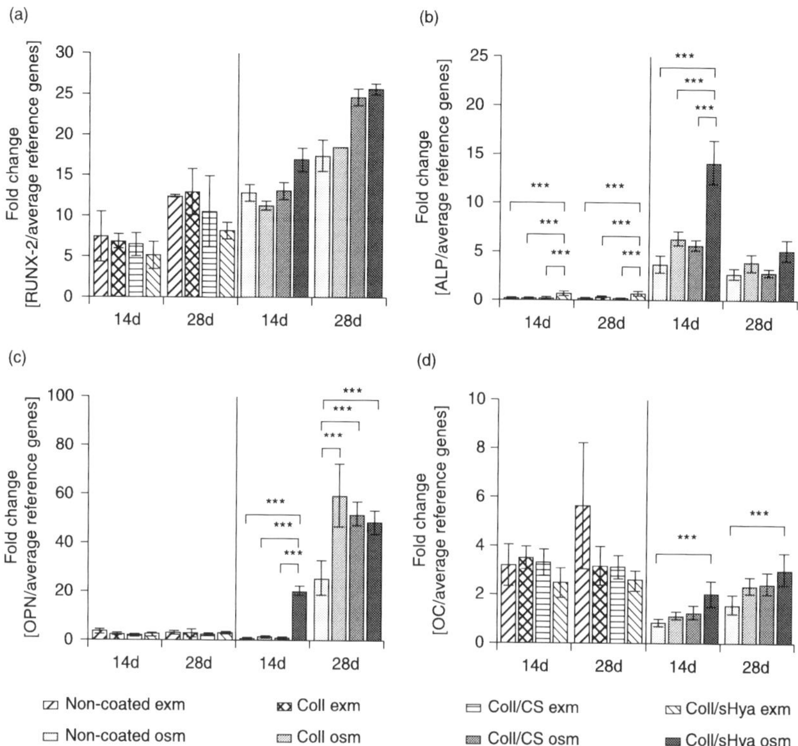

7.6 Gene-expression analysis of (a) RUNX-2, (b) ALP, (c) OPN and (d) OC of human mesenchymal stem cells cultivated on non-coated, collagen I-coated (coll), coll/chondroitin sulfate-coated (coll/CS) and coll/high-sulfated hyaluronan derivatives-coated (coll/sHya) scaffolds. Data points present the average ± SD. ($n=3$); *$p<0.05$, **$p<0.01$, ***$p<0.001$. (Reprinted from *Biomaterials*, 33, Hess *et al.*, 'Synergistic effect of defined artificial extracellular matrices and pulsed electric fields on osteogenic differentiation of human MSCs', 8975–85, 2012 with permission from Elsevier.)

sHya-coated scaffolds (Fig. 7.7(c)). In contrast, stimulation by electric field induced a five-fold higher expression than the control by day 28. Finally, OC expression (independent of coatings) obtained in osteogenic differentiation medium was retarded from relative to that in expansion medium, as shown in Fig. 7.7(d). Once the electric field was applied, the OC expression of coll/sHya-coated scaffolds achieved a two-fold higher level than coll/sHya-coated scaffold without electric field by day 28. From these results, it is seen that the combination of sHya coating, osteogenic differentiation medium and electric field can significantly promote the expression of these four proteins, indicating an excellent level of osteointegration (bone regeneration). This work successfully demonstrated the efficiency of biochemical and biophysical stimulation on the control of the hMSC differentiation.

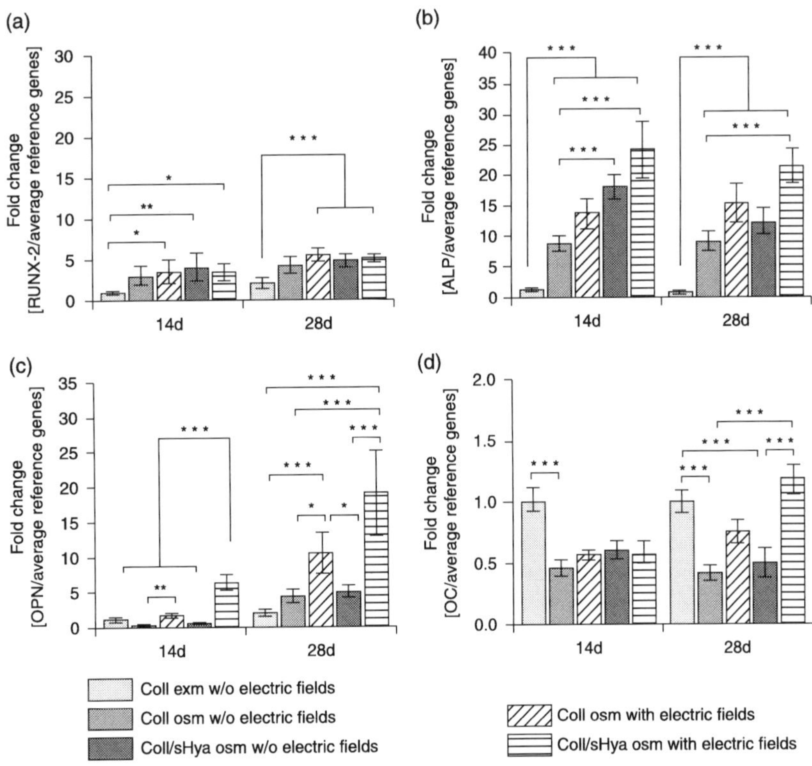

7.7 Gene-expression analysis of (a) RUNX-2, (b) ALP, (c) OPN and (d) OC of human mesenchymal stem cells cultivated either in exm or osm on coll and coll/sHya-coated scaffolds with and without electric fields. Data points present the average±SD. (n=3); *p<0.05, **p<0.01 ***p<0.001. (Reprinted from *Biomaterials*, 33, Hess *et al.*, 'Synergistic effect of defined artificial extracellular matrices and pulsed electric fields on osteogenic differentiation of human MSCs', 8975–85, 2012 with permission from Elsevier.)

7.4 SiNWs fabricated by the electroless metal deposition (EMD) method and their controllable spring constants

Nanostructured materials are preeminent stimulation sources to regulate the fates of cells because of their comparable dimensions to biomolecules (nucleic acids, amino acids and other macromolecules), simple chemical modification (functional group binding), controllable mechanical properties and various structures (nanowires, nanoholes, nanorods, etc.). All these characteristics obtained from nanostructured materials facilitate study of the fundamental information of cells.

This section will start with introducing the fabrication of SiNW by the EMD method, followed by engineering the spring constant of SiNW that aims to manipulate the fates of hMSC.

The mechanism of the metal-assisted etch method for SiNW formation, involving a galvanic reaction of Si oxidation and Ag^- reduction occurring on the surface of Si wafers, has been discussed elsewhere in detail (Yao *et al.*, 2005; Peng *et al.*, 2002; Peng *et al.*, 2006a, b; this volume, Chapter 2). Such a process leads to single crystalline, vertically aligned and large area (up to wafer scale) SiNW. Note that these fabricated SiNW exhibit preferential direction along <100> directions because this redox process is highly anisotropic, as shown in Fig. 7.8 (Chen *et al.*, 2008). Interestingly, the dopant conditions for intrinsic, *n*- and *p*-type Si wafer, do not affect SiNW fabrication.

According to the EMD method, we can readily fabricate *n*-SiNW arrays with the engineered dimensions, as shown in Fig. 7.9(a)–(d) (Kuo *et al.*, 2012). We carefully measured the correlation among *n*-SiNW diameter (▲), length (♦) and EMD processing time from 5 to 20 minutes, as shown in Fig. 7.9(e), and observed a linear tendency between the length of *n*-SiNW and processing time with the growing (etching) rate up to 1.06 μm/min. On the other hand, the diameter of *n*-SiNW remains constant in the range between 160 nm and 200 nm. This indicates that our employed parameters, tabulated in Table 7.4, would dominantly influence the *n*-SiNW length. This constant growing (etching) rate can further demonstrate a consistent linear behavior up to 2 hours, meaning about 120 μm-long SiNW can be achieved.

The reason why this linear behavior only stands for 2 hours is that the reaction is transformed from reaction control to diffusion control under the simple

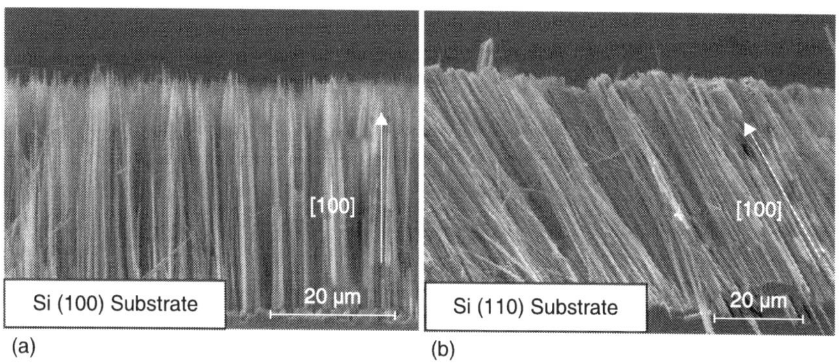

7.8 Cross-sectional scanning electron microscope images of silicon nanowires formed on a Si(100) substrate (a) and a Si(110) substrate (b). (Reprinted from *Advanced Materials*, 20, Chen *et al.*, 'Morphological control of single crystalline silicon nanowire arrays near-room temperatures', 3811–15, 2008, John Wiley and Sons.)

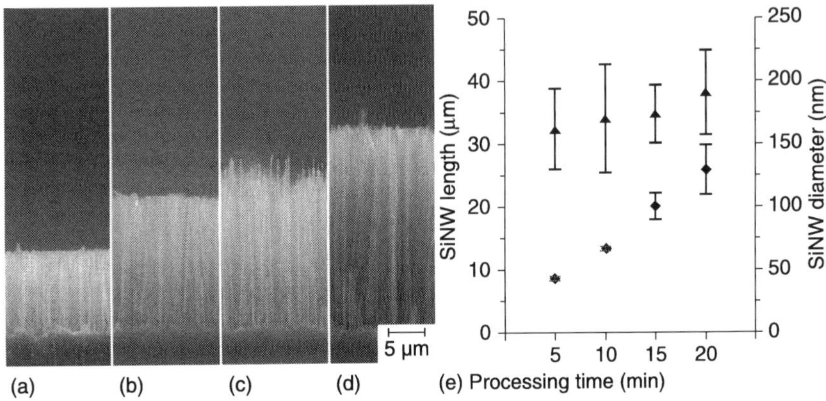

7.9 Scanning electron microscope cross-sectional view and the characteristics of the silicon nanowires (SiNW). The flat Si was treated by the electroless metal deposition (EMD) method for different processing periods, namely (a) 5 minutes, (b) 10 minutes, (c) 15 minutes and (d) 20 minutes. This gave rise to uniform and vertically aligned SiNW arrays with sequentially increased SiNW length. (e) Correlations between the length of the SiNW, the diameter of the SiNW and the processing period. (Reprinted from *Biomaterials*, 33, Kuo *et al.*, 'Regulation of the fate of human mesenchymal stem cells by mechanical and stereo-topographical cues provided by silicon nanowires', 5013–22, 2012, with permission from Elsevier.)

Table 7.4 Theoretical calculation of spring constants of Group I to IV silicon nanowires via the combination of Hook's law and Beam theory

Group		I	II	III	IV
SiNW dimension	Length (μm)	8.73±0.38	13.50±0.37	20.18±2.19	25.93±4.02
	Diameter (nm)	162.3±33.1	170.6±43.5	174.7±23.9	191.7±34.4
Spring constant, Kx (μN/m)		$(4.4\pm3.1)\times10^4$	$(1.5\pm1.3)\times10^4$	$(4.8\pm1.1)\times10^3$	$(3.3\pm0.8)\times10^3$

electrolyte ($AgNO_3$+HF). As shown in Fig. 7.10(a), the growing curve of SiNW clearly showed three distinct stages: (I) a reaction controlled region; (II) a diffusion controlled region; and (III) a saturated region, which further restricted the linear growth rate. The transformation between stages I and II is dominantly caused by the dramatic deposition of Ag dendrites covering the entire Si substrate. Because most of the Ag formed part of the dendrites instead of

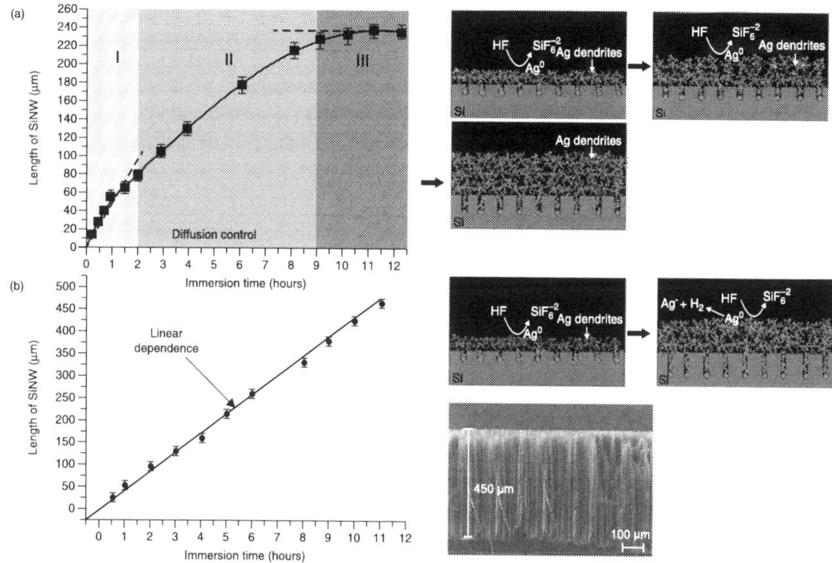

7.10 Growing curve for silicon nanowires by (a) traditional electroless metal deposition (EMD) method and (b) HNO_3 solution incorporated electrolyte. The right side illustrations indicate how the reaction proceeded. The scanning electron microscope image in the lower right corner shows the ability of HNO_3 addition electrolyte to grow ultralong SiNW (the wafer thickness here is 500 μm).

nanoparticles on the surface, the etching process was therefore slowed down and gradually reached a saturated region at the end. A solution was found to solve this issue; that is, the addition of HNO_3 solution into the original electrolyte ($AgNO_3$ + HF). With this simple correction, HNO_3 efficiently dissolves the over-deposited Ag dendrites into the desired Ag^-, which retains its role of depositing Ag nanoparticles as local cathodes. In this way, we can obtain a perfect linear dependence to fabricate an ultralong SiNW, as shown in Fig. 7.10(b). These ultralong SiNW possess a length of 450 μm on the Si wafer substrate with a thickness of 500 μm.

Figure 7.11(a),(b) presents TEM images of as-prepared SiNW from (100) and (110) Si substrates. Selected area electron diffraction (SAED) patterns characterize the corresponding axial directions and show single crystalline SiNW arrays in (100) and (110) Si substrate. The [100] direction is the preferential axial orientation of fabricated SiNW for both differently oriented wafers. The high-resolution TEM images in Fig. 7.11(c),(d) further reveal the high resolution images of single crystalline structure of fabricated SiNW. The noted axial crystallographic orientations of SiNW were found to be consistent with the results shown in Fig. 7.11(a),(b).

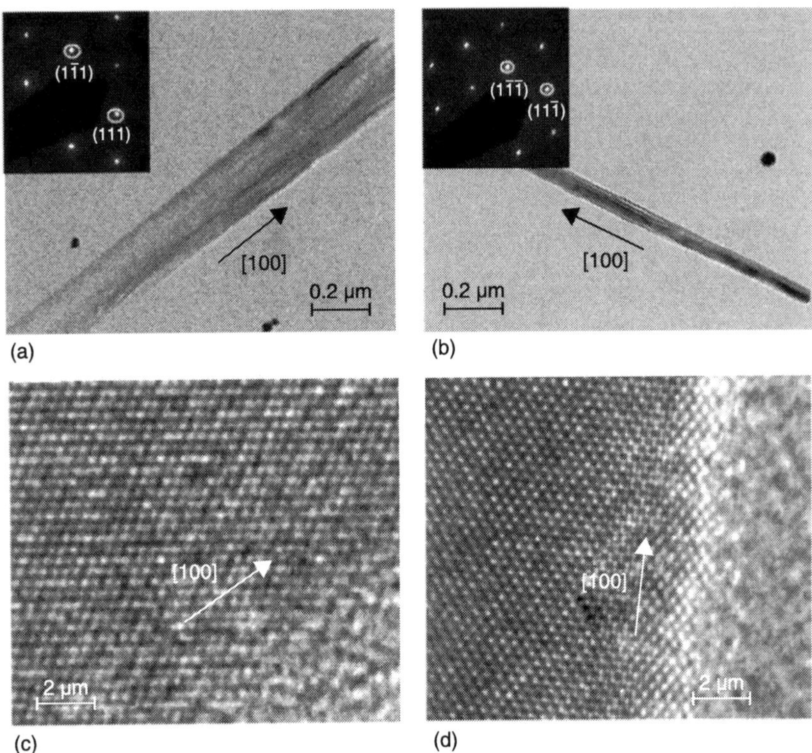

7.11 TEM images of as-prepared silicon nanowires (SiNW) Fabricated on (a) Si (100) substrates and (b) Si (110) substrates. Insets: the corresponding selected area electron diffraction (SAED) patterns taken from SiNW with the zone axis along [110] direction. The high resolution images of as-prepared SiNW fabricated on (c) Si (100) and (d) Si (110) substrates. (Reprinted from *Advanced Materials*, 20, Chen *et al.*, 'Morphological Control of Single-Crystalline Silicon Nanowire Arrays Near Room Temperatures', 3811–15, 2008, John Wiley and Sons.)

With respect to determining the mechanical properties of SiNW, we chose the parameter of spring constant (K_x) to determine the results on fate regulation of hMSC. First, we combined Hooke's law (Eq. 7.1) and Beam theory (Eq. 7.2) (Lam and Yang, 2003) into Eq. 7.3 to permit the theoretical calculation of spring constants of SiNW of various lengths. We assumed the thickness (t) and width (w) of Eq. 7.2 are comparable with diameter (d) of SiNW and transform Beam theory into the form of Eq. 7.3. The other notation here represents the spring constant of SiNW (K_x), the average length of SiNW (l) and Young's modulus for flat Si (100) (E = 170 GPa).

$$F = -K_x X \qquad [7.1]$$

$$K_x = \frac{wt^3 E}{4l^3} \quad [7.2]$$

$$K_x = \frac{d^4 E}{4l^3} \quad [7.3]$$

According to Eq. 7.3 and the measured data shown in Fig. 7.9(e), the calculation of the corresponding spring constants of the SiNW were tabulated in Table 7.4. Table 7.4 clearly shows that the length of SiNW significantly contributes to the spring constant and that shorter SiNW have a higher spring constant. Notice that the previous work evaluates the correlation between elasticity of a given planar substrate and the differentiation of hMSC. The levels of elasticity of planar substrates indeed correspond to specific differentiation results (Even-Ram *et al.*, 2006). The functions of spring constant may also be regarded as playing a corresponding role in regulating the fates of hMSC differentiation by nanowire structure. We expect to observe osteogenic differentiation of hMSC regulated by the SiNW with a stiffer spring constant and also assist in determining the associated differentiation pathway.

7.5 Mediated differentiation of stem cells by engineered SiNWs

In our work (Kuo *et al.*, 2012), the cell morphology of hMSC grown on flat Si substrates was found to be of a typical flat spindle-shaped morphology and greatly extended its cell body randomly. In contrast, the various SiNW groups all had similar cell morphologies with a spherical shape standing on the tips of SiNW bundles and showed less elongation and produced sturdy protrusions. As a result of capillarity effects, the well-aligned and dense SiNW array during the culture process easily forms obvious bundles, which supported the effective adhesive locations for hMSC. Different from hMSC grown on flat Si surfaces, hMSC on SiNW favored the growth of thick and numerous protrusions vertically (along the SiNW) instead of horizontally. Moreover, larger hMSC protrusions were formed for the cells adhering to the shorter SiNW (Groups I and II) than for the cells attached to the longer SiNW (Groups III and IV). We selected two early osteogenic lineage specific marker genes, COL1A1 and RUNX-2 to characterize the osteogenic differentiation of the hMSC (Fig. 7.12). An obvious promotion of COL1A1 expression was found on the shortest SiNW (Group I), which was about four-fold higher than that on flat Si surfaces. A comparable expression level of COL1A1 was also found across Groups II, III, IV and flat Si surfaces. Results of RUNX-2 expression on the Group I and II samples were also higher than that on flat Si surfaces. These results suggested that the level of osteogenic differentiation of hMSC on each SiNW group was different. The highest osteogenic differentiation

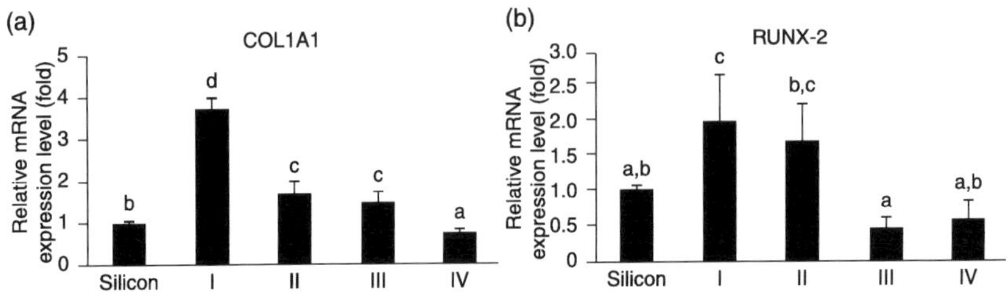

7.12 The various groups of silicon nanowires (SiNW) affect the fate commitment of human mesenchymal stem cells (hMSC) in a length-dependent manner. Relative quantitative gene expression of the osteogenic lineage marker genes (a) COL1A1 and (b) RUNX-2 in the hMSC, which were determined by semi-qPCR after 72 hours of culture. Data represent mean±S.D. $p<0.05$, $n=3$. (Reprinted from *Biomaterials*, 33, Kuo *et al.*, 'Regulation of the fate of human mesenchymal stem cells by mechanical and stereo-topographical cues provided by silicon nanowires', 5013–22, 2012, with permission from Elsevier.)

was promoted by the shortest SiNW, which exhibited the stiffest spring constant. hMSC differentiation was strongly correlated to SiNW length and the shorter nanowires demonstrate a greater potential for osteoblast differentiation.

As we knew that osteogenic differentiation induced by various mechanical stimulations would carry changes in morphology with it, the pathways to modulate cytoskeletal reorganization also posed a significant challenge. Therefore, F-actin cytoskeleton and two adhesion complex molecules, pFAK and vinculin, were investigated (Plate VI, see colour section between pp. 94 and 95). The elongated hMSC cultured on flat Si surfaces provided the denser and longer F-actin stress fibers observed by confocal microscopy (Plate VI). However, F-actin stress fibers showed a shorter and rod-like structure because of the spherical shape of hMSC grown on SiNW groups. By comparing the relative fluorescence intensity, the shorter SiNW (Group I) induced a higher expression level of F-actin than the others (Plate VI). Additionally, the expression level of pFAK and vinculin (Plate VI) also followed the consistent trend to that shown on Plate VI.

These focal adhesion complexes were located at the end of F-actin stress fibers of the cytoplasm of hMSC. Therefore one may observe a clear location of pFAK and vinculin at the end of F-actin (Plate VI). In contrast, the short and rod-like pFAK significantly gathered at the cytoplasm because of how the spherical morphology affects the SiNW. The results (Plate VI) strongly suggest that the nanowire lengths (and their associated spring constants) dominated the expression of these two focal adhesion molecules (vinculin and pFAK) in hMSC grown on SiNW.

The aforementioned findings informed us that an osteogenic differentiation of hMSC is regulated by shorter SiNW (with a stiffer spring constant). Therefore two important combinations of integrin heterodimer, such as beta 1/alpha 2 integrin and beta 1/alpha 5 integrin, perhaps dominate the entire differentiation mechanism. In Fig. 7.13(a) and (b), the expression of beta 1 and alpha 2 integrin on shorter SiNW performed at higher levels than that on flat Si surfaces. However, the expression of alpha 5 integrin demonstrated the opposite results, that the longer SiNW gave the higher level and achieved the highest level on flat Si surfaces (shown in Fig. 7.13(c)). Furthermore, expression of FAK in hMSC on SiNW follows a similar trend to the expression of beta 1 and alpha 2 integrin. Accordingly, we understood that combinations of integrin heterodimer: beta 1 integrin + alpha 2 integrin formed the major pathway to initiate the osteogenic differentiation when receiving the mechanical stimulation from SiNW. The distribution and expression of alpha 2 integrin in hMSC were found to locate near the interface between the hMSC and the SiNW (Plate VII). Quantitative expression of alpha 2 integrin in the hMSC grown on SiNW was higher for all SiNW compared with flat Si surfaces (Plate VII). The above findings showed that expression of alpha 2 integrin and FAK were higher on shorter SiNW.

Among the above findings, the uniform, well-aligned and dense SiNW were fabricated in various lengths (Group I to IV) and presented a series of spring

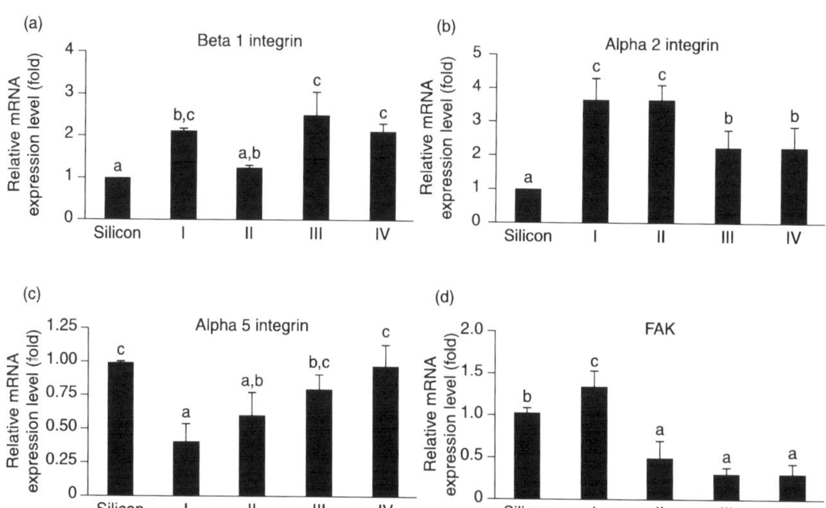

7.13 Relative quantitative gene expression of integrin and FAK of human mesenchymal stem cells (hMSC) grown on the various groups of silicon nanowires (SiNW). Relative quantitative gene expression of (a) beta 1 integrin, (b) alpha 2 integrin, (c) alpha 5 integrin and (d) total FAK in hMSC cultured on 2-D flat Si and on Group I, II, III and IV SiNW were detected by semi-qPCR after 72 hours of culture. Data represent mean ± S.D. $p < 0.05$, $n = 3$. (Reprinted from *Biomaterials*, 33, Kuo *et al.*, 'Regulation of the fate of human mesenchymal stem cells by mechanical and stereo-topographical cues provided by silicon nanowires', 5013–22, 2012, with permission from Elsevier.)

constants. The theoretical spring constant calculated by Eq. 7.3 indicated the shortest SiNW (Group I) possessed the stiffest spring constant, whereas the longer SiNW gave the softer one. This mechanical stimulation was first conducted by the combination of beta 1 and alpha 2 integrin (mechanical sensors) on the cell membrane. Then integrins transferred the stimulating signal through focal adhesion molecules, and nuclei of hMSC react with this stimulation to thereafter promote cytoskeletal rearrangement for changing the cell shape, (Cary *et al.*, 1999; Hynes, 2002). Consequently, hMSC morphology transformed from a normal fibroblast-like undifferentiated phenotype to a spherical shape, as we observed with hMSC cultured on SiNW. The spherical shape changes were led by a series of contractions of stress fibers generated by F-actin filament rearrangement, shown as short rods and a dotted form (see p. 0) (Chen, 2004; Kuo *et al.*, 2012). These results indicate that the shortest SiNW favors the formation of the highest amount of F-actin, which thereafter generates the largest cellular spread area and optimization of osteogenesis. The mechanical stimulation from spring constant of SiNW to regulate the fate of hMSC differentiation indeed presents a possible means to induce osteogenesis.

7.6 Conclusion

This chapter introduces basic information regarding stem cells and the means to effectively manipulate hMSC differentiation via biophysical and biochemical approaches. We subsequently utilized single crystal, well-aligned and dense SiNW fabricated by an EMD method to input a controllable spring constant naturally generated by dimensional differences, to stimulate hMSC. It was found that a specific combination of integrin heterodimer (beta 1/alpha 2 integrin) was responsible for receiving the mechanical signal from SiNW, transferring it through focal adhesion molecules, and with the nuclei of hMSC reacting with this stimulation to thereafter promote cytoskeletal rearrangement for changing of the cell shape. This reaction pathway is dominant in regulating the fates of hMSC. Moreover, this concept provides an alternative route to modulating different differentiation pathways of hMSC via the tuning of different spring constants.

7.7 Future trends

Great attention is currently being directed toward the use of nanotechnology and nanomaterials for their widespread applications in semiconductors, optoelectronics and biology that were previously inaccessible. This chapter has described how the use of the straightforward physical parameter of stiffer spring constant in a semiconductor nanowire can induce the osteogenesis of hMSC. In considering the clinical use of stem cell therapies, directing the osteogenesis of hMSC alone is far from practical employment. A significant challenge concerns the determination of a full picture of hMSC differentiation pathways. Therefore, nanotechnology-driven biophysical or biochemical stimulations to regulate the fates of hMSC are clearly worth additional study. Ideally, the tunable size-dependent properties of nanomaterials may permit new extraordinary stimulation methods, thereby leading to a different physiology of hMSC and other stem cell types. Overall, such efforts leading to clarification of the differentiation mechanism should accelerate their practical application to the clinic.

7.8 Acknowledgements

The authors would like to express their appreciation to Dr Oscar Kuang-Sheng Lee and Mr. Shu-Wen Kuo in Taipei Veterans General Hospital in Taiwan for their professional support in cell culture experiments and analysis.

7.9 References

Akhtar, S., Tanaka, A., Usami, K., Tsuchiya, Y. and Oda, S. (2008) 'Influence of the crystal orientation of substrate on low temperature synthesis of silicon nanowires from Si_2H_6', Thin Solid Films, 517, 317–19 (doi:10.1016/j.tsf.2008.08.155).

Ankey, R.A. (2008) *New Technologies: Ethics of Stem Cell Research*, International Encyclopedia of Public Health, 533–6.

Aubin, J.E. (2001) 'Regulation of osteoblast formation and function', *Reviews in Endocrine & Metabolic Disorders*, 2, 81–94 (doi:10.1023/A:1010011209064).

Bosnakovski, D., Mizuno, M., Kim, G., Takagi, S., Okumura, M. and Fujinaga, T. (2006) 'Chondrogenic differentiation of bovine bone marrow mesenchymal stem cells (MSC) in different hydrogels: influence of collagen type II extracellular matrix on MSC chondrogenesis', *Biotechnology and Bioengineering*, 93, 1152–63 (doi:10.1002/bit.20828).

Brammer, K.S., Choi, C., Frandsen, C.J., Oh, S. and Jin, S. (2011) 'Hydrophobic nanopillars initiate mesenchymal stem cell aggregation and osteo-differentiation', *Acta Biomaterialia*, 7, 683–90 (doi:10.1016/j.actbio.2010.09.022).

Charbord, P. (2010) 'Bone marrow mesenchymal stem cells: historical overview and concepts', *Human Gene Therapy*, 21, 1045–56 (doi:10.1089/hum.2010.115).

Cary, L.A., Han, D.C. and Guan, J.L. (1999) 'Integrin-mediated signal transduction pathways', *Histology and Histopathology*, 14, 1001–9.

Chen, C.S., Tan, J. and Tien, J. (2004) 'Mechanotransduction at cell-matrix and cell-cell contacts', *The Annual Review of Biomedical Engineering*, 6, 275–302 (doi: 10.1146/annurev.bioeng.6.040803.140040).

Chen, C.C., Chen, Y.Z., Huang, Y.J. and Sheu, J.T. (2011) 'Using silicon nanowire devices to detect adenosine triphosphate liberated from electrically stimulated HeLa cells', *Biosensors and Bioelectronics*, 26, 2323–8 (doi:10.1016/j.bios.2010.10.003).

Chen, C.Y., Wu, C.S., Chou, C.J. and Yen, T.J. (2008) 'Morphological control of single crystalline silicon nanowire arrays near-room temperatures', *Advanced Materials*, 20, 3811–15 (doi: 10.1002/adma.200702788).

Chen, K.I, Li, B.R. and Chen, Y.T. (2011) 'Silicon nanowire field-effect transistor-based biosensors for biomedical diagnosis and cellular recording investigation', *Nano Today*, 6, 131–54 (doi:10.1016/j.nantod.2011.02.001).

Cheng, T., Yang, C., Weber, N., Kim, H.T. and Kuo, A.C. (2012) 'Fibroblast growth factor 2 enhances the kinetics of mesenchymal stem cell chondrogenesis', *Biochemical and Biophysical Research Communications*, 426, 544–50 (doi:10.1016/j.bbrc.2012.08.124).

Collins, M.N. and Birkinshaw, C. (2013) 'Hyaluronic acid based scaffolds for tissue engineering – a review', *Carbohydrate Polymers*, 92, 1262–79 (doi:10.1016/j.carbpol.2012.10.028).

Dai, L.J., Moniri, M.R., Zeng, Z.R., Zhou, J.X., Rayat, J. and Wamock, G.L. (2011) 'Potential implications of mesenchymal stem cells in cancer therapy', *Cancer Letters*, 305, 8–20 (doi:10.1016/j.canlet.2011.02.012).

Danišovič, L., Varga, I. and Polák, Š. (2012) 'Growth factors and chondrogenic differentiation of mesenchymal stem cells', *Tissue and Cell*, 44, 69–73 (doi:10.1016/j.tice.2011.11.005).

Dash, M., Chiellini, F., Ottenbrite, R.M. and Chiellini, E. (2011) 'Chitosan – a versatile semi-synthetic polymer in biomedical applications', *Progress in Polymer Science*, 36, 981–1014 (doi:10.1016/j.progpolymsci.2011.02.001).

Even-Ram, S., Artym, V. and Yamada, K.M. (2006) 'Matrix control of stem cell fate', *Cell*, 126, 645–7 (doi:10.1016/j.cell.2006.08.008).

Friedenstein, A.J., Chailakhyan, R.K. and Gerasimov, U.V. (1987) 'Bone marrow osteogenic stem cells: in vitro cultivation and transplantation in diffusion chambers', *Cell & Tissue Kinetics*, 20, 263–72 (doi: 10.1111/j.1365-2184.1987.tb01309.x).

Fuchs, E. and Segre, J.A. (2000) 'Stem cells: a new lease on life', *Cell*, 100, 143–55 (doi:10.1016/S0092-8674(00)81691-8).

Genovese, J.A., Spadaccio, C., Rivello, H.G., Toyoda, Y. and Patel, A.N. (2009) 'Electrostimulated bone marrow human mesenchymal stem cells produce follistatin', *Cytotherapy*, 11, 448–56 (doi:10.1080/14653240902960445).

Ghaedi, M., Tuleuova, N., Zern, M.A., Wu, J. and Revzin, A. (2011) 'Bottom-up signaling from HGF-containing surfaces promotes hepatic differentiation of mesenchymal stem cells', *Biochemical and Biophysical Research Communications*, 407, 295–300 (doi:10.1016/j.bbrc.2011.03.005).

Her, G.J., Wu, H.C., Chen, M.H., Chen, M.Y., Chang, S.C. and Wang, T.W. (2012) 'Control of three-dimensional substrate stiffness to manipulate mesenchymal stem cell fate toward neuronal or glial lineages', *Acta Biomaterialia*, (doi:10.1016/j.actbio.2012.10.012).

Hess, R., Jaeschke, A., Neubert, H., Hintze, V., Moeller, S., *et al.* (2012) 'Synergistic effect of defined artificial extracellular matrices and pulsed electric fields on osteogenic differentiation of human MSC', *Biomaterials*, 33, 8975–85 (doi:10.1016/j.biomaterials.2012.08.056).

Hu, Y., Cai, K., Luo, Z., Xu, D., Xie, D., *et al.* (2012) 'TiO_2 nanotubes as drug nanoreservoirs for the regulation of mobility and differentiation of mesenchymal stem cells', *Acta Biomaterialia*, 8, 439–48 (doi:10.1016/j.actbio.2011.10.021).

Hynes R.O. (2002) 'Integrins: bidirectional, allosteric signaling machines', *Cell*, 110, 673–87 (doi:10.1016/S0092-8674(02)00971-6).

Indrawattana, N., Chen, G., Tadokoro, M., Shann, L.H., Ohgushi, H., *et al.* (2004) 'Growth factor combination for chondrogenic induction from human mesenchymal stem cell', *Biochemical and Biophysical Research Communications*, 320, 914–19 (doi:10.1016/j.bbrc.2004.06.029).

Jiang, X., Cao, H.Q., Shi, L.Y., Ng, S.Y., Stanton, L.W. and Chew, S.Y. (2012) 'Nanofiber topography and sustained biochemical signaling enhance human mesenchymal stem cell neural commitment', *Acta Biomaterialia*, 8,1290–302 (doi:10.1016/j.actbio.2011.11.019).

Kim, M.S., Park, S.J., Gu, B.K. and Kim, C.H. (2012) 'Inter-connecting pores of chitosan scaffold with basic fibroblast growth factor modulate biological activity on human mesenchymal stem cells', *Carbohydrate Polymers*, 87, 2683–9 (doi:10.1016/j.carbpol.2011.11.060).

Kim, S.M., Jung, J.U., Ryu, J.S., Jin, J.W., Yang, H.J., *et al.* (2008) 'Effects of gangliosides on the differentiation of human mesenchymal stem cells into osteoblasts by modulating epidermal growth factor receptors', *Biochemical and Biophysical Research Communications*, 371, 866–71 (doi:10.1016/j.bbrc.2008.04.162).

Koide, T. (2005) 'Triple helical collagen-like peptides: engineering and applications in matrix biology', *Connective Tissue Research*, 46, 131–41 (doi:10.1080/03008200591008518).

Kuo, S.W., Lin, H.I, Ho, J.H., Shih, Y.R., Chen, H.F., *et al.* (2012) 'Regulation of the fate of human mesenchymal stem cells by mechanical and stereo-topographical cues provided by silicon nanowires', *Biomaterials*, 33, 5013–22 (doi:10.1016/j.biomaterials.2012.03.080).

Lam, D.C.C. and Yang, F. (2003) 'Experiments and theory in strain gradient elasticity', *Journal of the Mechanics and Physics of Solids*, 51, 1477–508 (doi:10.1016/S0022-5096(03)00053-X).

Lee, J., Kang, B.S., Hicks, B., Chancellor Jr, T.F., Chu, B.H., *et al.* (2008) 'The control of cell adhesion and viability by zinc oxide nanorods', *Biomaterials*, 29, 3743–9 (doi:10.1016/j.biomaterials.2008.05.029).

Lee, K.D., Kuo, T.K., Whang-Peng, J., Chung, Y.F., Lin, C.T., *et al.* (2004) 'In vitro hepatic differentiation of human mesenchymal stem cells', *Hepatology*, 40, 1275–84 (doi: 10.1002/hep.20469).

Lee, O.K., Ko, Y.C., Kuo, T.K., Chou, S.H., Li, H.J., *et al.* (2004) 'Fluvastatin and lovastatin but not pravastatin induce neuroglial differentiation in human mesenchymal stem cells', *Journal of Cellular Biochemistry*, 93, 917–28 (doi: 10.1002/jcb.20241).

Li, Y., Chu, J.S., Kurpinski, K., Li, X., Bautista, D.M., *et al.* (2011) 'Biophysical regulation of histone acetylation in mesenchymal stem cells', *Biophysical Journal*, 100, 1902–9 (doi:10.1016/j.bpj.2011.03.008).

Liu, S.Q., Tian, Q., Hedrick, J.L., Hui, J.H.P., Ee, P.L.R. and Yang, Y.Y. (2010) 'Biomimetic hydrogels for chondrogenic differentiation of human mesenchymal stem cells to neocartilage', *Biomaterials*, 31, 7298–307 (doi:10.1016/j.biomaterials.2010.06.001).

Mathews, S., Bhonde, R., Gupta, P.K. and Totey, S. (2012) 'Extracellular matrix protein mediated regulation of the osteoblast differentiation of bone marrow derived human mesenchymal stem cells', *Differentiation*, 84, 185–92 (doi:10.1016/j.diff.2012.05.001).

McLaren, A. (2000) 'A scientist's view of the ethics of human embryonic stem cell research', *Cell Stem Cell*, 1, 23–6 (doi:10.1016/j.stem.2007.05.003).

Mohsin, S., Shams, S., Nasir, G.A., Khan, M., Awan, S.J., *et al.* (2011) 'Enhanced hepatic differentiation of mesenchymal stem cells after pretreatment with injured liver tissue', *Differentiation*, 81, 42–8 (doi:10.1016/j.diff.2010.08.005).

Niu, J., Sha, J. and Yang, D. (2004) 'Sulfide-assisted growth of silicon nano-wires by thermal evaporation of sulfur powders', *Physica E*, 24, 278–81 (doi:10.1016/j.physe.2004.05.002).

Noth, U., Rackwitz, L., Heymer, A., Weber, M., Baumann, B., *et al.* (2007) 'Chondrogenic differentiation of human mesenchymal stem cells in collagen type I hydrogels', *Journal of Biomedical Materials Research Part A*, 83, 626–35 (doi: 10.1002/jbm.a.31254).

Ozawa, K., Sato, K., Oh, I., Ozaki, K., Uchibori, R., *et al.* (2008) 'Cell and gene therapy using mesenchymal stem cells (MSCs)', *Journal of Autoimmunity*, 30, 121–7 (doi:10.1016/j.jaut.2007.12.008).

Pan, L., Lew, K.K., Redwing, J.M. and Dickey, E.C. (2005) 'Effect of diborane on the microstructure of boron-doped silicon nanowires', *Journal of Crystal Growth*, 277, 428–36 (doi:10.1016/j.jcrysgro.2005.01.091).

Peng, K.Q., Yan, Y.J., Gao, S.P. and Zhu, J. (2002) 'Synthesis of large-area silicon nanowire arrays via self-assembling nanoelectrochemistry', *Advanced Materials*, 14, 1164–7 (doi:10.1002/1521-4095(20020816)).

Peng, K.Q., Hu, J.J., Yan, Y.J., Wu, Y., Fang, H., *et al.* (2006a) 'Fabrication of single-crystalline silicon nanowires by scratching a silicon surface with catalytic metal particles', *Advanced Functional Materials*, 16, 387–94 (doi:10.1002/adfm.200500392).

Peng, K.Q., Fang, H., Hu, J., Wu, Y., Zhu, J., *et al.* (2006b) 'Metal-particle-induced, highly localized site-specific etching of Si and formation of single-crystalline Si nanowires in aqueous fluoride solution', *Chemistry–A European Journal*, 12, 7942–7 (doi: 10.1002/chem.200600032).

Pui, T.S., Agarwal, A., Ye, F., Huang, Y. and Chen, P. (2011) 'Nanoelectronic detection of triggered secretion of proinflammatory cytokines using CMOS compatible silicon nanowires', *Biosensors and Bioelectronics*, 26, 2746–50 (doi:10.1016/j.bios.2010.09.059).

Rodrigues, A.A., Batista, N.A., Bavaresco, V.P., Baranauskas, V., Ceragioli, H.J., *et al.* (2012) 'Polyvinyl alcohol associated with carbon nanotube scaffolds for osteogenic differentiation of rat bone mesenchymal stem cells', *Carbon*, 50, 450–9 (doi:10.1016/j.carbon.2011.08.071).

Rogacs, A., Steinbrenner, J.E., Rowlette, J.A., Weisse, JM., Zheng, X.L. and Goodson, K.E. (2010) 'Characterization of the wettability of thin nanostructured films in the presence of evaporation', *Journal of Colloid and Interface Science*, 349, 354–60 (doi:10.1016/j.jcis.2010.05.063).

Rosaz, G., Salem, B., Pauc, N., Gentile, P., Potié, A. and Baron, T. (2011) 'Electrical characteristics of a vertically integrated field-effect transistor using non-intentionally doped Si nanowires', *Microelectronic Engineering*, 88, 3312–15 (doi:10.1016/j.mee.2011.07.009).

Shah, K. (2012) 'Mesenchymal stem cells engineered for cancer therapy', *Advanced Drug Delivery Reviews*, 64, 739–48 (doi:10.1016/j.addr.2011.06.010).

Shi, X., Wang, Y., Varshney, R.R., Ren, L., Gong, Y. and Wang, D.A. (2010) 'Microsphere-based drug releasing scaffolds for inducing osteogenesis of human mesenchymal stem cells in vitro', *European Journal of Pharmaceutical Sciences*, 39, 59–67 (doi:10.1016/j.ejps.2009.10.012).

Sinha, V.R., Singla, A.K., Wadhawan, S., Kaushik, R., Kumria, R., *et al.* (2004) 'Chitosan microspheres as a potential carrier for drugs', *International Journal of Pharmaceutics*, 274, 1–33 (doi:10.1016/j.ijpharm.2003.12.026).

Tay, C.Y., Gu, H., Leong, W.S., Yu, H., Li, H.Q., *et al.* (2010) 'Cellular behavior of human mesenchymal stem cells cultured on single-walled carbon nanotube film', *Carbon*, 48, 1095–104 (doi:10.1016/j.carbon.2009.11.031).

Vija, L., Farge, D., Gautier, J.F., Vexiau, P., Dumitrache, C., *et al.* (2009) 'Mesenchymal stem cells: Stem cell therapy perspectives for type 1 diabetes', *Diabetes & Metabolism*, 35, 85–93 (doi:10.1016/j.diabet.2008.10.003).

Wang, P.Y., Tsai, W.B. and Voelcker, N.H. (2012) 'Screening of rat mesenchymal stem cell behaviour on polydimethylsiloxane stiffness gradients', *Acta Biomaterialia*, 8, 519–30 (doi:10.1016/j.actbio.2011.09.030).

Wu, X.B. and Tao, R. (2012) 'Hepatocyte differentiation of mesenchymal stem cells', *Hepatobiliary & Pancreatic Diseases International*, 11, 360–71 (doi:10.1016/S1499-3872(12)60193-3).

Yan, H.F., Xing, Y.J., Hang, Q.L., Yu, D.P., Wang, Y.P., *et al.* (2000) 'Growth of amorphous silicon nanowires via a solid–liquid–solid mechanism', *Chemical Physics Letters*, 323, 224–8 (doi:10.1016/S0009-2614(00)00519-4).

Yang, Y., Li, Y., Lv, Y., Zhang, S., Chen, L., *et al.* (2008) 'NRSF silencing induces neuronal differentiation of human mesenchymal stem cells', *Experimental Cell Research*, 314, 2257–65 (doi:10.1016/j.yexcr.2008.04.008).

Yao, Y., Li, F. and Lee, S.T. (2005) 'Oriented silicon nanowires on silicon substrates from oxide-assisted growth and gold catalysts', *Chemical Physics Letters*, 406, 381–5 (doi:10.1016/j.cplett.2005.03.027).

Yu, D.P., Xing, Y.J., Hang, Q.L., Yan, H.F., Xu, J., *et al.* (2001) 'Controlled growth of oriented amorphous silicon nanowires via a solid–liquid–solid (SLS) mechanism', *Physica E*, 9, 305–9 (doi:10.1016/S1386-9477(00)00202-2).

Wang, N., Cai, Y. and Zhang, R.Q. (2008) 'Growth of nanowires', *Materials Science and Engineering R*, 60, 1–51 (doi:10.1016/j.mser.2008.01.001).

Wu, S.L., Zhang, T., Zheng, R.T. and Cheng, G.A. (2012) 'Facile morphological control of single-crystalline silicon nanowires', *Applied Surface Science*, 258, 9792–9 (doi:10.1016/j.apsusc.2012.06.031).

Zhao, L., Wang, H., Huo, K., Zhang, X., Wang, W., *et al.* (2013) 'The osteogenic activity of strontium loaded titania nanotube arrays on titanium substrates', *Biomaterials*, 34, 19–29. (doi:10.1016/j.biometerials.2012.09.041)

Zhao, L., Liu, L., Wu, Z., Zhang, Y. and Chu, P.K. (2012) 'Effects of micropitted/nanotubular titania topographies on bone mesenchymal stem cell osteogenic differentiation', *Biomaterials*, 33, 2629–41 (doi:10.1016/j.biomaterials.2011.12.024).

8

Silicon nanoneedles for drug delivery

C. CHIAPPINI and C. ALMEIDA,
Imperial College London, UK

DOI: 10.1533/9780857097712.2.144

Abstract: Silicon nanoneedles are emerging as a strategy to negotiate the cell membrane and deliver drugs intracellularly. This chapter discusses the different strategies for silicon nanoneedle fabrication, drug loading, cell interfacing and drug delivery. The chapter also overviews four exemplary systems for nanoneedle-mediated intracellular drug delivery.

Key words: drug delivery, intracellular delivery, microfabrication, nanoneedles, nanowires.

8.1 Introduction

The efficiency of a drug delivery method, in general, depends on the internalisation of the drug being administered. In eukaryotic cells, there is physical segregation between the interior of the cell and the external environment by the plasma membrane, a structure composed of a phospholipid bilayer with embedded proteins that prevents unchecked influx and efflux of solutes from cells. Thus membrane permeability is one of the main constraints for the delivery of drugs with intracellular targets. Very few molecules that yield to specific parameters regarding molecular size, net charge and polarity are able to cross the plasma membrane by passive diffusion (Fischer *et al.*, 2005). Indeed, and for good evolutionary reasons, the cell tightly controls the flux of the great majority of molecules (including biomolecules), preventing their unaided internalisation (Fischer *et al.*, 2001). Because of these limitations, the vast majority of drugs with intracellular targets that are currently available on the market fall within the small molecule category, which are fairly permeable to the cell membrane, and can be internalised simply by imposing a concentration gradient across the cell membrane (Di *et al.*, 2012; Sugano *et al.*, 2010).

Recently, the efficacy of the use of antibodies for oncotherapy (Weiner *et al.*, 2010; Adams and Weiner, 2005) and the potential of short interfering RNA (siRNA) in modulation of gene expression (Dykxhoorn *et al.*, 2003) highlighted the vastly superior therapeutic potential of biologicals and renewed interest for delivery strategies that efficiently negotiate the plasmalemma. Biologicals are large and charged molecules that cannot negotiate the plasma membrane, requiring appropriate delivery strategies. The impact of biologicals in therapy is expected to be revolutionary. Developing efficient and universal strategies to negotiate the cell membrane could benefit both currently available small molecule drugs, allowing use of lower dosages, and molecules with promising molecular interactions but which are currently unable to reach the cell at therapeutic concentrations.

Several methods of molecular transfer across the membrane into living cells have been developed, each with different characteristics in terms of cell viability and transfer efficiency (Stephens and Pepperkok, 2001). The methods can be broadly categorised into biochemical and physical delivery strategies.

8.1.1 Biochemical delivery strategies

Biochemical delivery strategies exploit cellular uptake and transport mechanisms to reach the intracellular environment and deliver their cargo. Cellular uptake results from chemical modification of the molecule's surface or through encapsulation within a carrier that can act as mediator of transport (Chou *et al.*, 2011). In the former, cell-permeable molecules are coupled to cell-impermeable drugs to shuttle them across the plasma membrane. Several modulators of cell permeability exist, some of which are commercially available (Stephens and Pepperkok, 2001). The broad category of cell-penetrating peptides (CPP) is extremely versatile and widely used to successfully translocate biomolecules, including plasmid DNA, oligonucleotides, siRNA, peptide nucleic acids, proteins and peptides, liposomes and nanoparticles both *in vitro* and *in vivo* (Morris *et al.*, 2008). CPP are water-soluble, partly hydrophobic and/or rich in basic residues (Madani *et al.*, 2011) consisting of approximately 10–30 amino acids (Fischer *et al.*, 2005) that bind to their cargo either covalently or non-covalently (Morris *et al.*, 2008). Over 30 CPP have been identified so far and their biological and biophysical characteristics are very different (Fischer *et al.*, 2005) suggesting that, many routes of internalisation may exist. Although still debatable, endocytosis and direct penetration have been appointed as the two main routes of uptake (Madani *et al.*, 2011). Although CPP are broadly employed to enhance cell permeability, there are restrictions to the size of the molecules they can transfer, and transporting proteins almost always causes their unfolding (Stephens and Pepperkok, 2001).

Liposome encapsulation is an alternative biochemical approach to membrane translocation. Particularly their cationic formulations, which overcome some limitations of CPP, can introduce with high efficiency a variety of molecules (e.g. DNA, RNA, proteins, etc.) inside cells. Liposomes can vary in size and morphology (Balazs and Godbey, 2011), enabling the encapsulation of molecules and even nanoparticles without significant size restrictions (Stephens and Pepperkok, 2001). Moreover, the positive charge of cationic liposomes favours interaction with negatively charged backbones. As the plasma membrane is also negatively charged, interaction between the liposome and the plasma membrane enhance delivery. The most accepted concept is that cationic liposomes enter cells through adsorptive endocytosis and fusion with endosomal membranes then occurs, leading to cargo release (Sharma and Sharma, 1997). Thus, this internalisation pathway avoids lysosomal degradation. Liposome fusion might, however, interfere with lipid metabolism (Stephens and Pepperkok, 2001). Other formulations of liposomes are not as efficient for delivery of biomolecules: neutral liposomes have limited interaction with cells while anionic liposomes are electrostatically repelled by the negatively charged cell membrane (Balazs and Godbey, 2011). However, when internalisation through endocytosis occurs, a liposome can be delivered by the endosome into the lysosome or, in the case of pH-sensitive liposomes, it can induce endosome destabilisation, which results in drug delivery into the cytoplasm (Torchilin, 2005).

An alternative biochemical strategy employs viruses as drug carriers. Viruses can encapsulate genetic material by self-assembly of their coat proteins into a capsid after recombinant expression. This unique characteristic has been exploited to develop drug delivery systems, which are generally named virus-like particles (VLP). The genome of a virus is enclosed in the capsid, stabilised by electrostatic interaction with the basic polypeptide domains of the coat proteins. In a similar fashion, suitably charged moieties can be entrapped within VLP (Garcea and Gissmann, 2004). Obvious limitations are then the charge of the cargo to be encapsulated and its size. Although most of the VLP being developed for human application are typically not human pathogens (Manchester and Singh, 2006), they still can potentially exhibit pathogenicity.

8.1.2 Physical delivery strategies

Physical methods of delivery comprise direct transfer methods, where there is direct access to the intracellular compartment, or plasma membrane permeabilisation methods, which uses detergents, UV light, pore-forming toxins or electrical pulses to enhance the permeability of the plasma membrane and therefore introduce the cargo by passive diffusion (Stephens and Pepperkok, 2001). Among permeabilisation methods, electroporation is the most widely adopted. Electroporation involves transient increase in the plasma membrane permeability resulting from application of an external electric field (Neumann *et al.*, 1982). Briefly, electroporation uses short high-voltage pulses that just surpass the capacitance of the plasma membrane, thus creating transient pores. This reversible state permits the diffusion of small molecules or electrophoretic transfection of other molecules (Gehl, 2003), including DNA, proteins, etc. Besides being able to address a large range of compounds to be internalised, it is available to a wide range of cell types. It has proven beneficial both *in vitro* and *in vivo* (Gehl, 2003) and exhibits high efficiency when cells are in suspension but requires specialised equipment and may lead to cell death as a result of the electrical pulse (Chou *et al.*, 2011; Stephens and Pepperkok, 2001). Moreover, when used *in vivo*, it may cause decreased blood flow in the pulsed areas (Gehl, 2003).

Direct access to the intracellular compartment is advantageous and results in most cases in high efficiency of transfection. Microinjection is a conceptually simple technique where a glass micropipette with a tip of less than 0.5 μm injects the sample into the cell with the guidance of an optical microscope. Because it is a direct approach, it can reach high transfer efficiencies and survival rates (Celis, 1984). Furthermore, it allows co-injection of several distinct compounds and has few restrictions in terms of cell or payload type. However, because of its sequential nature, it possesses low throughput, is technically challenging and expensive (Stephens and Pepperkok, 2001). Microinjection's unmet potential for universal, low-toxicity intracellular delivery of molecules has stimulated further research into direct injection strategies. Nanoneedles conceptually originate as a miniaturisation of microneedles. They enhance microneedles' superior ability for localised, painless, minimally invasive drug delivery. Further, nanoneedles grant direct physical access to the cytosol with minimal disruption to the activity of cells, while they enable interaction with cellular components at the nanoscale.

8.1.3 Nanoneedle platforms

Nanoneedles are broadly definable as high aspect ratio structures with diameter at the nanoscale. The basic requirements for nanoneedles are similar to, and to some extent less stringent than, those for vertically aligned nanowires. Hence established techniques for nanowires can generate nanoneedles. Alongside the approaches adopted from the semiconductor industry, specific fabrication techniques are emerging that facilitate engineering of application-specific nanoneedles.

8.2 Strategies for nanoneedle fabrication

8.2.1 Vapour–liquid–solid (VLS) growth of nanoneedles

Vapour–liquid–solid (VLS) growth of silicon is established and largely employed in the synthesis of vertically aligned nanowires (Levitt, 1970) (Fig. 8.1(a)). In VLS, metal nanoparticles (usually Au) are heated above the eutectic temperature for the metal-silicon system in the presence of a silicon gas source (usually SiH_4 or $SiCl_4$). In such conditions a liquid nanodrop of metal–silicon alloy forms. The gas feeding the eutectic nanodrop supersaturates it with silicon, inducing the nucleation of solid Si. Once formed, the solid–liquid interface acts as growth interface catalysing continued incorporation of Si into the newly formed lattice. The metal nanoparticle rides the tip of the nanowire while it forms. This synthetic process can form straight cylindrical nanowires of constant diameter with lengths of hundreds of μm. The nanowire can grow either through incorporation of silicon

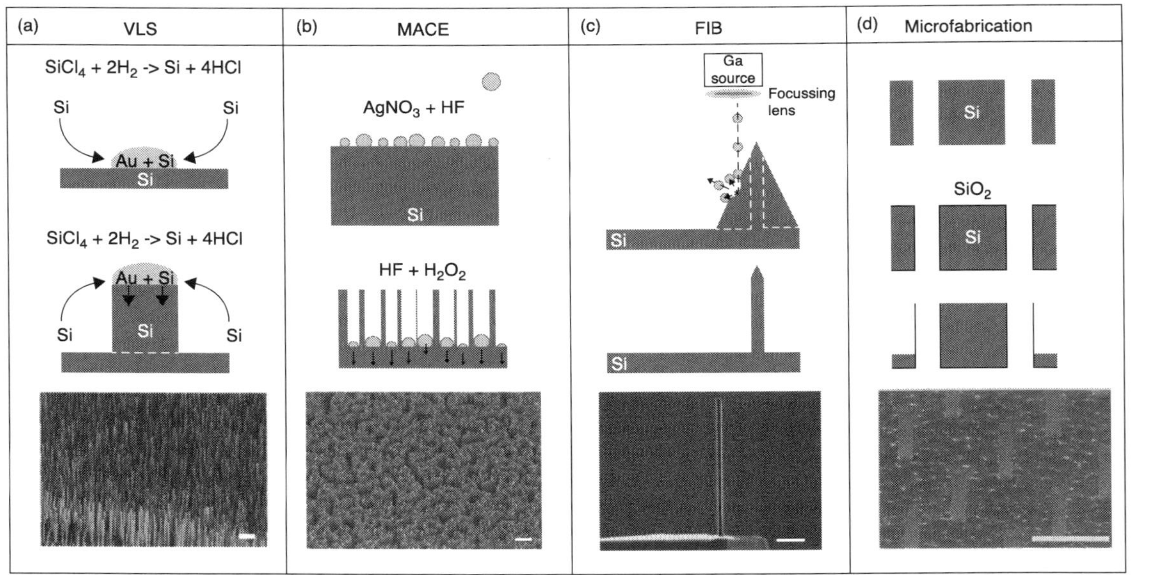

8.1 Fabrication strategies for nanoneedles. (a) In vapour–liquid–solid (VLS) synthesis, Si is incorporated from the vapour phase into a liquid eutectic Au-Si alloy nanodroplet. When the concentration of Si overcomes saturation for the eutectic, the Si nucleates in solid phase. This process is sustained as long as gaseous silicon is provided, forming silicon nanowires (pictured below). (b) In metal-assisted chemical etch (MACE), a noble metal mesh deposited on top of silicon catalyses the etching of silicon nanowires in an oxidizing solution of HF. Controlling the etching parameters results in solid or porous nanowires (pictured below). (c) Focused ion beam (FIB) can shape an already existing atomic force microscope (AFM) tip into a nanoneedle by milling the excess material (pictured below). (d) Microfabrication allows forming hollow nanoneedles (pictured below) by first etching deep nanopores into a Si membrane, conformably coating them in dielectric and then etching the top surface of the dielectric alongside the selective etch of silicon. Images reproduced with permission from Shalek *et al.*, 2010, Chiappini *et al.*, 2010, Han *et al.*, 2005b and Peer *et al.*, 2012 respectively. (All scale bars = 1 μm.)

at the interface between the droplet and the wire (liquid/solid interface), or between the gas and the wire (vapour/solid interface) (Lu and Lieber, 2006). Incorporation at the liquid/solid interface results in nanowire elongation, whereas incorporation at the vapour/solid interface results in its thickening. Which mechanism dominates during the growth process is determined by the conditions during synthesis. Pressure, gas flow rate, temperature and the nature of the reacting species and their gaseous byproducts all influence the growth (Kolasinski, 2006). With silanes, low temperatures favour nanowire elongation, reducing the thermal dissociation of the gas. Addition of hydrogen gas also promotes elongation either by passivating the nanowire surface through hydrogen termination (Wu *et al.*, 2004), or by preventing the dissociation of silane (Greytak *et al.*, 2004; Wang *et al.*, 2003). Careful control of the synthesis conditions allows the formation of uniform nanowires without diameter variation (see Chapter 2 for additional details).

VLS combined with lithography allows tight control over nanoneedle diameter (down to a few nanometres), their density and their arrangement over the carrier substrate (Fan *et al.*, 2006). Variants of the basic chemical vapour deposition (CVD) VLS principle generate the precursor gaseous Si by laser ablation (Morales, 1998), plasma generation (Hofmann *et al.*, 2003), and molecular beam epitaxy (Liu *et al.*, 1999). Other strategies employ Si precursor in solution (Heitsch *et al.*, 2008; Holmes, 2000) or in solid (Wong *et al.*, 2005) form. These alternatives were developed to address specific requirements such as growth temperature, integration, uniformity and doping, but share the same underlying formation/growth principle of VLS. Overall, VLS and associated techniques are extremely powerful tools to synthesize solid silicon nanowires. VLS integration with microfabrication allows for fine control of the diameter, length and arrangement of the nanowires. Even though VLS requires tightly controlled and harsh synthetic conditions, it is a highly reproducible, versatile, scalable and established technique.

8.2.2 Metal-assisted chemical etch of nanoneedles

Metal-assisted chemical etch (MACE) has risen to prominence in the past decade as a simple, low-investment strategy for wet anisotropic etch of silicon forming vertically aligned silicon nanowires (Fig. 8.1(b)) (Chiappini *et al.*, 2010; Hochbaum *et al.*, 2009; Huang *et al.*, 2007; Peng *et al.*, 2006). In MACE, a (noble) metal (usually silver or gold) deposited on a silicon substrate and immersed in an oxidizing solution of hydrofluoric acid, catalyses Si etching in the immediate vicinity of the metal itself, as a result of highly localised electrochemical dissolution of silicon (Chartier *et al.*, 2008; Li and Bohn, 2000). Silicon nanowires form by MACE following deposition of a random mesh of metal nanoparticles, obtained through electroless deposition from metal salt precursors. MACE is a versatile technique that integrates with conventional photolithography and microfabrication (Chiappini *et al.*, 2010), as well as with more specialised nanofabrication strategies such as nanosphere lithography (Chiappini *et al.*, 2010; Huang *et al.*, 2007), interference lithography (Choi *et al.*, 2008), anodised alumina (Huang *et al.*, 2008) and block copolymer templating (Chang *et al.*, 2009). This integration enables the synthesis of ordered arrays of high aspect ratio nanoneedles with a range of diameters comparable with those accessible to VLS: 10 nm to several microns. Lithographic patterning also permits arbitrary selection of the needles' pitch and arrangement, within the limitations in resolution dictated by the specific lithographic technique employed. Selection of the etching conditions and combination with post-synthesis dry etches can shape the needles along their major axis, determining the sidewall angle and enabling tunable shapes from cylindrical to conical, with extremely sharp tips.

A major feature of MACE is the possibility to directly form porous silicon (pSi) nanoneedles under appropriate etch conditions (Chiappini *et al.*, 2010). By selecting substrate resistivity, oxidant concentration and etch temperature, it is possible to finely control the porosity, pore size and crystalline orientation of the nanoneedles formed. By varying the oxidant concentration over time, it is possible to control the nanoneedle porosity along its axis, forming nanoneedles with multiple segments of different porosities. Thanks to their biodegradability (Chiappini *et al.*, 2010; Anderson *et al.*, 2003; Canham, 1995), biocompatibility (Piret *et al.*, 2011; Goh *et al.*, 2007; Low *et al.*, 2006; Chin *et al.*, 2001) and enhanced surface area (Herino, 1987), pSi nanoneedles are appealing for those biomedical applications where sustained release, delivery of large payloads, biocompatibility/biodegradability or protection of the payload from the external environment are concerned. These concerns are especially felt *in vivo* and for all those *in vitro* applications where nanoneedle treatment does not constitute an endpoint, such as the study of cellular pathways or cellular reprogramming for *in vivo* implantation. In these applications where pSi is favoured, and in all settings where VLS is impractical for integration or investment issues, MACE is strongly positioned as a reliable, versatile and low-cost strategy for the synthesis of nanoneedles.

8.2.3 Focused ion beam etch of nanoneedles

Focused ion beam (FIB) etching is a strategy that can sharpen existing microstructures into high aspect ratio nanostructure (Fig. 8.1(c)). With FIB, a beam of heavy ions (usually gallium) physically mills the surface of materials. The beam of ions can be focused and directed to desired areas on the sample with nanometric precision, to generate features of arbitrary geometry at the nanoscale. FIB controls the etch rate and etch resolution by tuning the ion current intensity. Higher currents mill faster but have lower resolution, and tend to increase the amount of re-deposited material and/or melt the surrounding material. FIB is a direct writing technique, where each needle must be milled individually. FIB is thus inherently low throughput and unsuitable for large-scale manufacturing, but combined with scanning electron microscopy (SEM) in a dual beam, SEM-FIB allows for high precision shaping of existing micro- and nano-structures that cannot be handled with conventional microfabrication (Han *et al.*, 2005b). FIB has been very successful for the sharpening of existing atomic force microscope (AFM) tips to form AFM-operated nanoneedles with diameters smaller than 50 nm. AFM-operated tips can be employed for the nuclear insertion of genetic material into hard-to-transfect, or rare cells owing to the low cytotoxicity and high transfection rate. Furthermore, the AFM-operated needle is suitable for delivery of drugs to certain specific cells within a culture, as it integrates with optical imaging of the area of interest.

8.2.4 Microfabrication of hollow nanoneedles

Conventional microfabrication strategies are at a disadvantage when trying to create high aspect ratio nanostructures into silicon, and cannot compete with the strategies outlined above for nanoneedle synthesis. Mostly the limitations arise from the challenges associated with the dry etching of silicon structures with nanoscale cross-sections and high aspect ratios (Woldering *et al.*, 2008). Microfabrication, however, provides simple strategies to form hollow needles, which have been successfully used to interface cells with a drug reservoir for extended periods of time, thus enabling sustained or repeated drug delivery (Fig. 8.1(d)) (Peer *et al.*, 2012; VanDersarl *et al.*, 2012). Hollow nanoneedle synthesis starts with an array of nanopores conformably covered with a thin dielectric film. The film is then etched away from the horizontal surfaces and remains as a lining shell over the walls of the pores. Finally, selectively etching the material around the pore forms a hollow needle. Although this strategy is quite effective to reach the desired scope, the diameter of the resulting needles is quite large compared with the other strategies outlined, and the needles thus formed require surfactants to mediate drug delivery. Furthermore, the aspect ratio achieved thus far is quite limited, and the large needle size and requirements for saponification limit the potential applications of these needles and negatively impact biocompatibility.

8.3 Drug loading of nanoneedles and release patterns

The strategies for loading nanoneedles with drugs are mostly determined by the nature of the nanoneedles. Solid, porous and hollow needles are amenable to different loading mechanisms, resulting into markedly different release patterns.

8.3.1 Solid nanoneedles

Solid silicon nanoneedles, which have been fabricated by VLS or FIB, are loaded with drugs by physisorption on their surface (Shalek *et al.*, 2010; Han *et al.*,

2008). Electrostatic interaction is conventionally used to improve physisorption, by ensuring the needles' surface charge is opposite to that of the molecules being adsorbed. Common strategies include initial oxidation of silicon to provide a hydrophilic and negatively charged surface at physiological pH (Tasciotti *et al.*, 2008). Surface functionalisation with an amine-terminated silane (the most common being 3-(aminopropyl)triethoxysilane, APTES) provides positively charged surfaces, which electrostatically favour adsorption of nucleic acids and most proteins, and are thus favoured over negatively charged ones (Shalek *et al.*, 2010; Tasciotti *et al.*, 2008). Further, the instability of the silane layer towards hydrolysis can potentially favour desorption of molecules from the surface and their release into cells, as suggested by the marked improvement in delivery efficiency when employing APTES compared with polyethylene imine (Shalek *et al.*, 2010; Kim *et al.*, 2007). Chemisorption of molecules on the walls of nanoneedles has been attempted, but although they were successfully loaded, they did not effectively mediate delivery (Han *et al.*, 2005b; Obataya *et al.*, 2005a; McKnight *et al.*, 2004).

This loading strategy has successfully delivered a wide variety of bioactive compounds, including some, like nucleic acids, that do not easily transverse the cell and nuclear membranes. The surface physisorption loading readily exposes the drug to solution, causing it to rapidly release it away from the needles, limiting the timeframe for successful delivery by nanoneedle application (Han *et al.*, 2008). To overcome this issue, a large excess of dissolved drug is spotted on the nanoneedles and often left to dry, in order to form thick coating layers that prevent immediate desorption of all the drug. Nonetheless, AFM-operated nanoneedles loaded in this fashion have a maximum of 3 minutes in solution before they become unable to deliver the drug intracellularly, as a result of its diffusion away from the needle. Similarly for nanoneedle arrays, it was never demonstrated that the drug delivery occurs from the nanoneedles into the intracellular compartment. The localisation of the delivered moieties in the perinuclear region suggests trafficking through the endolysosomal system instead of intracellular presentation (Shalek *et al.*, 2010). Alternative explanations for the observed delivery include cell poration by nanoneedles in a fashion similar to electroporation, or uptake caused by the significant concentration in solution close to cells. As delivery of nucleic acids, known for the complexity of their delivery, can be achieved by appropriately coated flat surfaces, the needles array need not play a different role from the flat surfaces in this instance, and thus may not actively insert the drug payload into cells.

8.3.2 Porous nanoneedles

Porous nanoneedles possess certain advantages with respect to their solid counterparts. Their high surface area and their pore volume provide a large reservoir for the loading of drugs, improving payload density by several orders of magnitude over solid structures (Salonen *et al.*, 2005). The ability of porous structures to harvest molecules from solution allows attainment of loading concentrations several orders of magnitude higher than the equilibrium in solution. Moreover, alongside the electrostatic loading method, porous structures can be loaded with drugs from melt powders, to achieve higher loading concentrations and loading in the amorphous phase (Ambrogi *et al.*, 2010; Riikonen *et al.*, 2009). Further, it is possible to cap the pore openings and protect the payload from the external environment, modulate its release and prevent its premature release. Capping the pores with agarose prevents the degradation of protein payloads by proteases without influencing the release profile (De Rosa *et al.*, 2011). Pore openings capped with environmentally responsive molecular valves that open at low pH prevent extracellular leakage of the payload, maximizing intracellular delivery (Xue *et al.*, 2011).

Differently from solid nanoneedles, mesoporous needles are well positioned for loading nanoparticles that can penetrate and accumulate within the porous structure, from which they are slowly released (Tasciotti *et al.*, 2008). Several different classes of nanoparticles were loaded successfully within pSi enhancing their therapeutic properties. Liposomes loaded with siRNA were loaded into pSi to extend their efficacy at silencing target genes for over 21 days (Tanaka *et al.*, 2010), while pSi-loaded Gd nanoparticles enhanced their MRI contrast potential (Ananta *et al.*, 2010) and Au nanoshells improved their photothermal effects (Shen *et al.*, 2012).

Porous silicon also modulates the solubility of the payload, both by limiting its diffusion from within the pores and by progressively dissolving and desorbing payload (Salonen *et al.*, 2005). Modulating solubility allows controlling the release rate of the payload, enables sustained release, mitigating the limitations of solid nanoneedles that rapidly release their payload, and capturing the advantages of hollow nanoneedles, without the need for a stable transmembrane opening that exposes the inside of the cell to the outer environment. Although porous nanoneedles can sustain drug delivery, their reservoir of drug is limited to what can be loaded within their small volume, whereas hollow nanoneedles can feed off an arbitrarily large external reservoir, acting as conduits.

8.3.3 Hollow nanoneedles

Hollow needles put in communication a drug reservoir with the cell cytosol (Peer *et al.*, 2012; VanDersarl *et al.*, 2012). In this fashion the needles are not loaded with drugs in the conventional interpretation of the term, but simply act as a conduit for drug delivery. This type of nanoneedle is the most strictly analogous to conventional needles and microneedles that act as conduits to carry drugs from large reservoirs where they are present in solution at high concentrations. Whereas needles and microneedles tend to be used to actively inject drugs into the target, nanoneedles for the most part rely on passive diffusion through nanochannels.

The passive diffusion delivery reduces the needles' cytotoxicity, as part of the toxicity associated with microinjection is caused by intracellular pressure buildup following injection (Zhang and Yu, 2008). By relying on diffusion the delivery is slower, but nanoneedles are parallelised into arrays feeding off a common reservoir, which grant a higher throughput than manually operated microneedles. However, similar to what occurs in microneedles, those nanochannels could be easily blocked by proteinaceous material depositing across their opening, preventing their use over long periods of time. In this regard, it is as yet unknown whether hollow nanoneedles can sustain intracellular delivery for longer than porous ones, which are limited by the size of their drug reservoir.

8.4 Drug delivery using nanoneedles

Typically, two strategies allow nanoneedles to deliver drugs to the cell cytosol: either the cells internalise the needles, or a force is applied to the needles in order to cross the cell membrane.

8.4.1 Internalisation

Internalisation is the strategy most often employed with arrays of nanoneedles supported on a substrate (Peer *et al.*, 2012; Shalek *et al.*, 2010). Whether solid, porous or hollow, cells seeded over substrates decorated over nanoneedles can internalise a wide range of bioactive payloads, which are able to alter cell phenotype as expected. Small molecules, nucleic acids, proteins and nanoparticles can all be localised in the cytosol following growth on appropriately loaded nanoneedle substrates. DNA and siRNA delivered to cells correctly alter the gene expression pattern of target cells. This delivery strategy requires cells to play an active role and can be sensitive to cell type and environmental conditions (i.e. temperature, inhibitors, medium). The density of needles is an important parameter that affects their ability to penetrate cells (Qi *et al.*, 2009). Cells grow on top of needles with a density of several needles per μm^2, whereas they grow at the bottom of needles at densities lower than 1 needle/μm^2. It is as yet unproven whether this strategy can mediate *in vivo* delivery, as it relies on cell growth on a needle decorated substrate, which is not a process known to occur *in vivo*.

8.4.2 Forcible insertion

The AFM-operated needle instead relies on forcible insertion to present the payload intracellularly. A force between 0.5 and 2 nN is required to insert a needle of less than 200 nm diameter across the cell membrane, and there is indirect evidence that penetration across the nuclear membrane is also possible (Han *et al.*, 2008; Obataya *et al.*, 2005b). As the needle diameter increases, its insertion requires a stronger force and results in bulging of the membrane, with associated cytotoxicity (detailed in the following section). DNA plasmids were successfully delivered by AFM-operated needles and expressed target genes (Han *et al.*, 2008). Two examples exist where supported nanoneedles were forcibly inserted into cells, in one instance using the force originated by a centrifuge, in another simply relying on the substrate weight (McKnight *et al.*, 2004). In both instances the estimated force applied was greater than that required by the AFM experiments and needles successfully localised within the cell cytosol. The capillary interfacing was successful in delivering the drug; efficacy could not be evaluated in the centrifugation strategy, as the DNA plasmid payload was covalently attached to the needles and failed to express the target gene.

The forcible insertion strategy closely resembles the usual mechanism of delivery through needles, at a smaller scale. This delivery strategy does not depend on the interaction of the needles with cells and is clearly more amenable to *in vivo* applications than internalisation. Additionally, the reduced diameter of nanoneedles improves their cytocompatibility, as will be clear in the next section.

8.5 Toxicity of nanoneedles

To date the toxicity of nanoneedles has only been investigated *in vitro*, and is limited to cytotoxicity studies. AFM-operated needles of different diameter showed that cytotoxicity decreases with decreasing diameter, and is proportional to the insertion force and the degree of membrane bulging (Yum *et al.*, 2010). Membrane bulging is thought to correlate with intracellular fluid leaking on needle insertion. When minimal bulging occurs for minimal insertion forces, none to minimal leaking occurs and the cells are likely to survive the nanoneedle penetration. Conversely, if the membrane is bulged by several microns, a significant portion of intracellular fluid is thought to leak out when the needle finally pierces the cell, resulting in likely cell death, or major disruption to cell function. A nanoneedle diameter of around or less than 200 nm has little impact on cell viability and proliferation following needle insertion, whereas diameters larger than 400 nm results in the death of the large majority of cells that underwent interfacing with nanoneedles. The forcible insertion of arrays of nanoneedles with diameter smaller than 200 nm also do not induce significant cell death, affect cells' proliferation or their metabolism in cell culture (McKnight *et al.*, 2004).

A similar trend occurs when cells internalise nanoneedles, with needles up to 200 to 300 nm in diameter having little effect on cell viability while still being capable of drug delivery. The cells grown on nanoneedles survive for several days on the silicon substrate. Further cells grown on arrays of nanowires with diameters of 30, 90 and 400 nm show that cell longevity is directly dependent on the diameter, with cell death occurring within a day for the 400 nm diameter wires and cells lasting for 5 days when 30 nm nanowires were used. Cells internalising nanoneedles express all 300 immune response genes investigated at levels comparable with those of untreated cells (Shalek *et al.*, 2012). Their metabolic activity, proliferation and viability, as measured in multiple studies, with multiple types of needles, and by multiple assays appear unaffected by the presence of the needles, if needle diameter is below the 200 to 300 nm threshold. The density of the needles also does not affect their viability, while it determines the vertical location of cell growth as mentioned above. These data indicate that forcibly inserted nanoneedles can be optimised to cause such minimal disruption to the cell integrity that they do not elicit toxicity, while cells that are grown on arrays of nanoneedles can perceive them as a standard cell culture substrate.

8.6 Overview of nanoneedle applications

8.6.1 Atomic force microscope (AFM)-operated nanoneedles

AFM actuation is the first strategy implemented for the use of silicon nanoneedles for drug delivery (Fig. 8.2). The force measurements with an AFM instrument can be used to study cellular events in individual cells with great sensitivity (Lamontagne *et al.*, 2008). Among the applications is the injection at the nanoscale of specific molecular entities in individual living cells. Solid silicon nanoneedles optimised for harmless cell penetration (Obataya *et al.*, 2005b) are developed with 6 μm length and 200 nm width by FIB from Si AFM tips (Han *et al.*, 2005a). Subsequent treatment of the surface of the newly formed nanoneedles with

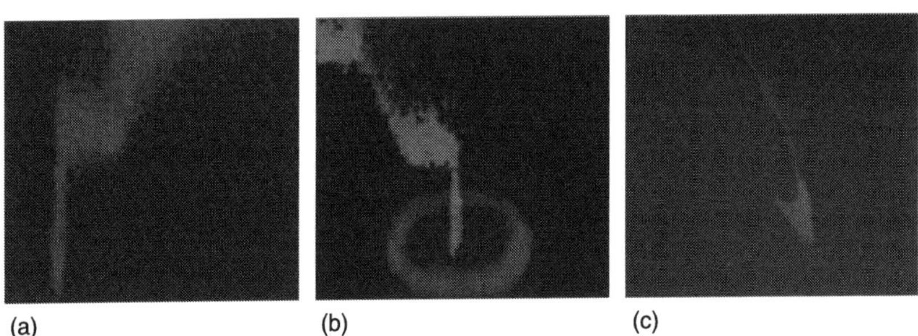

8.2 The atomic force microscope (AFM)-operated nanoneedle induces gene expression *in vitro*. AFM-operated nanoneedles can load fluorescently labelled DNA plasmids by drying a drop of plasmid solution on their surface (a); they cross the cell membrane and potentially the nuclear membrane displaying their payload in the cytosolic and nuclear environments as shown by the cross-sectional view obtained by confocal microscopy of fluorescently labelled needles interfaced with cells genetically engineered to attain a fluorescently labelled membrane (b).The AFM nanoneedle can deliver a GFP plasmid that is efficiently expressed by cells (c). Images reproduced with permission from Han *et al.*, 2008.

3-mercaptopropyltrimethoxysilane (MPTS) followed by incubation with N-(6-maleimidocaproyloxy)succinimide (EMCS) provides the means for further functionalisation. DNA is bound to the needle by first soaking the succinimidyl nanoneedle in an avidin solution and then incubating with a solution of biotinylated green fluorescent protein (GFP) DNA fragment. Force–distance measurements indicate that the nanoneedles can penetrate human embryonic kidney cells with a reproducible profile that outlines the different stages of the needle penetration. The penetration force profile does depend on DNA immobilisation on the nanoneedle's surface, suggesting that the DNA molecules and cell components do not interact. Furthermore, calculations of friction force applied to one molecule of DNA in the penetration process indicate that the DNA does not detach from the surface of the nanoneedle during insertion. The cells manipulated by AFM can proliferate after repeated penetrations with a DNA-functionalised nanoneedle (Han *et al.*, 2005a).

The silicon AFM-based nanoneedle system can also mediate intracellular presentation proteins, allowing insertion of two different His-tagged, fluorescently labelled proteins into HeLa cells. The nanoneedle surface is chemically modified with nitrilotriaceticacid (NTA) groups, and then chelated with $NiCl_2$ to conjugate poly-histidine-modified proteins. The protein–nanoneedle hybrid inserted into HeLa cells shows constant fluorescence intensity at the surface of the device while kept inside the cell, to indicate a stable conjugation of the protein to the needle (Obataya *et al.*, 2005a).

Although neitheir of these examples demonstrate drug delivery, they indicate an avenue to use silicon nanoneedles for manipulation of living cells and intracellular access, without inflicting critical cell damage. Indeed an AFM nanoneedle can successfully transport electrostatically bound GFP plasmid DNA into cells. The transfection efficiency of over 50% is sufficient to prove molecular delivery through AFM-operated nanoneedles (Han *et al.*, 2005a). Similarly AFM nanoneedles mediate efficient (above 70%) transfection of GFP plasmid DNA into the nucleus of mesenchymal stem cells, known hard-to-transfect cells using microinjection because of their flat shape (Han *et al.*, 2008). When the plasmid DNA is only non-specifically bound to the surface of the nanoneedle, the system releases its load inside the target cell but also in the surrounding media. If the cell penetration is not rapid enough, the delivery fails.

Using a nanoneedle actuated by AFM surmounts the limitation of whole-cell population studies of typical cell biology methods by being able to manipulate single cells (Lamontagne *et al.*, 2008) with minimal invasiveness. Moreover, these nanoneedles can also access specific regions (e.g. nuclei) inside living cells and deliver to target areas, a feature not available with conventional delivery methods. However, it is a time-consuming technique because of the need to manipulate each cell individually, limited by the availability of the nanoneedles, and requires a highly specialised setup, trained operators and costly consumables.

8.6.2 Vapour–liquid–solid nanoneedles for universal intracellular delivery

Large arrays of nanoneedles, employed in parallel, allow for high throughput delivery to multiple cells (Fig. 8.3). Nanowire arrays fabricated by VLS can act as a substrate for mouse embryonic stem cells and human embryonic kidney cells (Kim *et al.*, 2007). Electrostatic forces can immobilise DNA encoding GFP on a nanowire array for gene delivery in the HEK293 cell line. The fluorescence emitted by GFP allows the estimation of the transfection efficiency to the cells, with less than 1% of cells expressing the protein. The limited transfection efficiency observed is attributed to the chemistry used to retain the DNA and this could be improved with other conjugations.

Similarly, an array fabricated by VLS can mediate the delivery of virtually any type of molecule desired into both immortalised and primary cells (Shalek *et al.*, 2010). Molecules adsorb on the surface by electrostatic interaction through the simple modification of the surface with an aminosilane (3-aminopropyltrimethoxysilane) to attain non-covalent and non-specific binding of the molecules. The incubation time affects the position of HeLa cells with respect to the needles: whereas after 15 minutes the cells sit on top of the silicon nanowires, after 1 hour most of the cells are completely penetrated, irrespective of the nanowire density (always lower than 1 needle/μm^2) or the molecule

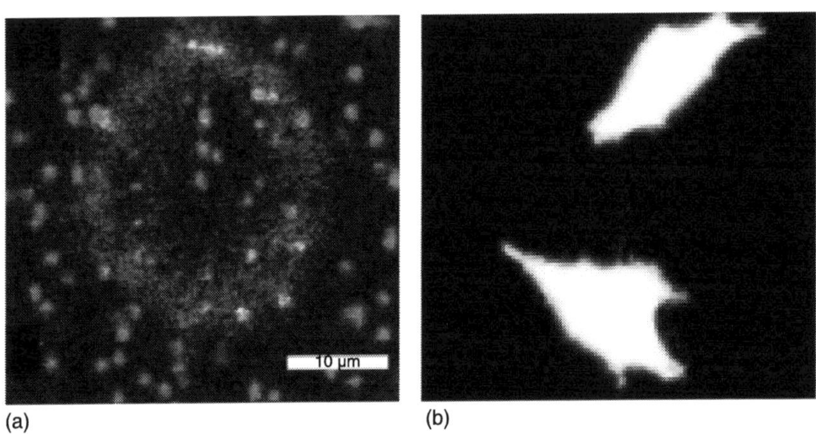

8.3 The vapour–liquid–solid (VLS) nanoneedle in action induces gene expression *in vitro*. VLS nanoneedles effectively cross the cell membrane displaying their payload in the cytsol, as indicated by confocal microscopy with cell membrane stain and fluorescent labelling of the Au seed at the tip of the nanoneedles (a). The VLS nanoneedles can deliver a YFP plasmid inducing its expression (b), alongside many other classes of bioactive molecules. Images reproduced with permission from Shalek *et al.*, 2010.

immobilised on the nanowire surface. The impaled cells retain normal metabolism and growth, although the initial growth rate is slightly reduced when compared with glass coverslips. Further, the integrity of the cell membrane appears to be conserved as rat hippocampal neuron cells can retain the intracellular ionic concentrations required to fire action potentials.

Delivery and co-delivery of various biomolecules (fluorescently tagged or encoding fluorescent proteins) are possible in HeLa cells and primary cells. Expression of fluorescent proteins occurs when plasmid DNA is administered through the silicon nanowires or following delivery of fluorescently labelled biomolecules (siRNA, DNA, proteins, peptides). Further, molecules co-deposited on the surface of the nanowires can be co-delivered. This system is compatible with microarray technology because of the direct mediation of biomolecule delivery by the surfaces of the nanowires. Also, by arraying biomolecules on a nanowire surface, parallel live-cell screening of diverse biological effectors is possible.

8.6.3 Hollow nanoneedles for sustained delivery

Hollow nanoneedles are synthesised both from alumina and silicon dioxide, and employed for repeated delivery of biomolecules from an effectively infinite reservoir (Peer *et al.*, 2012; VanDersarl *et al.*, 2012) (Fig. 8.4). The silicon dioxide needles are realised by first forming a 10 μm silicon membrane through KOH etch of a silicon on insulator wafer, which also defines the drug reservoir. Pores with

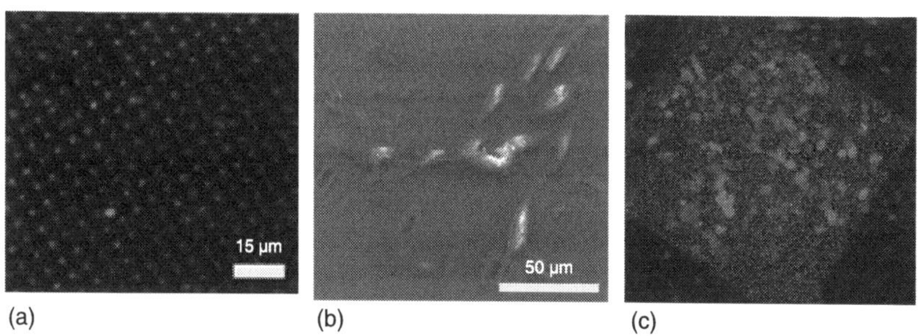

8.4 The hollow nanoneedle induces gene expression *in vitro*. Hollow nanoneedles act as a conduit between a reservoir of fluorescent dextran and the cell culture environment enabling multiple exposure of cells to bioactive agents, as shown by the highly fluorescent spots associated with the needles (a); they allow interfaced cells to grow successfully as evidenced through scanning electron microscope (SEM) micrographs (b). The hollow nanoneedles can deliver red fluorescent protein expressing plasmids to cells cultured over them and in the neighbouring area with the adjuvant effect of surfactants in solution(c). Images reproduced with permission from Peer *et al.*, 2012.

500 nm diameter and 5 μm pitch are patterned by e-beam lithography and etched through the membrane by deep reactive ion etch. Thermal oxide is grown on the whole system to a thickness of 100 nm. The top oxide surface is removed by reactive ion etching and the hollow needles formed by thinning the silicon membrane through deep silicon etching.

This fabrication strategy requires multiple deposition, growth and etch steps on a thin membrane, which is very fragile, and easily destroyed by interfacial stresses. Additionally, the overall thickness of the membrane is limited by the depth of the pores that can be etched by deep reactive ion etch (DRIE), which in turn is determined by the pore diameter. The DRIE step fundamentally limits the aspect ratio of the needles, and is inefficient at forming nanopores. Smaller pore diameter would mean more shallow etches, and in turn shorter needles. Even for these very large nanoneedles with 500 nm diameter, their length is limited to approximately 3 μm in order to preserve the integrity of the 10 μm membrane.

These nanoneedles can transport fluorescently labelled dextran and DNA plasmids across the membrane during a 2-hour period. Two human cell lines (HEK 293 and NIH3T3) can be seeded over the needles and grow for up to 48 hours, without noticeable effect on cell proliferation. Cells tend to grow on the upper section of the needles without being able to reach the bottom substrate. It is unclear whether the needles are able to cross the cell membrane. Loading a solution containing fluorescently tagged dextran mixed with the permeation enhancer saponin at concentrations of 3 and 4 μg/mL allows repeated delivery of dextran molecules following multiple sessions of 10 minutes' incubation. Similarly, the hollow nanoneedles deliver DNA plasmids for the expression of red fluorescent proteins, also in the presence of saponin. The delivery is not limited to the cells growing on the needles, but extended to neighbouring cells grown on flat surfaces.

Although the platform appears interesting, and can regulate sustained diffusion across two reservoirs, the need for saponin to mediate the delivery shows that the needles alone cannot penetrate cells and mediate intracellular delivery. To this extent, the advantage of this platform over a porous membrane is unclear.

8.6.4 Biodegradable porous silicon nanoneedles

Combining MACE with conventional microfabrication allows formation of arrays of vertically aligned nanoneedles (Fig. 8.5). The simplest form of needles, as vertically aligned porous nanowires, results from MACE of a silicon substrate covered with a noble metal mesh from electroless deposition. Ordered and shape-defined nanoneedles result by first depositing a thin film of low stress silicon nitride over a silicon substrate (Chiappini *et al.*, 2010). The nitride is then patterned with a large-scale array of nano-sized dots with desired pitch, and metal is selectively deposited in the field. The silicon then undergoes MACE to form high aspect ratio pillars. The pillars can then be shaped into conical needles by reactive ion etch, to form porous silicon nanoneedles.

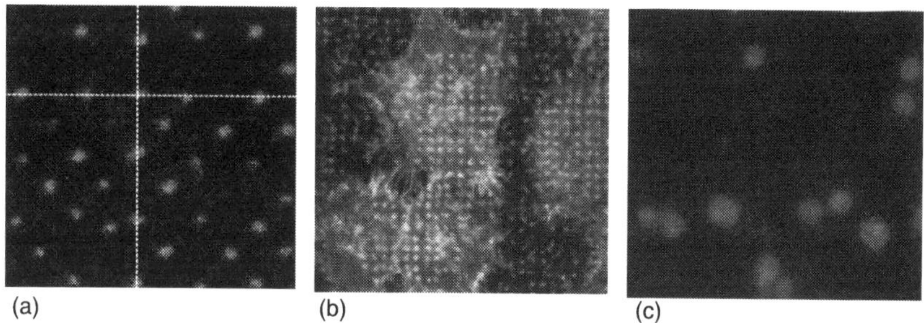

8.5 The porous silicon nanoneedle induces gene expression *in vitro*. Porous silicon nanoneedles load fluorescently labelled molecules throughout their volume as shown by the uniform intensity through each needle of an array in confocal micrographs (a); and can penetrate the cell cytosol as shown by confocal microscopy of dye-labelled cells and nanoneedles (b). The porous nanoneedles can deliver green fluorescent protein plasmids to cells with high efficiency (c).

Controlling MACE parameters allows for control of the porosity and pore size of the needles over a wide range, which in turn influences their biodegradation kinetics, and the loading and release kinetics of molecules and nanoparticles. Biodegradation is faster for higher porosity materials, which results in faster release. Further, nanoparticles can be size excluded through pore size, allowing for selective nanoparticle harvesting from solution. MACE also allows for forming needles with segments of different porosity along their axis, enabling differential loading of nanoparticles (Chiappini *et al.*, 2010). Quantum dots of 3.3 nm and 6.3 nm can be segregated into different segments of the same nanoneedle, allowing orthogonal loading of multiple nanoparticles. Also, the porous structure allows loading of multiple drugs simultaneously.

Cells grown on MACE nanoneedles maintain their normal viability and metabolism, and appear to internalise them. Molecules loaded on the nanoneedles are effectively delivered to the cell, and nucleic acid delivery can either induce (DNA) or silence gene expression (siRNA). The porous needles degrade over a few days, leaving behind a smooth substrate that resembles and behaves similarly to a standard cell culture substrate.

Porous needles are versatile, in that they allow co-loading, segregated loading and sustained delivery of multiple molecules, as well as efficient loading of nanoparticles. Porous needles are potentially less toxic than the solid ones in the long term, owing to their biodegradability. The elevated biocompatibility and biodegradability of pSi make porous nanoneedle structures amenable for all applications where silicon remaining *in situ* poses concerns. All applications where the nanoneedles are directly employed *in vivo* or where cells are treated *in vitro* with nanoneedles for further use *in vivo* fall into this category.

Although porous needles may appear more fragile than their solid counterparts, their mechanical properties can be optimised, controlling their porosity. Mechanical properties can range from values close to those of solid silicon to significantly lower values that could better match those of cells or the tissue with which to interface. The dissolution of the needles limits the timeframe of their usability, as the degradation process shortens, dulls and makes the needles more flexible, thus less capable of performing their function. They may thus be at a disadvantage wherever repeated applications from the same needles are required over extended periods of time. Solid or hollow needles would be advantageous in comparison for such applications.

8.7 Conclusion

All of the different nanoneedle-based approaches developed thus far can effectively deliver drugs inside the cell. They combine low cytotoxicity with broad applicability in a manner that surpasses most of the currently available strategies to negotiate the cell membrane. Nanoneedles also do not appear to disrupt or alter any cellular function, an important aspect when trying to investigate the change in cellular phenotype as a consequence of drug delivery.

Nanoneedles deliver many classes of bioactive molecules inside cells, and can deliver them simultaneously. Drug co-delivery, together with cytocompatibility, makes nanoneedles an extremely promising material for intracellular drug delivery *in vitro*. The cytosolic drug delivery approach also requires lower overall amounts of drugs and minimises unwanted off-target interactions in the intracellular space.

Both when operated by AFM and when used as large supported arrays, nanoneedles can deliver their payload in patterns of choice. Patterned delivery can modulate cellular phenotype of a subset of cells within a culture to form defined structures, and allows for designing co-cultures with arbitrary arrangements of differently behaving cells. The patterned delivery, combined with the minimal disruption to cellular activity, makes nanoneedles an extremely promising approach for high-throughput microarray setups that investigate cell response to cytosolic presentation of drugs. Further, patterned delivery enables complex organotypic cultures that contribute to investigation of cell interaction in tissue structures. The information from such cultures would provide important design guidelines when developing topographical cues in scaffolds for tissue engineering.

The low impact, minimally invasive approach of nanoneedles is well amenable for *in vivo* applications. Combining stimuli-responsive strategies with nanoneedle delivery would enable targeted intracellular drug delivery, contributing to a novel strategy to the practice of personalised medicine, where the therapeutic index is maximised by both the lower amount of drug required and its selective delivery. Further *in vivo* applications can include localised patterned delivery of genetic material for the small scale, precision engineering of tissue. In such vision, the local alterations to cell phenotype will form nano- to microscale functional tissue structures, which are essential for tissue function. However, significant barriers still exist to the efficient use of nanoneedles *in vivo*, requiring further innovation before their implementation. Presently, nanoneedle arrays rely on cell seeding and internalisation to deliver drugs, and this would not occur *in vivo*. The major obstacle towards using needles to actively pierce cells is the large amount of cells that detach from the original substrate in this case. Strategies to alleviate this concern are being developed, by increasing the interaction between the substrate and cells, but still rely on cell behaviour that is not observed *in vivo*.

Nanoneedle length and overall mechanical stability are also not suitable at present for transdermal delivery, presently requiring surgical placement of the needles to the site of interest. Further non-degradable needles would not be easily allowed *in vivo*, especially because of the known adverse effects of high aspect ratio silica structures. In this regard, porous silicon nanoneedles possess an advantage for *in vitro* applications as they biodegrade completely into harmless byproducts.

Overall, we are still at the infancy stage in the development of nanoneedle technology for drug delivery, which nonetheless promises to efficiently negotiate the major biological barrier constituted by the cell membrane without altering cell behaviour. Whether these promises will be transformed into technological advances that can see their way into laboratories and the clinic still depends on the ability to overcome what currently appear as strict design limitations of this technology.

8.8 References

Adams, GP and Weiner, LM (2005), Monoclonal antibody therapy of cancer, *Nat. Biotechnol.*, 23(9), 1147–57. [10.1038/nbt1137]

Ambrogi, V, Perioli, L, Pagano, C, Marmottini, F, Moretti, M, *et al.* (2010), Econazole nitrate-loaded MCM-41 for an antifungal topical powder formulation, *J. Pharm. Sci.*, 99(11), 4738–45. [10.1002/jps.22183]

Ananta, JS, Godin, B, Sethi, R, Moriggi, L, Liu, X, *et al.* (2010), Geometrical confinement of gadolinium-based contrast agents in nanoporous particles enhances T_1 contrast, *Nat. Nanotechnol.*, 5(11), 815–21. [10.1038/nnano.2010.203]

Anderson, SHC, Elliott, H, Wallis, DJ, Canham, LT and Powell, JJ (2003), Dissolution of different forms of partially porous silicon wafers under simulated physiological conditions, *Physica Status Solidi (a)*, 197(2), 331–5. [10.1002/pssa.200306519]

Balazs, DA and Godbey, W (2011), Liposomes for use in gene delivery, *J. Drug Deliv.*, 2011, 326497. [10.1155/2011/326497]

Canham, LT (1995), Bioactive silicon structure fabrication through nanoetching techniques, *Adv. Mater.*, 7(12), 1033–7. [10.1002/adma.19950071215]

Celis, JE (1984), Microinjection of somatic cells with micropipettes: comparison with other transfer techniques, *Biochem. J.*, 223(2), 281–91.

Chang, S-W, Chuang, VP, Boles, ST, Ross, CA and Thompson, CV (2009), Densely packed arrays of ultra-high-aspect-ratio silicon nanowires fabricated using block-copolymer lithography and metal-assisted etching, *Adv. Funct. Mater.*, 19(15), 2495–500. [10.1002/adfm.200900181]

Chartier, C, Bastide, S and Lévy-Clément, C (2008), Metal-assisted chemical etching of silicon in HF–H_2O_2, *Electrochim. Acta*, 53(17), 5509–16. [10.1016/j.electacta.2008.03.009]

Chiappini, C, Liu, X, Fakhoury, JR and Ferrari, M (2010), Biodegradable porous silicon barcode nanowires with defined geometry, *Adv. Funct. Mater.*, 20(14), 2231–9. [10.1002/adfm.201000360]

Chin, V, Collins, BE, Sailor, MJ and Bhatia, SN (2001), Compatibility of primary hepatocytes with oxidized nanoporous silicon, *Adv. Mater.*, 13(24), 1877–80. [10.1002/1521–4095(200112)13:24<1877::AID-ADMA1877>3.0.CO;2–4]

Choi, WK, Liew, TH, Dawood, MK, Smith, HI, Thompson, CV and Hong, MH (2008), Synthesis of silicon nanowires and nanofin arrays using interference lithography and catalytic etching, *Nano Lett.*, 8(11), 3799–802. [10.1021/nl802129f]

Chou, LY, Ming, K and Chan, WC (2011), Strategies for the intracellular delivery of nanoparticles, *Chem. Soc. Rev.*, 40(1), 233–45. [10.1039/c0cs00003e]

De Rosa, E, Chiappini, C, Fan, D, Liu, X, Ferrari, M and Tasciotti, E (2011), Agarose surface coating influences intracellular accumulation and enhances payload stability of a nano-delivery system, *Pharmacol. Res.*, 28(7), 1520–30. [10.1007/s11095–011–0453–2]

Di, L, Artursson, P, Avdeef, A, Ecker, GF, Faller, B, *et al.* (2012), Evidence-based approach to assess passive diffusion and carrier-mediated drug transport, *Drug Discov. Today*, 17(15–16), 905–12. [10.1016/j.drudis.2012.03.015]

Dykxhoorn, DM, Novina, CD and Sharp, PA (2003), Killing the messenger: short RNAs that silence gene expression, *Nat. Rev. Mol. Cell Bio.*, 4(6), 457–67. [10.1038/nrm1129]

Fan, HJ, Werner, P and Zacharias, M (2006), Semiconductor nanowires: from self-organization to patterned growth, *Small*, 2(6), 700–17. [10.1002/smll.200500495]

Fischer, PM, Krausz, E and Lane, DP (2001), Cellular delivery of impermeable effector molecules in the form of conjugates with peptides capable of mediating membrane translocation, *Bioconjugate Chem.*, 12(6), 825–41. [10.1021/bc0155115]

Fischer, R, Fotin-Mleczek, M, Hufnagel, H and Brock, R (2005), Break on through to the other side—biophysics and cell biology shed light on cell-penetrating peptides, *ChemBioChem*, 6(12), 2126–42. [10.1002/cbic.200500044]

Garcea, RL and Gissmann, L (2004), Virus-like particles as vaccines and vessels for the delivery of small molecules, *Curr. Opin. Biotechnol.*, 15(6), 513–17. [10.1016/j.copbio.2004.10.002]

Gehl, J (2003), Electroporation: theory and methods, perspectives for drug delivery, gene therapy and research, *Acta Physiol. Scand.*, 177(4), 437–47. [10.1046/j.1365–201X.2003.01093.x]

Goh, AS, Chung, AY, Lo, RH, Lau, TN, Yu, SW, *et al.* (2007), A novel approach to brachytherapy in hepatocellular carcinoma using a phosphorous^{32} (^{32}P) brachytherapy delivery device – a first-in-man study, *Int. J. Radiat. Oncol. Biol. Phys.*, 67(3), 786–92. [10.1016/j.ijrobp.2006.09.011]

Greytak, AB, Lauhon, LJ, Gudiksen, MS and Lieber, CM (2004), Growth and transport properties of complementary germanium nanowire field-effect transistors, *Appl. Phys. Lett.*, 84(21), 4176–8. [10.1063/1.1755846]

Han, SW, Nakamura, C, Kotobuki, N, Obataya, I, Ohgushi, H, *et al.* (2008), High-efficiency DNA injection into a single human mesenchymal stem cell using a nanoneedle and atomic force microscopy, *Nanomedicine*, 4(3), 215–25. [10.1016/j.nano.2008.03.005]

Han, SW, Nakamura, C, Obataya, I, Nakamura, N and Miyake, J (2005a), Gene expression using an ultrathin needle enabling accurate displacement and low invasiveness, *Biochem. Biophys. Res. Commun.*, 332(3), 633–9. [10.1016/j.bbrc.2005.04.059]

Han, SW, Nakamura, C, Obataya, I, Nakamura, N and Miyake, J (2005b), A molecular delivery system by using AFM and nanoneedle, *Biosens. Bioelectron.*, 20(10), 2120–5. [10.1016/j.bios.2004.08.023]

Heitsch, AT, Fanfair, DD, Tuan, H-Y and Korgel, BA (2008), Solution–liquid–solid (SLS) growth of silicon nanowires, *J. Am. Chem. Soc.*, 130(16), 5436–7. [10.1021/ja8011353]

Herino, R (1987), Porosity and pore size distributions of porous silicon layers, *J. Electrochem. Soc.*, 134(8), 1994–2000. [10.1149/1.2100805]

Hochbaum, AI, Gargas, D, Hwang, YJ and Yang, P (2009), Single crystalline mesoporous silicon nanowires, *Nano Lett.*, 9(10), 3550–4. [10.1021/nl9017594]

Hofmann, S, Ducati, C, Neill, RJ, Piscanec, S, Ferrari, AC, *et al.* (2003), Gold catalyzed growth of silicon nanowires by plasma enhanced chemical vapor deposition, *J. Appl. Phys.*, 94(9), 6005–12. [10.1063/1.1614432]

Holmes, JD (2000), Control of thickness and orientation of solution-grown silicon nanowires, *Science*, 287(5457), 1471–3. [10.1126/science.287.5457.1471]

Huang, Z, Fang, H and Zhu, J (2007), Fabrication of silicon nanowire arrays with controlled diameter, length, and density, *Adv. Mater.*, 19(5), 744–8. [10.1002/adma.200600892]

Huang, Z, Zhang, X, Reiche, M, Liu, L, Lee, W, *et al.* (2008), Extended arrays of vertically aligned sub-10 nm diameter [100] Si nanowires by metal-assisted chemical etching, *Nano Lett.*, 8(9), 3046–51. [10.1021/nl802324y]

Kim, W, Ng, JK, Kunitake, ME, Conklin, BR and Yang, P (2007), Interfacing silicon nanowires with mammalian cells, *J. Am. Chem. Soc.*, 129(23), 7228–9. [10.1021/ja071456k]

Kolasinski, K (2006), Catalytic growth of nanowires: vapor–liquid–solid, vapor–solid–solid, solution–liquid–solid and solid–liquid–solid growth, *Curr. Opin. Solid State Mater. Sci.*, 10(3–4), 182–91. [10.1016/j.cossms.2007.03.002]

Lamontagne, CA, Cuerrier, CM and Grandbois, M (2008), AFM as a tool to probe and manipulate cellular processes, *Pflugers Arch*, 456(1), 61–70. [10.1007/s00424–007–0414–0]

Levitt, AP (1970), *Whisker technology*, Wiley-Interscience.

Li, X and Bohn, PW (2000), Metal-assisted chemical etching in HF/H_2O_2 produces porous silicon, *Appl. Phys. Lett.*, 77(16), 2572–4. [10.1063/1.1319191]

Liu, JL, Cai, SJ, Jin, GL, Thomas, SG and Wang, KL (1999), Growth of Si whiskers on Au/Si(1 1 1) substrate by gas source molecular beam epitaxy (MBE), *J. Cryst. Growth*, 200(1–2), 106–11. [10.1016/S0022–0248(98)01408–0]

Low, SP, Williams, KA, Canham, LT and Voelcker, NH (2006), Evaluation of mammalian cell adhesion on surface-modified porous silicon, *Biomaterials*, 27(26), 4538–46. [10.1016/j.biomaterials.2006.04.015]

Lu, W and Lieber, CM (2006), Semiconductor nanowires, *J. Phys. D: Appl. Phys.*, 39(21), R387–R406. [10.1088/0022–3727/39/21/r01]

Madani, F, Lindberg, S, Langel, U, Futaki, S and Graslund, A (2011), Mechanisms of cellular uptake of cell-penetrating peptides, *J. Biophys.*, 2011, 414729. [10.1155/2011/414729]

Manchester, M and Singh, P (2006), Virus-based nanoparticles (VNPs): platform technologies for diagnostic imaging, *Adv. Drug Deliv. Rev.*, 58(14), 1505–22. [10.1016/j.addr.2006.09.014]

McKnight, TE, Melechko, AV, Hensley, DK, Mann, DGJ, Griffin, GD and Simpson, ML (2004), Tracking gene expression after DNA delivery using spatially indexed nanofiber arrays, *Nano Lett.*, 4(7), 1213–19. [10.1021/nl049504b]

Morales, AM (1998), A laser ablation method for the synthesis of crystalline semiconductor nanowires, *Science*, 279(5348), 208–11. [10.1126/science.279.5348.208]

Morris, MC, Deshayes, S, Heitz, F and Divita, G (2008), Cell-penetrating peptides: from molecular mechanisms to therapeutics, *Biol. Cell*, 100(4), 201–17. [10.1042/BC20070116]

Neumann, E, Schaefer-Ridder, M, Wang, Y and Hofschneider, PH (1982), Gene transfer into mouse lyoma cells by electroporation in high electric fields, *EMBO J.*, 1(7), 841–5.

Obataya, I, Nakamura, C, Han, S, Nakamura, N and Miyake, J (2005a), Direct insertion of proteins into a living cell using an atomic force microscope with a nanoneedle, *NanoBiotechnology*, 1(4), 347–52. [10.1385/nbt:1:4:347]

Obataya, I, Nakamura, C, Han, S, Nakamura, N and Miyake, J (2005b), Mechanical sensing of the penetration of various nanoneedles into a living cell using atomic force microscopy, *Biosens. Bioelectron.*, 20(8), 1652–5. [10.1016/j.bios.2004.07.020]

Peer, E, Artzy-Schnirman, A, Gepstein, L and Sivan, U (2012), Hollow nanoneedle array and its utilization for repeated administration of biomolecules to the same cells, *ACS Nano*, 6(6), 4940–6. [10.1021/nn300443h]

Peng, KQ, Hu, JJ, Yan, YJ, Wu, Y, Fang, H, *et al.* (2006), Fabrication of single-crystalline silicon nanowires by scratching a silicon surface with catalytic metal particles, *Adv. Funct. Mater.*, 16(3), 387–94. [10.1002/adfm.200500392]

Piret, G, Galopin, E, Coffinier, Y, Boukherroub, R, Legrand, D and Slomianny, C (2011), Culture of mammalian cells on patterned superhydrophilic/superhydrophobic silicon nanowire arrays, *Soft Matter*, 7(18), 8642–9. [10.1039/c1sm05838j]

Qi, S, Yi, C, Ji, S, Fong, CC and Yang, M (2009), Cell adhesion and spreading behavior on vertically aligned silicon nanowire arrays, *ACS Appl. Mater. Interfaces*, 1(1), 30–4. [10.1021/am800027d]

Riikonen, J, Makila, E, Salonen, J and Lehto, VP (2009), Determination of the physical state of drug molecules in mesoporous silicon with different surface chemistries, *Langmuir*, 25(11), 6137–42. [10.1021/la804055s]

Salonen, J, Laitinen, L, Kaukonen, AM, Tuura, J, Bjorkqvist, M, *et al.* (2005), Mesoporous silicon microparticles for oral drug delivery: loading and release of five model drugs, *J. Controlled Release*, 108(2–3), 362–74. [10.1016/j.jconrel.2005.08.017]

Shalek, AK, Gaublomme, JT, Wang, L, Yosef, N, Chevrier, N, *et al.* (2012), Nanowire-mediated delivery enables functional interrogation of primary immune cells: application to the analysis of chronic lymphocytic leukemia, *Nano Lett.*, 12(12), 6498–504. [10.1021/nl3042917]

Shalek, AK, Robinson, JT, Karp, ES, Lee, JS, Ahn, DR, *et al.* (2010), Vertical silicon nanowires as a universal platform for delivering biomolecules into living cells, *Proc. Natl. Acad. Sci. USA*, 107(5), 1870–5. [10.1073/pnas.0909350107]

Sharma, A and Sharma, US (1997), Liposomes in drug delivery: progress and limitations, *Int. J. Pharm.*, 154(2), 123–40. [10.1016/S0378-5173(97)00135-X]

Shen, H, You, J, Zhang, G, Ziemys, A, Li, Q, *et al.* (2012), Cooperative, nanoparticle-enabled thermal therapy of breast cancer, *Adv. Healthc. Mater.*, 1(1), 84–9. [10.1002/adhm.201100005]

Stephens, DJ and Pepperkok, R (2001), The many ways to cross the plasma membrane, *Proc. Natl. Acad. Sci. USA*, 98(8), 4295–8. [10.1073/pnas.081065198]

Sugano, K, Kansy, M, Artursson, P, Avdeef, A, Bendels, S, *et al.* (2010), Coexistence of passive and carrier-mediated processes in drug transport, *Nat. Rev. Drug Discov.*, 9(8), 597–614. [10.1038/nrd3187]

Tanaka, T, Mangala, LS, Vivas-Mejia, PE, Nieves-Alicea, R, Mann, AP, *et al.* (2010), Sustained small interfering RNA delivery by mesoporous silicon particles, *Cancer Res.*, 70(9), 3687–96. [10.1158/0008-5472.CAN-09-3931]

Tasciotti, E, Liu, X, Bhavane, R, Plant, K, Leonard, AD, *et al.* (2008), Mesoporous silicon particles as a multistage delivery system for imaging and therapeutic applications, *Nat. Nanotechnol.*, 3(3), 151–7. [10.1038/nnano.2008.34]

Torchilin, VP (2005), Recent advances with liposomes as pharmaceutical carriers, *Nat. Rev. Drug Discov.*, 4(2), 145–60. [10.1038/nrd1632]

VanDersarl, JJ, Xu, AM and Melosh, NA (2012), Nanostraws for direct fluidic intracellular access, *Nano Lett.*, 12(8), 3881–6. [10.1021/nl204051v]

Wang, D, Wang, Q, Javey, A, Tu, R, Dai, H, *et al.* (2003), Germanium nanowire field-effect transistors with SiO_2 and high-κ HfO_2 gate dielectrics, *Appl. Phys. Lett.*, 83(12), 2432–4. [10.1063/1.1611644]

Weiner, LM, Surana, R and Wang, S (2010), Monoclonal antibodies: versatile platforms for cancer immunotherapy, *Nat. Rev. Immunol.*, 10(5), 317–27. [10.1038/nri2744]

Woldering, LA, Willem Tjerkstra, R, Jansen, HV, Setija, ID and Vos, WL (2008), Periodic arrays of deep nanopores made in silicon with reactive ion etching and deep UV lithography, *Nanotechnology*, 19(14), 145304. [10.1088/0957-4484/19/14/145304]

Wong, YY, Yahaya, M, Mat Salleh, M and Yeop Majlis, B (2005), Controlled growth of silicon nanowires synthesized via solid–liquid–solid mechanism, *Sci. Technol. Adv. Mat.*, 6(3–4), 330–4. [10.1016/j.stam.2005.02.011]

Wu, Y, Cui, Y, Huynh, L, Barrelet, CJ, Bell, DC and Lieber, CM (2004), Controlled growth and structures of molecular-scale silicon nanowires, *Nano Lett.*, 4(3), 433–6. [10.1021/nl035162i]

Xue, M, Zhong, X, Shaposhnik, Z, Qu, Y, Tamanoi, F, *et al.* (2011), pH-Operated mechanized porous silicon nanoparticles, *J. Am. Chem. Soc.*, 133(23), 8798–801. [10.1021/ja201252e]

Yum, K, Wang, N and Yu, MF (2010), Nanoneedle: a multifunctional tool for biological studies in living cells, *Nanoscale*, 2(3), 363–72. [10.1039/b9nr00231f]

Zhang, Y and Yu, LC (2008), Single-cell microinjection technology in cell biology, *Bioessays*, 30(6), 606–10. [10.1002/bies.20759]

Part III

Silicon nanowires for detection and sensing

9
Semiconducting silicon nanowire array fabrication for high throughput screening in the biosciences

J. WU, Georgia Southern University, USA

DOI: 10.1533/9780857097712.3.171

Abstract: This chapter discusses various strategies to fabricate silicon nanowire field effect transistor arrays for high throughput screening in the biosciences. At the beginning, three major strategies to fabricate addressable nanowire field effect transistor arrays are reviewed, as well as their interfacing to microfluidic devices and other complementary characterization instruments for high throughput screening. Then the underlying mechanisms of screening and practical applications in the field of DNA hybridization, virus detection, cancer diagnosis and recording of cell signals are discussed. Finally, future research trends and important reference resources are given.

Key words: nanowire arrays, silicon, high throughput screening (HTS), field effect transistor, cancer diagnosis, DNA hybridization, cell signals recording, virus detection, protein interactions.

9.1 Introduction

Since its invention in the 1990s, high throughput screening (HTS) has been playing an increasingly important role in biosciences by assisting pharmaceutical and biotechnological companies, as well as academic researchers, in the diagnosis of diseases at an early stage and the discovery of new and efficient drugs within a much shortened timeframe (Hertzberg and Pope, 2000; Kojima *et al.*, 2012; Mayr and Bojanic, 2009). HTS methods utilize automated instrumentation and miniaturized assays combined with large-scale data analysis to screen large numbers of chemical and biological species for their activity against biological targets (Mayr and Bojanic, 2009; Collins and Franzblau, 1997; Janzen, 2009). A typical HTS device consists of 96-well microtiter plate arrays and a nanoliter pipette operated by a robotic system and integrated with characterization instruments like fluorescence spectroscopy (Bleicher *et al.*, 2003). HTS methods can be categorized into label-based and label-free approaches. Label-based approaches include fluorescent dyes (Smilkstein *et al.*, 2004), surface enhanced Raman spectroscopy (SERS) (Kim *et al.*, 2006), gold nanoparticles (NP) (Han *et al.*, 2006), quantum dots (Han *et al.*, 2001), dye and rare-earth doped NP (Wang and Liu, 2009) and magnetic glass bead-based methods (Sathe *et al.*, 2006).

Label-free methods use atomic force microscopy (AFM), surface plasmon resonance, field effect transistors (FET), scanning tunneling microscopy (STM), etc., to detect the presence of analyte molecules and investigate the efficacy of drug species without changing their biochemical properties (Chandra *et al.*, 2011; Mayr and Bojanic, 2009). Compared with other HTS methods, silicon nanowire FET arrays possess the advantages of high sensitivity, high reproducibility, real-time measurements and the potential of integrating high density addressable units on a single chip for high throughput and valid screening (Ray *et al.*, 2010). In this chapter, recent dynamic advances using silicon nanowire arrays for high throughput screening in the biosciences are discussed, with a special focus on their fabrication, disease diagnostics and their potential applications in drug discovery. Key sources of further information and advice, including important books and research groups, etc., are provided. The challenges and future research trends using silicon nanowires for high throughput screening will also be commented on.

9.1.1 Background to high-throughput screening methods

Quantitative and valid analyses of biological processes with high throughput are fundamentally important to disease diagnosis and drug discovery. Therefore, various HTS methods have been developed in the past decade, utilizing technologies ranging from classical dye-labeling and spectroscopic methods to nanomaterial-based approaches using quantum dots (Wang *et al.*, 2008; Sheng *et al.*, 2009), nanofilms (Schmatloch *et al.*, 2004), nanogaps (Im *et al.*, 2010), nanowires (Cheng *et al.*, 2006; Stoevesandt *et al.*, 2009) and nanotubes (Byon *et al.*, 2008, Allen *et al.*, 2007). Compared with their bulk counterparts, nanomaterials possess the advantage of extremely high surface area to volume ratios, thus allowing surface chemistry to play a dominant role in sensing biological processes of interest. Once the surface atoms bind to certain biological or chemical molecules selectively (via appropriate surface modifications), the physical and chemical properties of these nanomaterials would be dramatically changed resulting in the detection of diseases, viruses, DNA, cell signals, etc. Such a phenomenon is unlikely in the case of bulk materials like thin film FET. Secondly, biological species like DNA, RNA, enzymes, viruses and cells have dimensions comparable with nanoscale materials whose one or more external dimensions range from 1 to 100 nm, making them ideal platforms for biological sensors (Chen *et al.*, 2011). Quantum dots like CdSe, CdS and rare-earth doped nanoparticles have been used as efficient labeling agents for multiplexed optical coding of biomolecules (Chan *et al.*, 2002; Gao *et al.*, 2004; Wang and Liu, 2009). However, they have to be coupled to expensive and non-portable optical microscopic instruments to detect surface binding events. Recently, nanotubes and nanowires have been employed as building blocks to fabricate FET for highly sensitive and selective sensing in biosciences with the advantages of real-time measurements and being label-free, making the detection more precise and reliable. The measured signals can be easily read out as the changes in voltage, electrical current, conductance or frequency. Although carbon nanotube (CNT)-based FET have demonstrated very promising properties in sensing biological processes (facile chemical modification routes, high selectivity and sensitivity (So *et al.*, 2005)), they also have several untenable disadvantages in light of device fabrications: it is difficult and also expensive to separate semiconducting nanotubes from metallic ones (So *et al.*, 2005); they are not compatible with the current semiconducting industries based on silicon and germanium, so the device fabrication process is much more complex and costly. On the other hand, silicon nanowire (NW) FET are completely compatible with the semiconducting industries, and thus many conventional semiconductor device fabrication techniques can be employed such as photolithography, rapid thermal processing, magnetron sputtering, and e-beam evaporation, etc.; and multiple well-established methods are available to modify their surface with appropriate chemistry for highly selective biological and chemical recognition events (Puzder *et al.*, 2002).

In general, to realize high throughput screening in biosciences using silicon nanowire FET arrays, the following three steps have to be performed: fabrication of silicon nanowire FET arrays; subsequent surface modification of nanowire FET with DNA, RNA, enzyme, antibodies, etc., for biological recognition. Depending on whether *n*- or *p*-type silicon NW FET are used, the accumulation of negative charges on the NW surface will either reduce or increase the conductance of the NW, respectively; 3 integration of silicon nanowire FET with micro-fluidic devices for real-time measurements and coupling FET to complementary characterization instruments to cross-check the accuracy of the diagnosis and screening. Detailed mechanistic information in this regard is provided by Carlen and co-workers in Chapter 12.

9.2 Fabrication of silicon nanowire (SiNW) field effect transistor (FET) arrays for high throughput screening (HTS) in the biosciences

Generally speaking, silicon nanowire FET can be fabricated using three different strategies: 1 'top-down' methods that are based on the conventional photolithography, thin film deposition and other standard semiconductor processing techniques; 2 'bottom-up' methods that start with the synthesis of free-standing nanowires using a chemical vapor deposition method; these nanowires are then assembled and patterned using either classical photolithography or advanced e-beam lithography in order to obtain FET arrays. Using bottom-up methods, sub-10 nm silicon nanowires can be fabricated, which is a formidable challenge for 'top-down' methods. 3 'Superlattice nanowire pattern transfer' (SNAP) methods that lie between bottom-up and top-down methods.

9.2.1 Fabrication of SiNW FETS via top-down methods

As mentioned above, top-down methods are based on conventional photolithography, and other standard semiconductor processing techniques as shown in Fig. 9.1. Usually the starting materials are silicon on insulator (SOI) wafers that are commercially available. The top Si layer has a thickness ranging from tens to 100 nm, while the silicon dioxide insulator layer is about 200 nm. The SOI wafer must be strictly cleaned using proper wet chemical techniques to remove impurities before it is used for photolithography. In the next step, optical lithography and reactive ion etching (RIE) are employed to define the active layer. Ion implantation is then used to dope the active layer with boron or phosphorous to obtain *p*- or *n*-type Si, respectively. Next, the source and drain segments are defined using a combination of standard photolithography, thin film deposition techniques (like thermal or e-beam evaporation), and RIE (Pui *et al.*, 2009; Stern *et al.*, 2008). To obtain SiNW FETs with channel widths around 100 nm, e-beam lithography and RIE treatment are mandatory. Finally, thermal evaporation and rapid thermal annealing are used to make the top gate and contact leads (Stern *et al.*, 2007). The total fabrication process is shown in Fig. 9.1. Notably, expensive e-beam lithography is required to fabricate SiNW FET with NW widths less than 100 nm because it is beyond the resolution limit of optical lithography. The minimum resolution R achievable with projection lithography is determined by Eq. 9.1, where λ is the light wavelength, k_1 is a process-dependent parameter in the range of 0.4 to 1, and NA stands for numerical aperture whose value is less than 1 if the surrounding medium is air (Wallraff and Hinsberg, 1999).

$$R = k_1 \lambda / NA \qquad [9.1]$$

According to this equation, a light source with a wavelength shorter than 250 nm, for example X-ray, electron beam or ultra-deep UV, has to be used to obtain a resolution smaller than 100 nm. In addition, it is extremely challenging even for e-beam lithography to fabricate SiNW FET with NW widths less than 10 nm (Hu *et al.*, 2004). However, top-down methods possess the advantage of high reproducibility because of the performance of mature semiconductor processing techniques. Furthermore, the density and position of NW FET can be easily controlled and addressed.

9.2.2 Fabrication of SiNW FET via bottom-up methods

In general, 'bottom-up' methods typically consist of three major steps as shown in Fig. 9.2. In the first step, free-standing SiNW of controllable diameters and lengths are synthesized using a chemical vapor deposition method via a well-known vapor–liquid–solid (VLS) mechanism that was first discovered by Wagner and Ellis (1964) in the1960s and extrapolated to the nanoscale by Lieber *et al.*

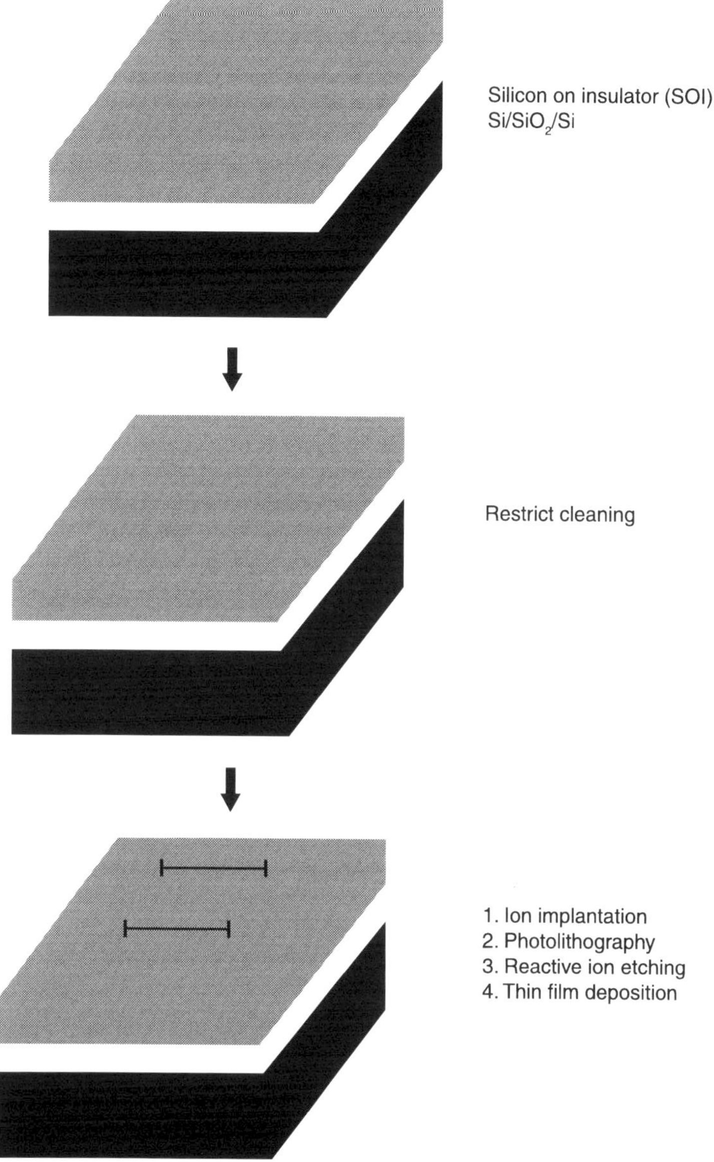

9.1 Diagram of silicon nanowire (SiNW) field effect transistor (FET) arrays fabrication via top-down method. First the starting material, silicon-on-insulator wafer ($Si/SiO_2/Si$) is strictly cleaned before it is used for photo- or electron-beam lithography. Secondly, it is doped with either P or B to make *n*- or *p*-type silicon through ion implantation. Photolithography and reactive ion etching are used to define NW structure, drain, source and gate for the FET device. Finally the junctions between NW, drain and source are covered with thin insulator thin film to protect from corrosion during biological sample measurements.

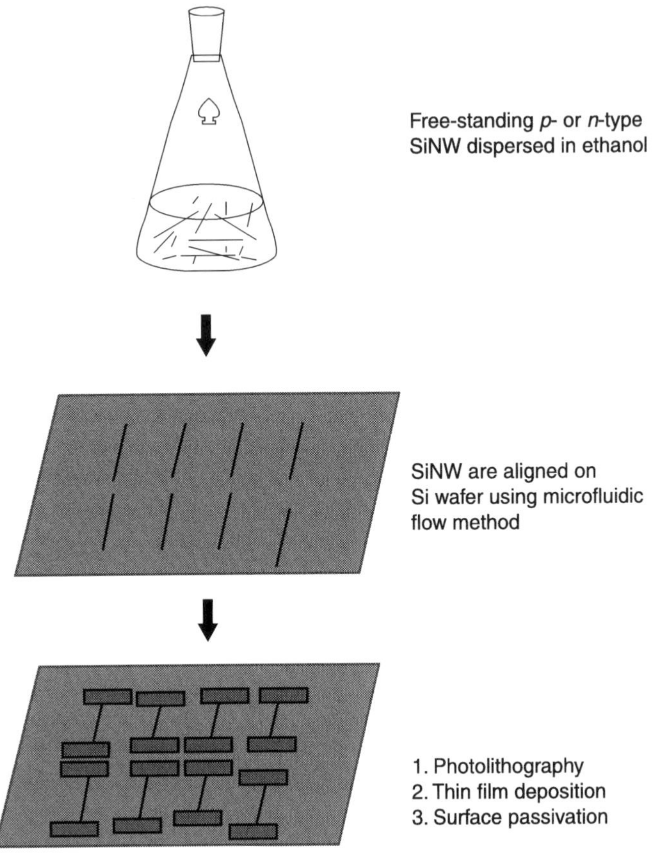

9.2 Diagram of silicon nanowire (SiNW) field effect transistor (FET) arrays fabrication via bottom-up method. First, free-standing *p*- or *n*-type SiNW prepared using a chemical vapor deposition method via a vapor–liquid–solid (VLS) mechanism are dispersed in ethanol under sonication. Then SiNW are aligned on a piece of silicon wafer by passing the ethanol solution containing SiNW through a microfluidic channel on the top of Si wafer. Finally, photolithography, thin film deposition and surface passivation techniques are employed to make drain, source and gate for FET devices.

(Cui *et al.*, 2001). It is noteworthy to point out that SiNW can also be synthesized below the eutectic point of Si-metal catalysts via a vapor-solid-solid mechanism, but the NW growth rate is much reduced compared with the one produced via a VLS route (Sunkara *et al.*, 2001). In the second step, these NW are further assembled and patterned through different strategies as will be discussed later in this section. Finally, conventional optical or e-beam lithography is utilized to fabricate the source, drain, gate and the leads to obtain addressable FET arrays.

In the synthesis of SiNW via a VLS mechanism, colloidal gold nanoparticles and silane (SiH_4) are commonly used as the metal catalysts and Si source respectively. Au nanoparticles can be uniformly adsorbed on the poly-L-lysine coated Si wafer because of a strong electrostatic interaction between negatively charged Au colloids and positively charged poly-L-lysine. The growth temperature should be above the eutectic point of Au-Si alloys. The diameter of NW is mainly determined by the size of gold nanocrystals, and the length of NW can be affected by the growth temperature and reaction time. Typically, the growth temperature is 460°C, and reaction time ranges from 10 to 20 minutes. To synthesize *p*-type SiNW, silane (SiH_4) and diborane (B_2H_6) are introduced into the chemical vapor deposition reactor simultaneously, while phosphine (PH_3) and silane are used to obtain *n*-type SiNW (Patolsky *et al.*, 2006).

As-prepared SiNW are then removed from the substrate and dispersed into ethanol solution under sonication. Aliquots of an ethanol solution containing SiNW are added to the surface of a Si chip dropwise using a microliter pipette, allowing each ethanol drop to evaporate completely before another is added. An optimal density of NW is approximately 1–2 NW per 100 μm^2, which can be monitored using dark field optical microscopy. Although this pipette method is straightforward and very simple, it is critically important to align as-prepared SiNW into a certain direction and precisely control the density of NW on the Si chip for the purpose of obtaining reproducible FET devices (Cui *et al.*, 2001). Otherwise their practical applications would be seriously limited. Recently, enormous amounts of creative efforts have been dedicated to the alignment of SiNW and controlling NW density per unit area. In 2001, Lieber's research group reported a breakthrough method to align NW by passing NW suspended in ethanol solution through micro-fluidic channels formed between a poly(dimethylsiloxane) (PDMS) mold and a silicon wafer (Huang *et al.*, 2001). It was found that a faster fluidic flow through the channels benefits a better NW alignment with a narrower angular distribution ($< \pm 5$ °C). The shear force generated by the fluidic flow is the underlying driving force for the NW alignment (Huang *et al.*, 2001). The density of aligned NW can be controlled by either the flow duration or the surface chemistry of the silicon wafer. If the Si wafer is modified with positively charged amine groups, NW can be adsorbed on the wafer more rapidly compared with the bare or methyl-terminated wafer. Given that the lengths of NW can be precisely controlled, the spacing between aligned NW can also be precisely manipulated.

There are also other approaches to assemble NW into ordered structures, including a Langmuir-Blodgett film method (Tao *et al.*, 2003), an electric-field driven method (Fan *et al.*, 2004), a roll-printing method (Yerushalmi *et al.*, 2007), a polydimethylsiloxane (PDMS) stamping method (Yi-Kuei and Franklin Chau-Nan, 2009), etc. In the case of the Langmuir-Blodgett film method, NW were first dispersed in a monolayer surfactant Langmuir film, which was then compressed in a Langmuir trough to form aligned NW arrays, similar to logs in a river (Whang *et al.*, 2003; Yang, 2003). These aligned NW were then transferred to a solid substrate for FET device fabrication. This method requires a relatively large amount of NW and high density NW arrays can be obtained. The electric-field driven method utilizes a strong electric field to align the NW. It was reported that nearly 16 000 FET units can be successfully assembled on a single chip using this method (Freer *et al.*, 2010).

As with other methods, following SiNW assembly, standard optical or e-beam lithography, thin film deposition techniques, rapid thermal annealing, RIE, etc., are employed to pattern the source, drain and gate pads, as well as the metal leads. Notably, the metal electrodes and the junctions between SiNW and metal electrodes should be protected with insulating film to prevent corrosion and current leakage during the HTS measurements as corrosive solutions are used. Compared with top-down methods, bottom-up methods can be used to fabricate NW FET with extremely small dimensions (<10 nm). Secondly, expensive E-beam lithographic techniques are not necessarily required to fabricate NW FET arrays. However, the density of SiNW FET cannot be easily controlled and further development in the precise assembly of SiNW is essential for practical applications.

9.2.3 Fabrication of SiNW FET arrays via superlattice nanowire pattern transfer (SNAP) method

SiNW FET arrays can also be fabricated using a SNAP method as shown in Fig. 9.3. In this process, first the SOI wafer is doped via a unique 'spin-on-dopant' method to avoid generating high density defects such as those generated using conventional ion-implantation methods. It is noteworthy that these defects can significantly scatter electrons and thus greatly reduce the conductance of silicon NW (Heath, 2008). To make *p*-type Si, the SOI wafer is first coated with Boron A, a spin-on dopant which is then diffused into the SOI thin film using rapid thermal processing (RTP). A doping level of $10^{18}/cm^3$ can be achieved using this method (McAlpine *et al.*, 2008). Meanwhile, a GaAs layer of a $Al_xGa_{1-x}As$/GaAs superlattice is selectively etched away using a dilute $NH_4OH/H_2O_2/H_2O$ solution, forming a comb-like structure. Then a thin platinum film is deposited onto the $Al_xGa_{1-x}As$ edges at certain angles using electron beam evaporation to form PtNW. These PtNW, along with the superlattice, are deposited onto a SOI wafer coated with a heat-curable epoxy. After curing and wet etching using a $H_3PO_4/H_2O_2/H_2O$ solution, these PtNW are detached from the superlattice and transferred to the surface of a SOI wafer. After RIE treatment, patterned SiNW will be formed underneath the PtNW that function as a photomask. The NW can be millimeters in length, with SiNW widths on the order of tens of nanometers (Heath, 2008). Finally, as before, standard optical or electron beam lithography methods and other semiconductor processing techniques are employed to make drain, source and gate pads, as well as the metal leads for FET devices.

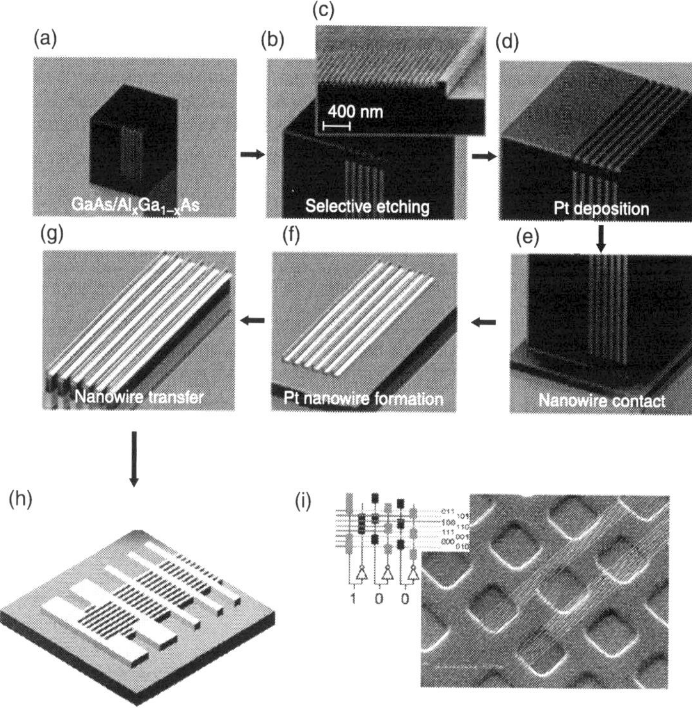

9.3 Diagram of silicon nanowires (SiNW) field effect transistor (FET) arrays fabrication via superlattice nanowire pattern transfer (SNAP) method. (a) A piece of GaAs/Al$_x$Ga$_{1-x}$As superlattice template. (b) GaAs layer is selectively etched to form a comb-like structure. (c) Scanning electron microscopy image of the comb-like structure. (d) Pt is evaporated onto the comb-like structure. (e) The superlattice is dropped onto silicon on insulator (SOI) substrate coated with epoxy thin film. (f) The interface between Pt and superlattice is etched way to leave Pt on the substrate. (g) Dry etching translates the metal nanowire array into the supporting substrate to form SiNW. (h) E-beam lithography and reactive ion etching are used to fabricate the NW FET arrays. (i) Scanning electron microscope image of the NW FET arrays prepared using the SNAP method and patterned at 33 nm pitch. Inset is the schematic of demultiplexing readouts for these NW FET arrays (adapted with permission from Heath, J. R. (2008) 'Superlattice Nanowire Pattern Transfer (SNAP)'. *Accounts of Chemical Research*, 41, 1609–17. Copyright 2008, American Chemical Society).

The SiNW widths and spacing fabricated using the SNAP method can be as small as several nanometers, which is difficult to obtain using the other two fabrication methods. In addition, this fabrication method is compatible with current semiconductor processing techniques and can be applied to any other electronic materials besides silicon. Expensive E-beam lithography is not necessarily required to fabricate this type of NW FET array. However, the doping density and uniformity cannot be as easily controlled as in the case of bottom-up methods using a VLS mechanism. Furthermore, the superlattice has to be custom-grown and other expensive metals (e.g. Pt) are needed in a sacrificial context to fabricate the FET arrays.

9.3 Surface modification of SiNW FETs for HTS in the biosciences

To realize a highly selective and sensitive biological recognition for disease diagnosis, sensing of cell signals and the investigation of protein–protein interaction kinetics, the surface of NW are usually modified with various kinds of chemical or biological species. Usually, the first step involves modifying the surface of SiNW with a small linker molecule that possesses a silanol group at one end and a primary amine group at another end. The silanol groups (Si-OR) can react with the dangling hydroxyl (OH) groups on the surface of SiNW to form Si-O-Si bonds, while the primary amine groups can be further coupled to other chemical or biological species like DNA, RNA or peptide nucleic acid (PNA) to recognize the complementary DNA, RNA and PNA sequences. The primary amine group can also be coupled to enzymes, antibodies, biomarkers, etc., to diagnose cancers, investigate cell activity and study the kinetics of protein–protein interaction, for example measurements of K, the equilibrium constant. Routes to this type of surface modification are outlined in more detail in Chapter 3.

Besides the effect of surface chemistry on the selectivity and sensitivity of SiNW FET for HTS in bioscience, the diameter of NW and the ionic strength of the electrolyte solution also play critical roles. For instance, it was found that the SiNW FET modified with primary amine groups are actually insensitive to the pH changes when the diameter of the NW is larger than 150 nm. In contrast, the conductance of SiNW FET is dramatically increased by nearly seven times when the NW diameter is 50 nm and exposed to a buffer solution of pH 3.0 (Elfström *et al.*, 2007). Under acidic conditions, the primary amine is protonated and positively charged, which functions like a positive gate voltage on SiNW FET, resulting in the increase of electrical conductance of NW (Elfström *et al.*, 2007). The smaller the NW diameter, the higher the surface to volume ratio and thus more atoms are located on the outer surface. As a result, smaller NW are more sensitive to the surface chemical binding (Gao *et al.*, 2009). For large and micrometer silicon wires, most atoms are inside the wires and not sensitive to the external chemical and physical changes, which is very similar to planar FET arrays (Pui *et al.*, 2010).

The sensitivity and selectivity are also affected by the ionic strength of the electrolyte solution. In the label-free SiNW FET-based HTS for biosciences, the underlying mechanism is the conductance change of FET when the NW are selectively bound to targeting biological or chemical species and thus result in the accumulation of either positive or negative charges on the NW FET gate surface. However, the effect of accumulated charges on the FET conductance can be significantly screened under a physiological environment during biological sensing. The typical ionic strength of the buffered solutions used for biosciences is ~0.2 M, at which ionic concentration the Debye-screening length is smaller than 1 nm as calculated using Eq. 9.2 (Kulkarni and Zhong, 2012).

$$\lambda_D = (\varepsilon k_B T/q^2 c)^{1/2}, \qquad [9.2]$$

where λ_D stands for Debye screening length, ε is the dielectric permittivity of the media, k_B is Boltzmann's constant, T is the temperature, q is the charge of the ion and c is the ionic strength of the electrolyte solution (Kulkarni and Zhong, 2012). When big protein molecules are bound to the SiNW surface, the short Debye-screening length will significantly reduce the changes in FET conductance. Therefore, it is sometimes necessary to desalinate the biological samples before they are used for HTS in order to improve the sensitivity (Patolsky *et al.*, 2006). These screening issues are discussed in greater detail in Chapter 12.

9.4 Integration of SiNW FETs with microfluidic devices for HTS in real-time measurements

As-fabricated SiNW FET arrays are commonly integrated with a microfluidic channel device for rapid, real-time and multiplexing analysis. A microfluidic channel device can be made from a master, silicone elastomer and elastomer curing agent (Whitesides *et al.*, 2001). First, the mixture of elastomer and elastomer curing agent is poured over the master, which is then cured, dried and peeled off from the master to form the microfluidic channel. Next, the polymeric microfluidic channel is clamped to the SiNW FET arrays chip such that the channel is overlapping the central active region of the chip, where the SiNW FET arrays are located. In the microfluidic channel, there is one inlet to inject samples using a syringe pump, and a channel outlet to balance internal pressure. The SiNW FET chip has to be mounted onto a chip carrier so that the contact pads of SiNW FET can be wire-bonded to the output pads of the chip carrier for facile electrical measurements. As the electrical current through these FET is very low (nA to pA), very sensitive lock-in amplifiers and electrical recording devices are required, which might seriously limit their practical applications (Patolsky *et al.*, 2007). To cross-check the accuracy of the measurements obtained from the SiNW FET arrays, mature complementary characterization instruments are commonly employed such as the simultaneous usage of cell-attached-voltage-clamp techniques and fluorescence spectroscopy to record/verify the cell potential signals (Foth *et al.*, 2006).

9.5 Examples/applications of SiNW FETs

9.5.1 DNA hybridization

Bunimovich *et al.* investigated the effects of surface chemistry and the native oxidative layer of SiNW on the DNA hybridization using SiNW FET arrays (Bunimovich *et al.*, 2006). The SiNW FET arrays were fabricated using a SNAP method as described earlier. Two procedures were employed to functionalize the SiNW with and without the native oxide layer. For the SiNW with the oxide layer, one-step silanol chemistry was used to introduce primary amine groups to the SiNW surface. To obtain NW without native oxide, SiNW were treated with 2% HF for a short time, and then reacted with tert-butyl allylcarbamate. After removing the tert-butoxycarbonyl (t-Boc) protection group from tert-butyl allylcarbamate, the SiNW were functionalized with a primary amine moiety and lacking a native oxide layer. These amine-functionalized *p*-type SiNW FET arrays were coupled to single strand 16-mer DNA via electrostatic interaction, whose sequence is complementary to the targeting DNA. After being bound to the targeting DNA, the accumulation of negative charges on the *p*-type SiNW FET was significantly reduced, resulting in the decrease of the FET conductance. The calculated association constants (K_a) between primary DNA and complementary DNA were about $1–2 \times 10^7$ (M^{-1}), which are comparable to the value obtained from a more conventional method, surface plasmon resonance (SPR), which is about 0.5×10^7 (M^{-1}). Compared with the SiNW functionalized with primary amine and with a native oxide layer, the detection limit for the NW functionalized with primary amine and without oxide layer was two orders of magnitude higher (~ tens of pM). This method has a great potential for screening large numbers of DNA and RNA for diagnosis of diseases.

Compared with DNA, PNA is more suitable for DNA hybridization because PNA lacks the negatively charged phosphate backbone and thus avoids the electrostatic repulsions during DNA–DNA hybridization (Hahm and Lieber, 2003). SiNW FET functionalized with 22-mer PNA were used to detect 22-mer fully complementary DNA, and 19-mer, 16-mer, 13-mer, 10-mer and 7-mer complementary DNA, whose FET conductance showed an exponential decrease with decreasing number of complementary base pairs (Zhang *et al.*, 2008). When the number of complementary base pairs is lowered, the negative charges on the targeting DNA are displaced further away from the surface of NW, resulting in a much reduced gating effect on the SiNW FET. The detection limit of DNA can be pushed back to as low as 1 pM using SiNW FET modified with PNA.

Remarkably, reusable SiNW FET have been developed using the reversible surface binding chemistry, such as the glutathione (GSH)/glutathione S-transferase (GST) and Ni^{2+}/polyhistidine tag (Lin *et al.*, 2009; 2010; Lo *et al.*, 2009). GST bound to the surface of SiNW modified with GSH can be easily displaced by washing the NW using high concentrations of GSH solutions. In the case of

Ni^{2+}/polyhistidine tag reversible binding, the bound Ni^{2+} can be removed using EDTA, which has a much stronger association constant with nickel ions than polyhistidine.

9.5.2 Detection of multiple viruses and small molecule–protein interactions

Dengus serotype 2 virus can be detected rapidly and sensitively using SiNW FET arrays, which were modified with synthetic PNA complementary to the specific fragment of the virus (Zhang *et al.*, 2010). There was more than 20% change in the FET conductance when virus was present, whereas the control only showed an insignificant change (less than 2%). The detection limit can be as low as 10 fM, owing partially to the usage of reversed transcription polymerase chain reaction (RT-PCR) that can amplify the original virus DNA concentration dramatically. For another example, *p*-type SiNW FET modified with monoclonal antibody receptors specific for either influenza A or adenovirus can detect the presence of influenza A and adenovirus simultaneously when the mixture of those two viruses was delivered to the FET (Patolsky *et al.*, 2004). The binding of negatively charged adenovirus to *p*-type SiNW FET resulted in positive conductance changes, while the binding of positively charged influenza A led to the negative conductance changes. The binding events were also confirmed by labeling the virus with fluorescent dyes and thus directly observed using optical microscopy. It is noteworthy that a single virus can be detected without any ambiguity (Patolsky *et al.*, 2004). This technology could be further developed for screening multiple viruses in real time and simultaneously for fast preliminary disease detection.

Studies of the specific binding events between small molecules and proteins are critically important for drug discovery and development (Strausberg and Schreiber, 2003; Stockwell, 2000). For instance, tyrosine kinases can mediate signal transduction in mammalian cells through phosphorylation of a tyrosine residue of a substrate protein using adenosine triphosphate (ATP). However, the normal phosphorylation process can be disrupted for people with diseases like cancer (Wang *et al.*, 2005). SiNW FET arrays modified with tyrosine kinase can be employed to identify the molecular inhibitors to tyrosine kinases rapidly and precisely (Wang *et al.*, 2005). If there are no inhibitors, ATP can bind to the kinases and thus result in the accumulation of negatively charged ATP on the surface of *p*-type SiNW FET. As a consequence, the conductance of *p*-type SiNW FET will increase, whereas the conductance of *n*-type FET will decrease. If small molecule inhibitors, for example Gleevec, were added to the ATP solution, the conductance of *p*-type SiNW FET decreased because the negatively charged ATP binding was inhibited. Notably, the presence of inhibitors in nM concentration can be reproducibly detected using this technology. Furthermore, three other small molecule inhibitors were also tested, indicating that SiNW

FET arrays can be employed for HTS to discover new drug molecules. In addition, peptide-modified SiNW FET can also function as 'electronic noses' to selectively detect small molecules like acetic acid and ammonia for implantable and rapid medical diagnosis (McAlpine *et al.*, 2008).

9.5.3 Detection of multiple cancer biomarkers

In the diagnosis of complex diseases like cancer, a test of a single biomarker is not sufficient to make a conclusive diagnosis (Zheng *et al.*, 2005). Therefore, real-time multiplexed detection of multiple biomarkers is highly desirable in the HTS for cancer diagnosis. Zheng *et al.* modified SiNW FET with multiple antibodies against different antigens including prostate specific antigen (PSA), carcino-embryonic antigen (CEA) and mucin-1. These FET devices being integrated with a microfluidic channel can be used to detect extremely low concentrations of antigens (0.9 pg/mL for PSA, 100 fg/mL for CEA and 75 fg/mL for mucin-1). The complementary data collected from *p*- and *n*-type SiNW FET further confirms the specific binding between antibody and antigen, and validity of the diagnosis. It should be noted that the blood samples containing these antigens have to be desalted before being measured using antibody-modified SiNW FET, as a consequence of the very small Debye screening length in the physiological fluid sample. Stern *et al.* developed a smart microfluidic purification method to solve this problem (Stern *et al.*, 2010). The targeting biomarkers are first anchored on a surface modified with antibodies to the biomarkers through a photocleavable linker. After the desalination, the linker is cleaved using UV-irradiation so that the free-standing biomarkers are released into the SiNW FET devices for diagnostic measurements (Stern *et al.*, 2007).

9.5.4 Temporal and spatially resolved recording of extracellular and intracellular signals from cells

Recording the electrical signals from cells and tissues is fundamentally important to the understanding of basic biophysical phenomena (Hille, 2001; Rutten, 2002; Huang *et al.*, 1999). Micropipette electrodes and patch clamp electrodes are commonly used to measure electrical signal propagation through individual neuron cells (Hille, 2001). But these techniques do not allow for multiplexed measurements (Patolsky *et al.*, 2006; Cohen-Karni *et al.*, 2009). Micro-fabricated multi-electrode arrays and planar FET do allow for multiplexed detections; however, these techniques suffer from the limits of low signal-to-noise ratios and low spatial resolution in the cell detection area (Banks *et al.*, 2002; Prohaska *et al.*, 1986; Sekirnjak *et al.*, 2006). Recently, NW and CNT FET have been employed as nano-electrode arrays to detect nucleic acids, proteins and viruses, and to record cellular signals, demonstrating high signal-to-noise ratios and excellent sensitivity and selectivity (Patolsky *et al.*, 2006; Timko *et al.*, 2009; Gao *et al.*, 2011; Pui *et al.*, 2009; Cellot *et al.*, 2009; Stern *et al.*, 2007; Gruner, 2006; Zheng *et al.*, 2005). SiNW FET arrays can be used to record the electrical signals from multiple points of a single neuron (Patolsky *et al.*, 2006), cardiomyocyte cells (Cohen-Karni *et al.*, 2009; Timko *et al.*, 2009), acute brain slices (Qing *et al.*, 2010) and detect the released neuro-transmitter of CgA (Wang *et al.*, 2007).

Patolsky *et al.* integrated the SiNW FET arrays with the individual axons and dendrites of live mammalian neurons to study the neuronal signals under stimulations and inhibitions (Patolsky *et al.*, 2006). The SiNW FET arrays were fabricated via a bottom-up method as discussed previously in this chapter. Polylysine was then used as both the adhesion and growth factor to define the neuron cell growth with respect to the SiNW FET device elements. Using a photolithography method, square regions of a 30 × 60 μm polylysine pattern were generated to boost the cell body adhesion, and 2 μm wide lines are projected to define the subsequent neurite growth. To survive the harsh cell culture and actual sample measurements, the SiNW–metal contacts were passivated such that the survival rate of devices is greater than 90% after 10 days at 37 °C (Patolsky *et al.*, 2006). Using such a unique integrated system of SiNW FET-cell arrays, the propagation and back-propagation of the action potential spikes in axon and dendrites can be measured separately and simultaneously. It was found that there is a clear reduction in FET conductance and temporal spreading in the dendrites, whereas very little change was observed for the axon, which is consistent with passive and active propagation mechanisms for dendrites and axon, respectively. The signal propagation rates in axon and dendrites are 0.46 and 0.15 m/s, respectively, as calculated from the conductance data collected from different SiNW FET that are in contact with different positions of dendrites and axon. The signal propagation can be blocked by applying an input voltage of 0.9 V. They also demonstrated that as many as 50 addressable NW FET with a 10 μm inter-device spacing can be integrated with a single axon with an 86% yield of functional devices (Patolsky *et al.*, 2006; 2007). The spacing can be further reduced to only 400 nm, indicating their potential applications in high throughput screening for real-time cellular testing, drug testing and discovery.

Cohen-Karni *et al.* recently reported the direct recording of electrical signals from embryonic chicken cardiomyocytes using SiNW FET arrays that were also fabricated using a bottom-up method (Cohen-Karni *et al.*, 2009). The cultured cardiomyocyte cells were first transferred to the top of thin transparent polydimethylsiloxane (PDMS) sheets and then attached to the surface of SiNW FET for electrical measurements, which will greatly increase the HTS speed in a real practical application compared with culturing cells directly on FET devices. While water gate-voltage potentials range from −0.5 to +0.1 V, the corresponding conductance amplitudes change from 31 to 7 nS. The conductance is also sensitive to the distance between the cardiomyocyte cells and the FET. For example, if the PDMS film coated with cells is 9.8 μm closer to the FET, the NW FET device exhibited an increase in conductance from 44 to 77 nS. NW FET arrays are also suitable for the measurement of signal propagations and temporal shifts in the cells. The measured signal propagation in cardiomyocytes ranges from 0.07 to 0.21 m/s, which is close to the literature reports (Fast and Kléber, 1994). Similarly, Timko *et al.* reported the direct monitoring of electrical signals from different parts of embryonic hearts using flexible and transparent SiNW FET arrays (Timko *et al.*, 2009). The SiNW FET arrays are built on a flexible substrate, Kapton, making it more suitable for measuring multiple positions of bulk samples like tissues and organs (Timko *et al.*, 2009). In addition, SiNW FET arrays can also be employed to detect neurotransmitters released from living cells (Wang *et al.*, 2007).

9.6 Conclusion

In the past decade, significant efforts have been put forth to utilize SiNW-based FET arrays for multiplexing HTS in the biomedical sciences. Addressable SiNW FET arrays for HTS applications can be fabricated using three major methods: top-down, bottom-up and SNAP approaches. They have been successfully applied to the fields of DNA hybridization, virus detection, small molecule–protein interactions, cancer diagnosis and recording of cell activities, showing very impressive sensitivity, selectivity, versatility and reliability. Especially, the capability of recording electrical signals from cells, organs and tissues under external stimulations or inhibitions using SiNW FET arrays could have broad applications in drug discovery and disease diagnosis. However, more efforts are needed to reduce the fabrication cost of NW FET arrays, to improve the precision of NW assembly in terms of density and spacing, and to increase the multitasking capability of NW FET arrays to measure thousands of samples simultaneously. Furthermore, the sensitivity of SiNW FET conductance to ionic strength is a serious concern in terms of their practical applications and should be addressed appropriately. Leading research efforts in this field include the Lieber group at Harvard, the Heath group at Caltech, and the Reed group at Yale, as well as the Chen group at National Taiwan University. Key reference books include *High Throughput Screening: The Discovery of Bioactive Substances* by John P. Devlin (Devlin, 1997) and *High Throughput Screening: Methods and Protocols* by William P. Janzen (Janzen, 2009).

9.7 Future trends

Although huge efforts and impressive breakthroughs have been made in the past decade using SiNW arrays for HTS in the field of biosciences, there are still many barriers to overcome before they can be used for practical applications and compete with current dominant methods. For example, the fabrication methods for NW FET arrays need to be further addressed to obtain highly reproducible NW assembly in the case of the bottom-up method. The SNAP method would be a very promising approach if the usage of expensive heavy metals and custom-grown toxic superlattice can be avoided. In addition to microfluidic systems, interfacing NW FET with nanofluidic devices to deliver biological samples could dramatically reduce the volume required for biological screenings (see Chapter 12). How to bridge each NW FET to an individual nanofluidic channel reproducibly, however, is a formidable challenge but an area worthy of intense exploration. If successful, these NW array devices would be able to compete with current HTS methods capable of measuring thousands of samples each day. Another research direction is to overcome the Debye screening length barrier using such NW FET arrays without sample pre-desalination.

Furthermore, achieving quantitative HTS using SiNW arrays is clearly an important research focus for the future. For example, can the concentrations of cancer cells, DNA, RNA and viruses in blood samples be quantitatively determined as readily as a flow cytometer using silicon NW FET arrays. As very sensitive and non-portable instruments are required to measure the weak electrical signals from SiNW FET, the development of inexpensive, sensitive and portable electrical measurement units is a priority. Until now, few *in vivo* studies have been done using these NW FET arrays. It would be quite interesting and scientifically meaningful to study the bio-compatibility, reproducibility and reliability of these devices *in vivo* implanted for disease monitoring and diagnosis, as well as to investigate the treatment efficacy of new drugs, thereby attracting significant commercial attention. Besides measuring electrical signals from SiNW FET arrays, would it be possible to address every individual nanowire and then correlate with standard analysis tools like AFM?

In summary, the use of SiNW arrays for HTS in the biosciences field has been a burgeoning and very exciting research field in the past decade and will continue to be so in the decades ahead.

9.8 References

Allen, B. L., Kichambare, P. D. and Star, A. (2007) 'Carbon nanotube field-effect-transistor-based biosensors'. *Adv Mater*, 19, 1439–51.

Banks, D. J., Balachandran, W., Richards, P. R. and Ewins, D. (2002) 'Instrumentation to evaluate neural signal recording properties of micromachined microelectrodes inserted in invertebrate nerve'. *Physiol Meas*, 23, 437.

Bleicher, K. H., Bohm, H.-J., Muller, K. and Alanine, A. I. (2003) 'Hit and lead generation: beyond high-throughput screening'. *Nat Rev Drug Discov*, 2, 369–78.

Bunimovich, Y. L., Shin, Y. S., Yeo, W.-S., Amori, M., Kwong, G. and Heath, J. R. (2006) 'Quantitative real-time measurements of DNA hybridization with alkylated nonoxidized silicon nanowires in electrolyte solution'. *J Am Chem Soc*, 128, 16323–31.

Byon, H. R., Kim, S. and Choi, H. C. (2008) 'Label-free biomolecular detection uisng carbon nanotube field effect transistors'. *Nano*, 03, 415–31.

Cellot, G., Cilia, E., Cipollone, S., Rancic, V., Sucapane, A., *et al.* (2009) 'Carbon nanotubes might improve neuronal performance by favouring electrical shortcuts'. *Nat Nano*, 4, 126–33.

Chan, W. C. W., Maxwell, D. J., Gao, X., Bailey, R. E., Han, M. and Nie, S. (2002) 'Luminescent quantum dots for multiplexed biological detection and imaging'. *Curr Opin Biotech*, 13, 40–6.

Chandra, H., Reddy, P. J. and Srivastava, S. (2011) 'Protein microarrays and novel detection platforms'. *Expert Rev Proteomics*, 8, 61–79.

Chen, K.-I., Li, B.-R. and Chen, Y.-T. (2011) 'Silicon nanowire field-effect transistor-based biosensors for biomedical diagnosis and cellular recording investigation'. *Nano Today*, 6, 131–54.

Cheng, M. M.-C., Cuda, G., Bunimovich, Y. L., Gaspari, M., Heath, J. R., *et al.* (2006) 'Nanotechnologies for biomolecular detection and medical diagnostics'. *Curr Opin Chem Bio*, 10, 11–19.

Cohen-Karni, T., Timko, B. P., Weiss, L. E. and Lieber, C. M. (2009) 'Flexible electrical recording from cells using nanowire transistor arrays'. *Proc Natl Acad Sci USA*, 106, 7309–13.

Collins, L. and Franzblau, S. G. (1997) 'Microplate alamar blue assay versus BACTEC 460 system for high-throughput screening of compounds against *Mycobacterium tuberculosis* and *Mycobacterium avium*'. *Antimicrob Agents Chemother*, 41, 1004–9.

Cui, Y., Wei, Q., Park, H. and Lieber, C. M. (2001) 'Nanowire nanosensors for highly sensitive and selective detection of biological and chemical species'. *Science*, 293, 1289–92.

Devlin, J. P. (1997) *High Throughput Screening: The Discovery of Bioactive Substances*, New York: Marcel Dekker, Inc.

Elfström, N., Juhasz, R., Sychugov, I., Engfeldt, T., Karlström, A. E. and Linnros, J. (2007) 'Surface charge sensitivity of silicon nanowires: size dependence'. *Nano Lett*, 7, 2608–12.

Fan, D. L., Zhu, F. Q., Cammarata, R. C. and Chien, C. L. (2004) 'Manipulation of nanowires in suspension by ac electric fields'. *Appl Phys Lett*, 85, 4175–7.

Fast, V. G. and Kléber, A. G. (1994) 'Anisotropic conduction in monolayers of neonatal rat heart cells cultured on collagen substrate'. *Circ Res*, 75, 591–5.

Foth, B. J., Goedecke, M. C. and Soldati, D. (2006) 'New insights into myosin evolution and classification'. *Proc Natl Acad Sci USA*, 103, 3681–6.

Freer, E. M., Grachev, O., Duan, X., Martin, S. and Stumbo, D. P. (2010) 'High-yield self-limiting single-nanowire assembly with dielectrophoresis'. *Nat Nano*, 5, 625.

Gao, A., Lu, N., Dai, P., Li, T., Pei, H., *et al.* (2011) 'Silicon-nanowire-based CMOS-compatible field-effect transistor nanosensors for ultrasensitive electrical detection of nucleic acids'. *Nano Lett*, 11, 3974–8.

Gao, X., Cui, Y., Levenson, R. M., Chung, L. W. K. and Nie, S. (2004) 'In vivo cancer targeting and imaging with semiconductor quantum dots'. *Nat Biotech*, 22, 969–76.

Gao, X. P. A., Zheng, G. and Lieber, C. M. (2009) 'Subthreshold regime has the optimal sensitivity for nanowire FET biosensors'. *Nano Lett*, 10, 547–52.

Gruner, G. (2006) 'Carbon nanotube transistors for biosensing applications'. *Anal Bioanal Chem*, 384, 322–35.

Hahm, J.-i. and Lieber, C. M. (2003) 'Direct ultrasensitive electrical detection of DNA and DNA sequence variations using nanowire nanosensors'. *Nano Lett*, 4, 51–4.

Han, M., Gao, X., Su, J. Z. and Nie, S. (2001) 'Quantum-dot-tagged microbeads for multiplexed optical coding of biomolecules'. *Nat Biotech*, 19, 631–5.

Han, M. S., Lytton-Jean, A. K. R., Oh, B.-K., Heo, J. and Mirkin, C. A. (2006) 'Colorimetric screening of DNA-binding molecules with gold nanoparticle probes'. *Angew Chem Inter Ed*, 45, 1807–10.

Heath, J. R. (2008) 'Superlattice nanowire pattern transfer (SNAP)'. *Acc Chem Res*, 41, 1609–17.

Hertzberg, R. P. and Pope, A. J. (2000) 'High-throughput screening: new technology for the 21st century'. *Curr Opin Chem Bio*, 4, 445–51.

Hille, B. (2001) *Ion Channels of Excitable Membranes, Third Edition*, Sinauer Associates Inc.

Hu, W., Sarveswaran, K., Lieberman, M. and Bernstein, G. H. (2004) 'Sub-10 nm electron beam lithography using cold development of poly(methylmethacrylate)' *J Vac Sci Technol B*, 22, 1711–16.

Huang, C. Q., Shepherd, R. K., Center, P. M., Seligman, P. M. and Tabor, B. (1999) 'Electrical stimulation of the auditory nerve: direct current measurement in vivo'. *Biomed Eng, IEEE Trans*, 46, 461–9.

Huang, Y., Duan, X., Wei, Q. and Lieber, C. M. (2001) 'Directed assembly of one-dimensional nanostructures into functional networks'. *Science*, 291, 630–3.

Im, H., Bantz, K. C., Lindquist, N. C., Haynes, C. L. and Oh, S.-H. (2010) 'Vertically oriented sub-10-nm plasmonic nanogap arrays'. *Nano Lett*, 10, 2231–6.

Janzen, W. P. (2009) *High Throughput Screening: Methods and Protocols (Methods in Molecular Biology)*, Totowa: Humana Press Inc.

Kim, J.-H., Kim, J.-S., Choi, H., Lee, S.-M., Jun, B.-H., *et al.* (2006) 'Nanoparticle probes with surface enhanced raman spectroscopic tags for cellular cancer targeting'. *Anal Chem*, 78, 6967–73.

Kojima, T., Ohuchi, S., Ito, Y. and Nakano, H. (2012) 'High-throughput screening method for promoter activity using bead display and a ligase ribozyme'. *J Biosci Bioeng*, 114, 671–6.

Kulkarni, G. S. and Zhong, Z. (2012) 'Detection beyond the Debye screening length in a high-frequency nanoelectronic biosensor'. *Nano Lett*, 12, 719–23.

Lin, S.-P., Pan, C.-Y., Tseng, K.-C., Lin, M.-C., Chen, C.-D., *et al.* (2009) 'A reversible surface functionalized nanowire transistor to study protein–protein interactions'. *Nano Today*, 4, 235–43.

Lin, T.-W., Hsieh, P.-J., Lin, C.-L., Fang, Y.-Y., Yang, J.-X., *et al.* (2010) 'Label-free detection of protein-protein interactions using a calmodulin-modified nanowire transistor'. *Proc Natl Acad Sci USA*, 107, 1047–52.

Lo, Y.-S., Nam, D. H., So, H.-M., Chang, H., Kim, J.-J., *et al.* (2009) 'Oriented immobilization of antibody fragments on Ni-decorated single-walled carbon nanotube devices'. *ACS Nano*, 3, 3649–55.

Mayr, L. M. and Bojanic, D. (2009) 'Novel trends in high-throughput screening'. *Curr Opin Pharm*, 9, 580–8.

McAlpine, M. C., Agnew, H. D., Rohde, R. D., Blanco, M., Ahmad, H., *et al.* (2008) 'Peptide–nanowire hybrid materials for selective sensing of small molecules'. *J Am Chem Soc*, 130, 9583–9.

Patolsky, F., Timko, B. P., Zheng, G. and Lieber, C. M. (2007) 'Nanowire-based nanoelectronic devices in the life sciences'. *MRS Bulletin*, 32, 142–9.

Patolsky, F., Zheng, G., Hayden, O., Lakadamyali, M., Zhuang, X. and Lieber, C. M. (2004) 'Electrical detection of single viruses'. *Proc Natl Acad Sci USA*, 101, 14017–22.

Patolsky, F., Zheng, G. and Lieber, C. M. (2006) 'Fabrication of silicon nanowire devices for ultrasensitive, label-free, real-time detection of biological and chemical species'. *Nat Protocols*, 1, 1711–24.

Prohaska, O. J., Olcaytug, F., Pfundner, P. and Dragaun, H. (1986) 'Thin-film multiple electrode probes: possibilities and limitations'. *Biomed Eng, IEEE Trans*, BME-33, 223–9.

Pui, T.-S., Agarwal, A., Ye, F., Balasubramanian, N. and Chen, P. (2009) 'CMOS-compatible nanowire sensor arrays for detection of cellular bioelectricity'. *Small*, 5, 208–12.

Pui, T.-S., Sudibya, H. G., Luan, X., Zhang, Q., Ye, F., *et al.* (2010) 'Non-invasive detection of cellular bioelectricity based on carbon nanotube devices for high-throughput drug screening'. *Adv Mater*, 22, 3199–203.

Puzder, A., Williamson, A. J., Grossman, J. C. and Galli, G. (2002) 'Surface chemistry of silicon nanoclusters'. *Phys Rev Lett*, 88, 097401.

Qing, Q., Pal, S. K., Tian, B., Duan, X., Timko, B. P., *et al.* (2010) 'Nanowire transistor arrays for mapping neural circuits in acute brain slices'. *Proc Natl Acad Sci USA*, 107, 1882–7.

Ray, S., Chandra, H. and Srivastava, S. (2010) 'Nanotechniques in proteomics: current status, promises and challenges'. *Biosens Bioelectron*, 25, 2389–401.

Rutten, W. L. C. (2002) 'Selective electrical interfaces with the nervous system'. *Annu Rev Biomed Eng*, 4, 407–52.

Sathe, T. R., Agrawal, A. and Nie, S. (2006) 'Mesoporous silica beads embedded with semiconductor quantum dots and iron oxide nanocrystals: dual-function microcarriers for optical encoding and magnetic separation'. *Anal Chem*, 78, 5627–32.

Schmatloch, S., Bach, H., van Benthem, R. A. T. M. and Schubert, U. S. (2004) 'High-throughput experimentation in organic coating and thin film research: state-of-the-art and future perspectives'. *Macromol Rapid Comm*, 25, 95–107.

Sekirnjak, C., Hottowy, P., Sher, A., Dabrowski, W., Litke, A. M. and Chichilnisky, E. J. (2006) 'Electrical stimulation of mammalian retinal ganglion cells with multielectrode arrays'. *J Neurophysi* 95, 3311–27.

Sheng, Z., Han, H., Hu, D., Liang, J., He, Q., *et al.* (2009) 'Quantum dots-gold(iii)-based indirect fluorescence immunoassay for high-throughput screening of APP'. *Chem Comm*, 2009, 2559–61.

Smilkstein, M., Sriwilaijaroen, N., Kelly, J. X., Wilairat, P. and Riscoe, M. (2004) 'Simple and inexpensive fluorescence-based technique for high-throughput antimalarial drug screening'. *Antimicrob Agents Chemother*, 48, 1803–6.

So, H.-M., Won, K., Kim, Y. H., Kim, B.-K., Ryu, B. H., *et al.* (2005) 'Single-walled carbon nanotube biosensors using aptamers as molecular recognition elements'. *J Am Chem Soc*, 127, 11906–7.

Stern, E., Klemic, J. F., Routenberg, D. A., Wyrembak, P. N., Turner-Evans, D. B., *et al.* (2007) 'Label-free immunodetection with CMOS-compatible semiconducting nanowires'. *Nature*, 445, 519–22.

Stern, E., Vacic, A., Rajan, N. K., Criscione, J. M., Park, J., *et al.* (2010) 'Label-free biomarker detection from whole blood'. *Nat Nano*, 5, 138–42.

Stern, E., Vacic, A. and Reed, M. A. (2008) 'Semiconducting nanowire field-effect transistor biomolecular sensors'. *Electron Devices, IEEE Trans*, 55, 3119–30.

Stockwell, B. R. (2000) 'Frontiers in chemical genetics'. *Trends Biotechnol*, 18, 449–55.

Stoevesandt, O., Taussig, M. J. and He, M. (2009) 'Protein microarrays: high-throughput tools for proteomics'. *Exper Rev Proteomics*, 6, 145–57.

Strausberg, R. L. and Schreiber, S. L. (2003) 'From knowing to controlling: a path from genomics to drugs using small molecule probes'. *Science*, 300, 294–5.

Sunkara, M. K., Sharma, S., Miranda, R., Lian, G. and Dickey, E. C. (2001) 'Bulk synthesis of silicon nanowires using a low-temperature vapor–liquid–solid method'. *Appl Phys Lett*, 79, 1546–8.

Tao, A., Kim, F., Hess, C., Goldberger, J., He, R., *et al.* (2003) 'Langmuir-Blodgett silver nanowire monolayers for molecular sensing using surface-enhanced raman spectroscopy'. *Nano Lett*, 3, 1229–33.

Timko, B. P., Cohen-Karni, T., Yu, G., Qing, Q., Tian, B. and Lieber, C. M. (2009) 'Electrical recording from hearts with flexible nanowire device arrays'. *Nano Lett*, 9, 914–18.

Wagner, R. S. and Ellis, W. C. (1964) 'Vapor–liquid–solid mechanism of single crystal growth'. *Appl Phys Lett*, 4, 89–90.

Wallraff, G. M. and Hinsberg, W. D. (1999) 'Lithographic imaging techniques for the formation of nanoscopic features'. *Chem Rev*, 99, 1801–22.

Wang, C.-W., Pan, C.-Y., Wu, H.-C., Shih, P.-Y., Tsai, C.-C., *et al.* (2007) 'In situ detection of chromogranin a released from living neurons with a single-walled carbon-nanotube field-effect transistor'. *Small*, 3, 1350–5.

Wang, F. and Liu, X. (2009) 'Recent advances in the chemistry of lanthanide-doped upconversion nanocrystals'. *Chem Soc Rev*, 38, 976–89.

Wang, J., Liu, G., Wu, H. and Lin, Y. (2008) 'Quantum-dot-based electrochemical immunoassay for high-throughput screening of the prostate-specific antigen'. *Small*, 4, 82–6.

Wang, W. U., Chen, C., Lin, K.-h., Fang, Y. and Lieber, C. M. (2005) 'Label-free detection of small-molecule–protein interactions by using nanowire nanosensors'. *Proc Natl Acad Sci USA*, 102, 3208–12.

Whang, D., Jin, S., Wu, Y. and Lieber, C. M. (2003) 'Large-scale hierarchical organization of nanowire arrays for integrated nanosystems'. *Nano Lett*, 3, 1255–9.

Whitesides, G. M., Ostuni, E., Takayama, S., Jiang, X. and Ingber, D. E. (2001) 'Soft lithography in biology and biochemistry'. *Annu Rev Biomed Eng*, 3, 335–73.

Yang, P. (2003) 'Nanotechnology: wires on water'. *Nature*, 425, 243–4.

Yerushalmi, R., Jacobson, Z. A., Ho, J. C., Fan, Z. and Javey, A. (2007) 'Large scale, highly ordered assembly of nanowire parallel arrays by differential roll printing'. *Appl Phys Lett*, 91, 203104.

Yi-Kuei, C. and Franklin Chau-Nan, H. (2009) 'The fabrication of ZnO nanowire field-effect transistors by roll-transfer printing'. *Nanotechnology*, 20, 195302.

Zhang, G.-J., Zhang, G., Chua, J. H., Chee, R.-E., Wong, E. H., *et al.* (2008) 'DNA sensing by silicon nanowire: charge layer distance dependence'. *Nano Lett*, 8, 1066–70.

Zhang, G.-J., Zhang, L., Huang, M. J., Luo, Z. H. H., Tay, G. K. I., *et al.* (2010) 'Silicon nanowire biosensor for highly sensitive and rapid detection of Dengue virus'. *Sensor Actuat B-Chem*, 146, 138–44.

Zheng, G., Patolsky, F., Cui, Y., Wang, W. U. and Lieber, C. M. (2005) 'Multiplexed electrical detection of cancer markers with nanowire sensor arrays'. *Nat Biotech*, 23, 1294–301.

10
Neural cell pinning on surfaces by semiconducting silicon nanowire arrays

C. VILLARD, Institut Néel, CNRS and
Université Joseph Fourier, France

DOI: 10.1533/9780857097712.3.192

Abstract: Building elementary neuronal networks above sub-micrometric devices is a pertinent approach toward the understanding of neuronal computation at one of the lowest levels of complexity. This chapter first describes how shaping neurons by adhesive micropatterns allows control of both the network topography and connectivity on the basis of a surface chemistry compatible with the top surface of an electronic chip. Then, a second section presents the recent achievements in neuronal positioning, recording and stimulation. This chapter will end with an overview of the trends and future challenges in the instrumentation of living neuronal networks.

Key words: neuronal networks, micropatterns, axons, silicon nanowires, intracellular, extracellular.

10.1 Introduction

Neurons are electrically polarized and excitable cells. Therefore, electrical recording can give privileged access to neuronal activity. Building *in vitro* neuronal networks while being able to detect and stimulate activity is an approach that has been extensively developed in the literature over the past two decades. It has allowed, with the advantages of accessibility and reduction of complexity, quantitative analysis of neuronal firing patterns (Robinson *et al.*, 1993) similar to those observed *in vivo* (Meister *et al.*, 1991). Moreover, the use of multiplexed extra-cellular probes like micro-electrode-arrays (MEA) has generated a vast bibliography devoted to various aspects of neuronal activity like the evolution of neuronal firing patterns during development (Wagenaar *et al.*, 2006), the mechanisms of synchronized firing (Maeda *et al.*, 1995) or learning (Shahaf and Marom 2001).

To address other fundamental issues in neuroscience like the mechanisms underlying synaptic plasticity or the relation between a neuronal architecture and its function, the next step is to record and stimulate, with a cellular or even a sub-cellular resolution, *in vitro* neuronal assemblies that reproduce in some extend the controlled architectures of physiological neuronal connections. Moreover, access to individual synapses would constitute a considerable advantage as 80% of brain diseases affect synaptic transmission in their primary stages (Mattson 2004). To this end, experimental developments must integrate three challenging prerequisites:

1 control over both the geometry and the connectivity of the neuronal network, 2 development of arrays of local sensors, 'local' meaning here with at least one dimension similar to the characteristic length of a synapse (250–300 nm), and 3 alignment of both cellular and physical networks. Active devices such as a matrix of silicon nanowire field effect transistors (SiNW-FET) are the most promising candidates to achieve this sub-cellular resolution (Stern *et al.*, 2006), and this technology has recently offered the sensitivity, lengthscale and timescale required to detect, stimulate or inhibit neuronal activity at a sub-cellular level (Patolsky *et al.*, 2006a; Qinga *et al.*, 2010).

Two types of strategies can be chosen at this point for the sensor array. The first option would rely on a high density of probes, eventually combined with an underlying Complementary Metal Oxide semiconductor (CMOS), addressing circuit to get a reasonable chance to follow the flow of information within a single cell or across a synaptic junction. At the larger scale of the network rather than at the cellular level, this is the method that has been pursued by the group of P. Fromherz in Munich (Hutzler *et al.*, 2006) using high-density two-dimensional multitransistor arrays. The second strategy would deal with neuronal networks aligned with dedicated geometries of sensors.

In all cases, going beyond the state of the art at that stage would imply full control of axo-dendritic polarity and network connectivity above an array of nanometric electric probes like SiNW-FET.

This chapter describes the cellular aspects of this global challenge. It starts with a presentation of the methods that have been used to control both network topography and axonal polarization, continues with neuron positioning above sensors, rapidly focusing on silicon nanowires and signals. Finally, this chapter will give the future trends concerning the subject of neural cell pinning above silicon devices.

10.2 Toward control of neuronal topography and axo-dendritic polarity

10.2.1 Building *in vitro* controlled neuron networks: an overview

Developing neurons extend neurites that subsequently compete with each other, and one of them becomes the axon, which is the extension that conveys the action potential towards post-synaptic neurons. This process, named axonal differentiation and characterized by a rapid elongation of one single neurite among all others, occurs about 36 hours after plating for hippocampal neurons extracted from mice embryos (Dotti *et al.*, 1998) or after 2–4 DIV (days *in vitro*) for cortical neurons (Barnes and Polleux, 2009). The other neurites further differentiate into dendrites and form the highly ramified structure that collects, and partly computes, the electric signals coming from pre-synaptic neurons.

Micropatterns are common tools to guide neurite outgrowth, independently of their axonal or dendritic nature, and to locate neuron cellular bodies (or soma) at specific positions, thus allowing a global design of neuronal architectures (Fig. 10.1) (Wyart *et al.*, 2002). However, failures in long-term confinement can occur as a result of the mechanical pulling forces exerted by neurites, which can be strong enough to displace cellular bodies over micrometric distances (Fromhertz, 2003). This is why strategies based on mechanical confinement have been developed, leading to long-term soma positioning and therefore to a stable neuron–sensor coupling. Let us cite the use of either parylene neurocages on top of micro electrode arrays (Erickson *et al.*, 2008) or picket fences of polyimide aligned above silicon transistors (Zeck and Fromherz, 2001) (Fig. 10.2). However, we will see later in this chapter that the mechanical properties of neurites can be exploited to impose the axo-dendritic polarity of the cell and, doing so, to control the topography and to direct the flow of information within neuronal networks.

Various experimental strategies have been developed in order to gain control over polarity at the single neuron level or, in other terms, to get an *a priori* specification of axons and dendrites. However, some of these methods have resulted in a global growth of all the axons in one direction using electric fields (Britland and McCaig, 1996), microfluidic channels (Peyrin *et al.*, 2011) or chemical gradients (Dertinger *et al.*, 2002), ruling out the fabrication of looped networks. A second set of experimental methods has paved the way to an axon-by-axon positioning, but at the price of a complex surface topography (Shinoe *et al.*, 2010) or chemistry (Oliva *et al.*, 2003) that may be hardly compatible with the top surface of an electronic chip.

An alternative strategy would consist of exploitation of axonal characteristics at the microscopic rather than at the molecular level. At least three properties

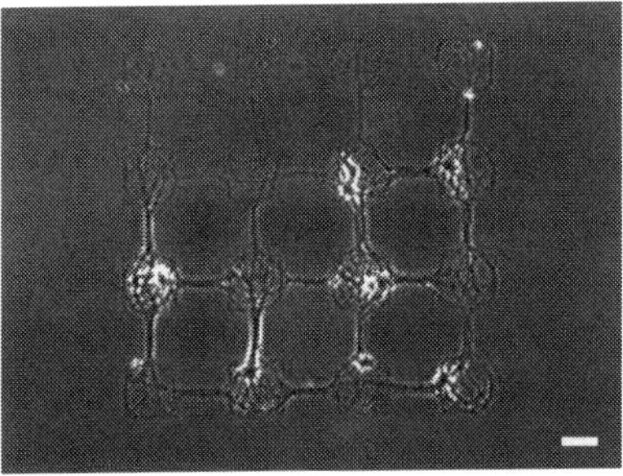

10.1 Image of neural networks of controlled architecture. Cell bodies of neurons are restricted to squares of 80 μm and neurites to lines (80 μm length, 2–4 μm wide). Scale bar is 50 μm. Reprinted with permission from *Journal of Neuroscience Methods* 117, 123 (2002). Copyright 2013 Elsevier Limited.

(a)

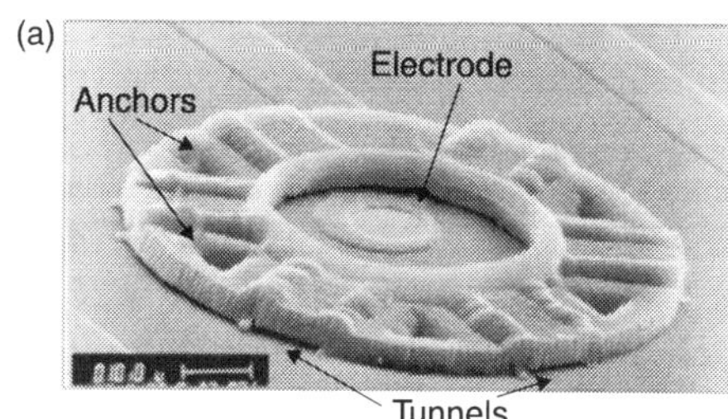

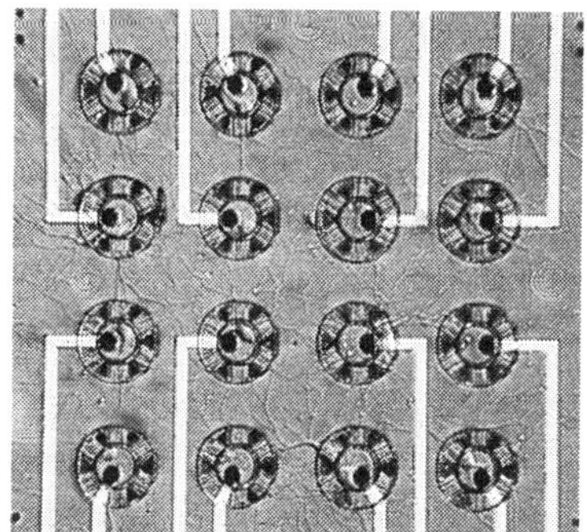

(b)

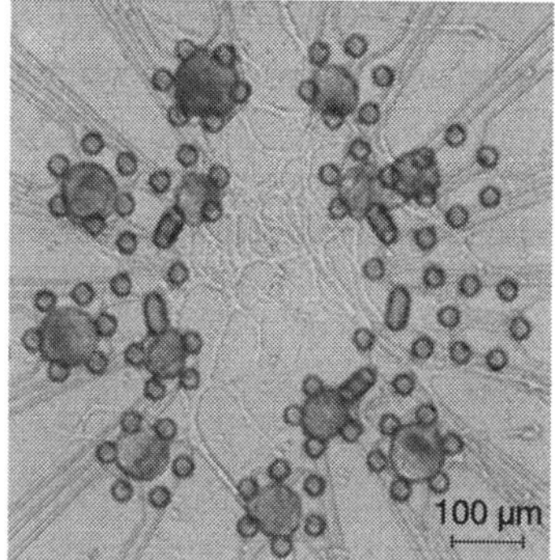

10.2 Mechanical confinement of the cellular body of neurons. (a) Parylene neuro-cages. Left: Electron micrograph of the final neurocage design. The main parts of the neurocage are labeled. The cage is made out of 4 μm thick parylene, a biocompatible polymer. Low-stress silicon nitride insulates the gold electrode and leads. Scale bar: 10 μm. Right: Neurochip culture at DIV10. Soma are trapped in the central chimney aegion of the cages spaced 110 μm apart. Process outgrowth through the tunnels is evident. A rich network has formed. The networking is even richer than shown in the photo as only the thickest processes are visible. Reprinted with permission from *Journal of Neuroscience Methods* 175, 1 (2008). Copyright 2012 Elsevier Limited. (b) Left: Electron micrograph of a snail neuron immobilized by picket fences after 3 days in culture. Scale bar 20 μm. Right: Micrograph of neuronal net with cell bodies (dark blobs) trapped within a double circle of circular fences with neurites grown in the central area (bright threads) after two days in culture. Scale bar 100 μm. Pairs of pickets in the inner circle are fused to bar-like structures. Reprinted with permission from *Proc. Natl. Acad. Sci. USA* 98, 10457 (2001). Copyright 2013 National Academy of Sciences of the United States of America.

specific to axons have been reported in the literature: resistance to bending associated with the naturally observed straightness of axons (Katz, 1985), the positioning of the centrosome at the base of the nascent axon (de Anda *et al.*, 2005) and the axonal preference for low adhesive conditions in contrast to the tendency of dendrites and soma to spread on surfaces (Prochiantz *et al.*, 1990). This is the strategy that we and others have followed, and each of these axonal properties has inspired various adhesive macro- or micropatterns, as detailed below.

10.2.2 Polarity control and axonal straightness

The cells used in the process detailed in the following paragraph are dissociated hippocampal neurons (a typical procedure for neuronal culture is given in the Appendix (see Section 10.7).

The tendency of axons to grow straight was recognized again and quantified well after the work of Katz (1985) by Feinerman *et al.* (2008) who explicitly wrote that 'axons keep their direction and make few turns, advancing in long stretches that are parallel to the pattern borders'. These authors used this property to impose an overall axo-dendritic polarity in neuronal networks constrained by means of asymmetric patterns of dimensions comparable with the measured axonal persistence length (420±50 μm) with the aim of building neuronal logic devices (Fig. 10.3). The basic idea here was to create thin (i.e. tens of micrometers wide) junctions between large triangular adhesive areas forming a ratchet geometry to create a 'funnel' effect that privilegiates straight neuronal extensions. In practice, three times more axons developed along one direction (i.e. crossed the thin bridges at the tip of the triangles) than along the opposite one. It would, however, be even more interesting to exploit the property of axonal straightness to control the position of the axon at the single cell level.

This challenge led us to the design of a set of patterns made of a 20 μm-diameter disk dedicated to soma adhesion and four branches for neurite outgrowth. More precisely, we used one straight (L1) and three curved (L2–L4) 2 μm-wide stripes (Fig. 10.4) (Roth *et al.*, 2012a). These geometric patterns were transferred on (3-glycidoxypropyl) trimethoxysilane silanized glass substrates (Nam *et al.*, 2006) to form adhesive poly-L-lysine (PLL) areas dedicated to cell adhesion. This protocol uses UV classical photolithography steps, including Shipley S1805 photoresist spinning (4000 rpm, 0.5 μm thickness, 115 °C annealing step for 1 minute), insulation through a mask, development (Microposit concentrate 1:1, Shipley), PLL deposition (1 mg/mL one night), multiple rinsing of the PLL and lift-off using an ultrasound pure ethanol bath.

The localization of axon specfication for neurons plated on these adhesive patterns was then quantified on fixed cells using the axonal marker Tau (more details in the Fixation and immunostaining section 10.7) after 3 days of *in vitro* differentiation (3DIV). The graph of Fig. 10.4 shows a growing preference of

(a)

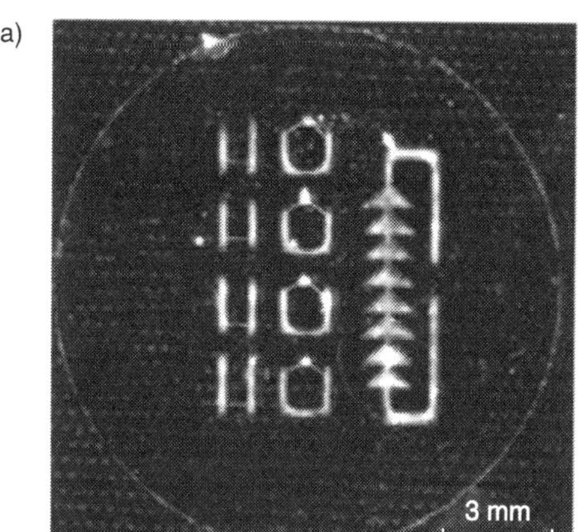

(c)

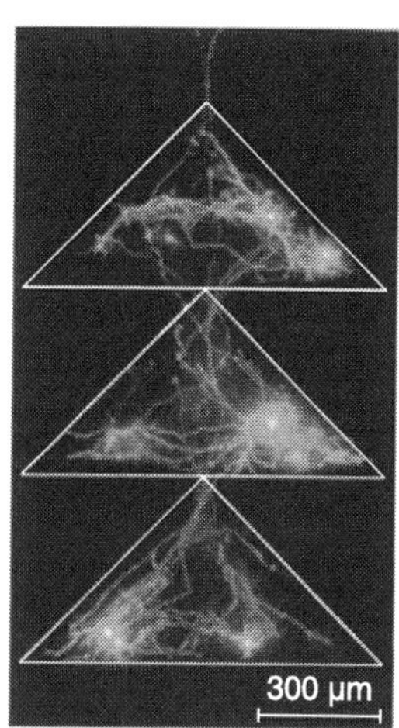

(b)

10.3 Patterning neuronal cultures to induce polarity at the network scale, adapted from Feinerman *et al.*, 2008. (a) Nine separate neuronal devices patterned on a single 13mm coverslip. The glass coverslip is first coated with Pluronic to form a cell-repellent surface. Then specific patterns are etched through this surface and the coverslip is recoated with fibronectin and laminin. Dark-field illumination; bright areas are concentrations of neurons. (b) Bright-field image of the device built from triangles (« diode » pattern). (c) Fluorescence images of devices transfected with non-specific green fluorescent protein (GFP). GFP transfection effciency was 1%. Long and uniformly thin processes are axons. Bright spots are cell bodies (somas) and thick processes near the somas are dendrites. On average, M_{fwd}=360 neurons send their axons forward across the apex and into the triangle beyond it, whereas M_{bwd}=120 neurons send their axons in the opposite direction. See Supplementary Information in Feinerman *et al.*, 2008. Reprinted with permission from *Nature Physics* 4, 967 (2008). Copyright 2012 Nature Publishing Group.

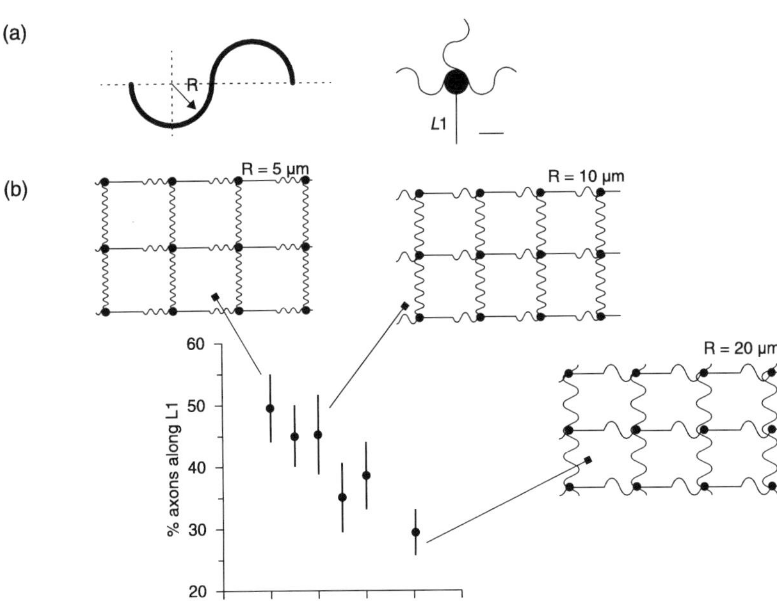

10.4 Curvature dependence of axonal inhibition on 2 µm wide stripes. (a) Drawing of wavy paths from half-circles of radius R. An elementary four branch-pattern with R = 10 µm is shown. Scale bar: 20 µm. (b) View of three square networks characterized by a 150 µm periodicity and different curvatures (R = 5, 10 and 20 µm). The percentage of axons along the straight direction L1 is maximum for R = 5 µm. Error bars denote 95% confidence intervals (n = 310, 341, 217, 359, 286 and 558 for R = 5, 7.5, 10, 12.5, 15 and 20 µm, respectively).

axonal localization along L1 for the lowest curvatures of the L2–L4 stripes. Below R = 10 µm, however, we observed that neurites almost systematically cut through the curved lines of the pattern, a feature resulting from neuritic tension during development (Roth *et al.*, 2012b). We therefore stopped our choice on selected value R = 10 µm for the design of a new set of patterns with only three branches dedicated to further exploration of the influence of the cellular shape on development and polarization (Fig. 10.5).

First, the control pattern named DC (for 'disk' and 'control') characterized by straight directions L1–L3 distributed according to a three-fold symmetry resulted as expected in axons randomly oriented along L1–L3 directions (35.8%, 33.2% and 31.1% along L1, L2 and L3, respectively; Fig. 10.5(a), (c)). Again, the bias introduced by the neuritic curvatures in the DW pattern (for 'disk' and 'wavy') led to a preferential axonal positioning along L1, reaching now a percentage of 69% in this three branch pattern (Fig. 10.5(a), (c)).

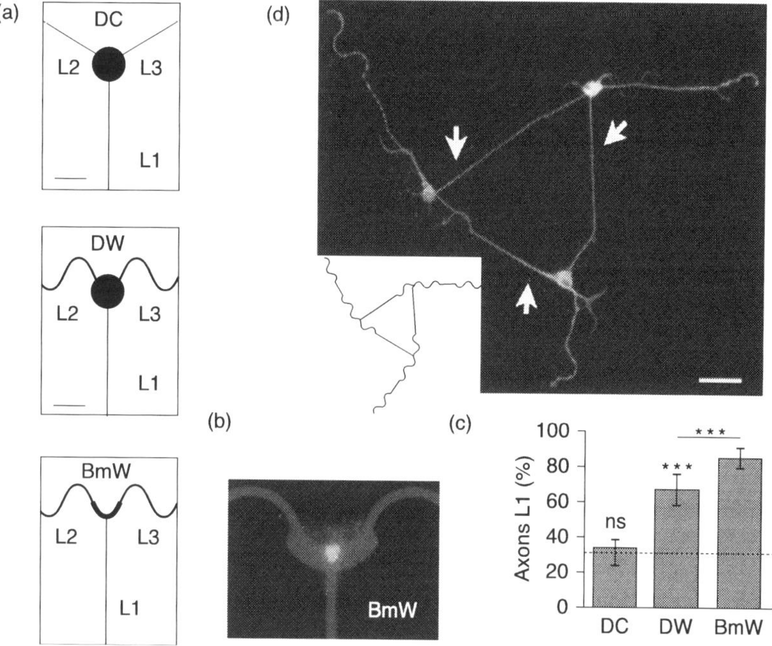

10.5 Effect of soma and neuritic constraints on axonal polarization. (a) Design of the DC, DW and BmW micropatterns; L1–L3 directions for neurite outgrowth are indicated. Scale bars: 20 μm. (b) Superimposition of centriole scatter plots on the BmW patterns for 1 DIV non-polarized neurons, n = 218. (c) Results in axonal polarization, i.e. percentage of 3 DIV neurons with their axon along the L1 direction. Error bars denote 95% confidence intervals and the ⋆⋆⋆ probability $p<0.001$ for these values to be significantly different from random; n = 194, 110, 161 for the DC, DW and BmW patterns, respectively. The dashed line corresponds to random polarization (33%). (d) Triangular looped circuit made of three connected neurons at DIV7. Arrows point to the three axons. Note that neuronal extensions often cut through the curved lines of the pattern, a feature resulting from neuritic tension during development (Roth *et al.*, 2012b). Inset: the poly-L-lysine adhesion pattern. Scale bar: 50 μm.

Although remarkably significant compared with DC, the results obtained on DW patterns were, however, not fully satisfactory for neuronal network engineering. Indeed, the mean percentage of correctly polarized *n*-neuronal networks built from elementary DW patterns (i.e. 0.69^{n} %) would statistically fall below 10% for only $n \geq 6$. Thus, a higher efficiency in axonal localization would be highly desirable to gain control in the fabrication of larger neuronal networks. In this aim, we turned our attention to the two other reported axonal specificities at the microscopic scale.

10.2.3 Centrosome and ultimately axonal localization

The cell polarity axis, in the case of epithelial cells, has been shown to respond to the cell adhesive micro-environment. In particular, cells forced to adopt a triangular shape on 'L-like' micropatterns were reproducibly polarized along the pattern axis of symmetry: that is, the internal organization of the cell was modified and the centrosome was aligned along this axis (Théry *et al.*, 2005). Therefore, in the aim to locate the centrosome at the axon base, we shaped the soma according to an axis of symmetry aligned with L1 by hollowing out the central disk of the DW pattern to form a 5 μm thick boomerang-like shape ('Bm', Fig. 10.5(a)). The centrosome distribution clearly localized along L1 in the BmW pattern (Fig. 10.5(b)) and, as a consequence, the percentage of axons along L1 in the BmW pattern reached the impressive rate of 85.7% (Fig. 10.5(c)). Note that this value is significantly different from both random and DW.

The high rate of axonal control obtained with the BmW pattern paved the way to the reliable fabrication of neuronal microcircuits of any geometry on the basis of an appropriate association of these elementary patterns. As a first example, we designed a looped triangular network with a connexion geometry that naturally includes a curve at the end of the L1 line (Fig. 10.5(d) inset). Figure 10.5(d) shows a real triangular three-neuron network at 7 DIV for which the axons were found exclusively along straight lines. As a second example, neuron chains were formed using elementary BmW patterns. Here, the presence of a curvature at the end of the L1 path of the pre-synaptic neuron proved unnecessary: an optimized 15 μm gap between the boomerang thick line dedicated to soma adhesion and the L1 direction was sufficient to avoid the formation of a neurite from the soma of the post-synaptic neuron while allowing the axon of the pre-synaptic neuron to cross that gap, eventually by branching as shown in Fig. 10.6. This example illustrates the generic capacity of neurites to explore non-adhesive areas on a maximum distance of 20 μm.

10.2.4 Differential adhesion and axonal localization

The joint exploitation of axonal straightness and centrosome positioning to select the position of the future axon in developing neurons has led to rather efficient, although complex, micropatterns. We suspected that the differential adhesiveness between axons and dendrites discussed by Prochiantz *et al.* (1990) might generate simpler designs.

Because of the constraint of a simple chemistry that we imposed at the start of this work, this differential adhesive condition should not result from the use

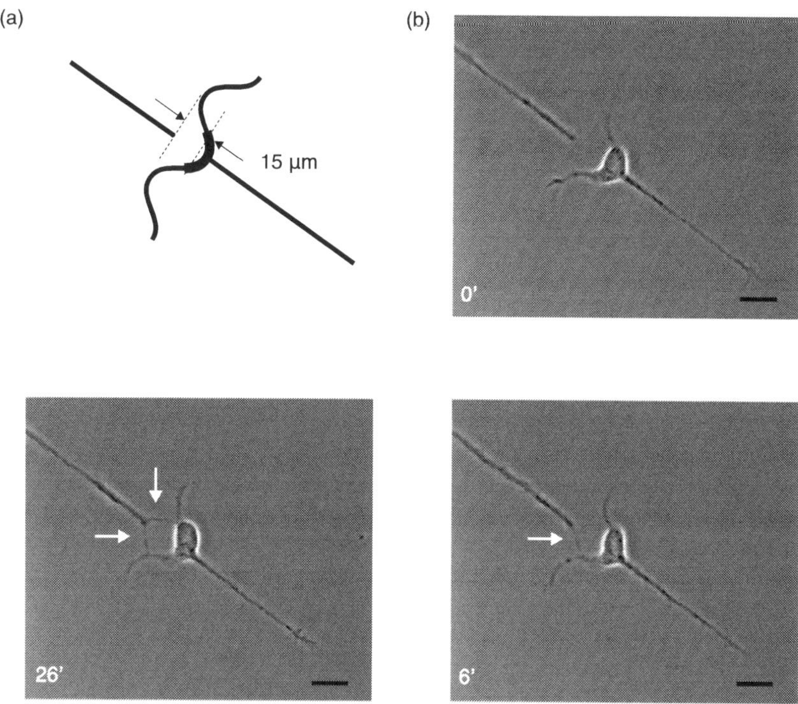

10.6 Building neuronal chains with BmW patterns. (a) A 15 µm gap separates two consecutive elementary BmW patterns. (b) Time-lapse experiment (time indicated in minutes) of a neuron developing on a BmW motif-based network. The neurite of the pre-synaptic cell is shown to cross the 15 µm gap and to connect the neurites of the post-synaptic (central) neuron (white arrows). Scale bars: 20 µm.

of different adhesive coatings. The alternative strategy we have developed consisted instead of the use of different poly-L-lysine stripe widths. We have thus designed a two branch, asymmetric pattern made of 2 µm and 6 µm wide stripes located on each side of a 20 µm diameter circular area dedicated to soma positioning (see Fig. 10.7(a) for an image of a bipolar neuron polarized on this pattern). The consequences effects of this design on axonal polarization were the most impressive, with the axon found exclusively along the thinnest 2 µm wide stripe (100% success for a population of 180 neurons; Fig. 10.7(b)). This success in imposing the localization of the axon plausibly resulted from an interplay between neuritic length and width. Indeed, we found that DIV2 neurites grown on 2 µm uniform wide stripes were on average more than three times longer than neurites that have been forced to grow on 6 µm uniform

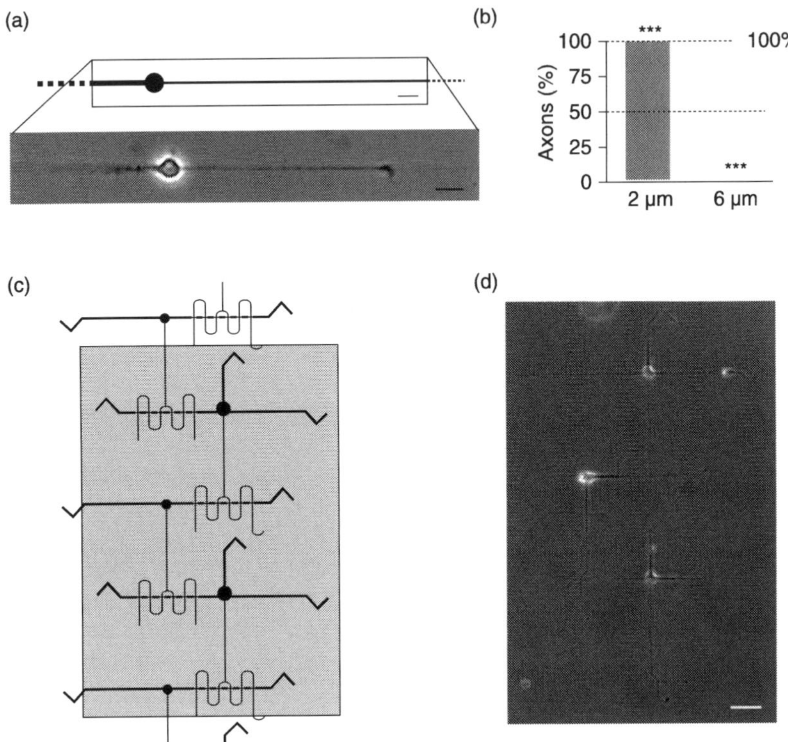

10.7 Effect of differential adhesion on axonal polarization. (a) The asymmetric pattern made of a 14 μm disk and two different stripe widths (left: 6 μm, right: 2 μm (poly-L-lysine areas in gray)), and one example of neuron development on this pattern. Scale bars: 20 μm. (b) Results on axonal polarization, i.e. percentage of 3DIV neurons with their axon on the 2 μm- and 6 μm-wide stripes of the asymmetric pattern ($n = 180$). The second dashed line corresponds to random polarization (50%). The ★★★ denote the probability $p < 0.001$ for these values to be significantly different from random. (c) The dark gray areas represent the poly-L-lysine adhesive patterns. One of the three thick, 6 μm wide, lines is periodically thinned to avoid its possible invasion by the axon located on the thin vertical line ended by curvatures and to form synaptic junctions. The light gray rectangle defines a region of interest (see (d)). (d) A real neuronal network at 3DIV grown according to the pattern shown in (c) within the region delimited by the light gray rectangle. Scale bar: 50 μm. Courtesy of C. Tomba.

wide stripes ($n = 180$ neurons for each condition). Considering the propensity of the axon to grow faster than the other neurites, this asymmetry in length is plausibly correlated to the axonal preference we observed for the thinnest stripe.

10.2.5 Existing methods of control and positioning

All these studies provide evidence that physical constraints directly influence neuronal dynamics of growth and polarization. However, our results demonstrate that it is possible to act separately on the fate of a given neurite, while a global regulation at the cell level ensures the formation of only one axon among beside dendrites.

From a practical point the view, this basic, physical approach to neuronal biology gives a panel of available strategies for the design of controlled neuron networks at a single cell resolution. Figure 10.7(c) gives an example of a design that integrates the use of differential adhesion and curvatures. Note here that, apart from the zig-zag that ends the wide stripes dedicated to dendritic growth, curvatures were put on the thinnest line used to locate the axon. Indeed, curvatures inhibit axonal differentiation if imposed near the soma. But once the axon is formed (Fig. 10.7(d)), that is after its length reaches a few tenths of microns, its growth is not disrupted by bending. Curvatures can thus be used to guide the axons perpendicularly to dendrites in order to locate synaptic connections.

10.3 Neuron networks on top of silicon nanowires (SiNWs)

The feasibility of aligning organized neuronal networks (however, without any control of axo-dendritic polarity) on top of electrical devices was demonstrated as early as in 2007 (Jun *et al.*, 2007) using MEA. Interestingly, these authors succeeded in measuring the propagation of an action potential from one micro-electrode to the other through poly-synaptic pathways. Concerning stimulation, it was demonstrated that cathodic electrical current pulses (50 μs, current pulse 200 μA) robustly produced evoked potential, whereas application of anodic pulses did not induce any activity. However, the microelectrodes were too large (surface 100 μm^2) to provide single cell resolution: a single electrode stimulates multiple neurites, possibly coming from different neurons.

The step forward in the recording and stimulation of organized networks at the single cell level was reported by Lieber's group in 2006 using a single neuron configuration (Patolski *et al.*, 2006). Instead of passive micro-devices like MEA, a linear matrix of polarized silicon nanowires was used to record, with a high signal to noise ratio, the propagation speed of an action potential (AP) at a sub-cellular scale in both axons and dendrites (back AP propagation in this latter case) after an intra-cellular stimulation at the soma level (15 ms, 0.5 nA current pulses). Figure 10.8 displays the reconstituted propagation of an action potential: each trace represents an experiment where intra-cellular stimulation occurs at the soma level and sets the origin of time for each SiNW-FET recording. Extra-cellular stimulation was also achieved by applying biphasic square wave pulses (amplitude 0–1 V) to drain and source electrodes simultaneously. Interestingly, the propagation of an action potential could be blocked when encountering a nanowire polarized at about 1 V.

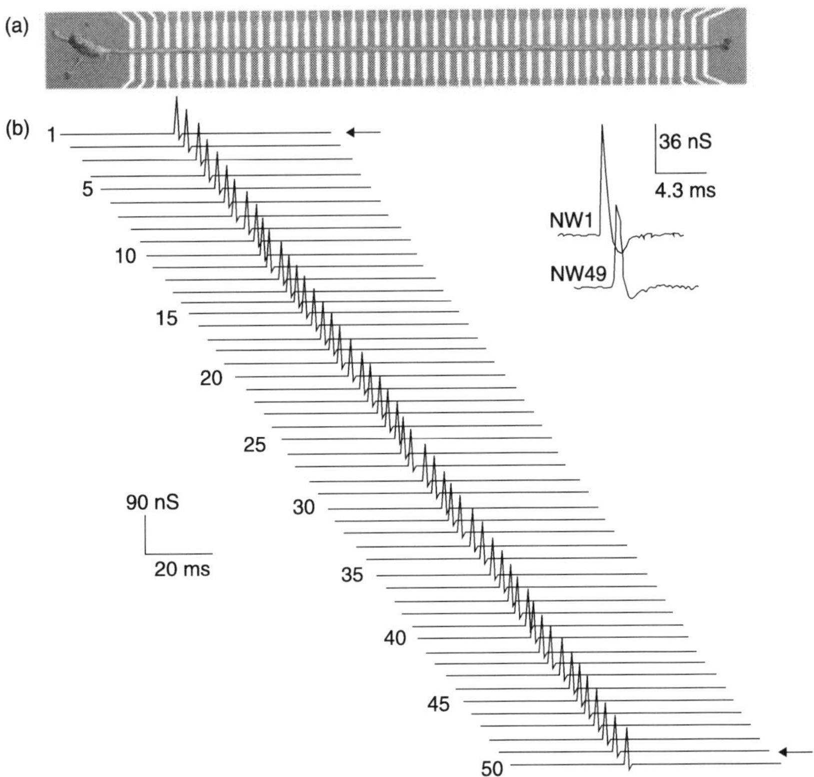

10.8 Propagation of an action potential on a linear array of 50 silicon nanowire (SiNW)-field effect transistors (FET). (a) Optical image of an aligned axon crossing an array of 50 NW devices with a 10-mm interdevice spacing. (b) Electrical data from the 50-device array shown above. The yield of functional devices is 86%. The peak latency from NW1 (top arrow) to NW49 (bottom arrow) was 1060 ms. Reprinted with permission from *Science* 313, 1100 (2006). Copyright 2013 The American Association for the Advancement of Science.

This first demonstration of neuronal recording and stimulation at a sub-cellular scale based on a field effect detection was a breakthrough that paved the way to a new generation of neuron-electronic interfacing.

10.3.1 Silicon nanowire fabrication: from bottom-up to top-down strategies

In the work of Patolsky *et al.* (2006a), SiNW-FET devices were built from a bottom-up approach: silicon nanowires were grown in a plasma reactor under a flow of SiH_4 (Cui *et al.*, 2001), then deposited and oriented from a dispersion within a liquid (Patolsky *et al.*, 2006b) and finally connected to the outside world by multiple photolithographic steps. This bottom-up approach allows the design of linear SiNW-FET, eventually according two perpendicular orientations. However, these restricted geometries may limit the range of possible SiNW-FET matrix designs, which would in turn restrict the diversity of the neuronal architectures that could be instrumented. Top-down technological approaches using Silicon On Insulator (SOI) substrates hardly produces nanowires of sizes and shapes as small as the ones used by Lieber's group (20 nm in diameter), but SOI devices have demonstrated their ability to detect biological signals like DNA hybridization and pH variations (Kim *et al.*, 2007). These results have suggested that these devices might also be used for monitoring of excitable cells.

We present here a top-down fabrication approach, suitable for looped neuron network architectures like the one shown in Fig. 10.5(d). NanoFET arrays were fabricated on uniformly boron implanted (10^{17}–10^{19} at/cm^3) SOI wafers (SOITEC) using the following technological steps: deep UV photolithography for the lift-off of gold contact pads, electron beam lithography and reactive ion etching for the definition of 100 nm wide SiNW, and two last steps of deep UV photolithography to deposit a 10 nm thick HfO_2 encapsulation layer above the SiNW by atomic layer deposition, and a micron thick oxide (or a parylene layer) for metal insulation (Fig. 10.9) (Delacour *et al.*, 2012). At the end of the fabrication

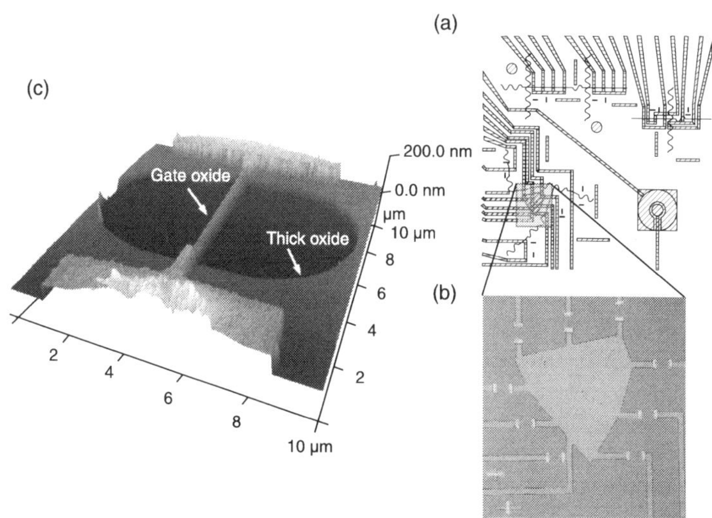

10.9 The neuro field effect transistor (FET) chip. (a) Partial view of the neuroFET chip. Gold metallic lines are represented by dashed surfaces and the poly-L-lysine adhesive patterns in solid lines. Scale bar: 100 μm. (b) Detail of the chip before the deposition of the insolating layers. (c) Atomic force microscopy image of a silicon nanowire (SiNW)-FET showing the final structure after deposition of the insulators. Courtesy of C. Delacour.

process, the SiNW-FET device (i.e. the NeuroFET chip) contains 89 silicon nanowires, 100 nm in width and a few microns in length organized into specific arrays dedicated to the recording and stimulation of neuronal networks of various architectures including triangles (Roth *et al.*, 2012a) similar to the one presented in Fig. 10.5(d).

10.3.2 Neuron positioning above SiNWs

A final step leading to the alignment of neuronal networks onto the NeuroFET chip is required. It consists of a lift-off step of poly-L-lysine, aligned with the underlying SiNW-FET. Figure 10.10(a) shows the aligned photoresist pattern (Shipley S1805, 0.5 μm thick) before the incubation step of poly-L-lysine, designed to position a neuron shaped using the BmW pattern above SiNW. Large zones of poly-L-lysine are moreover implemented outside the central part of the chip (not shown) to provide a reservoir of non-constrained cells for long-term survival through their release of diffusible factors (Goslin and Banker, 1991). Figure 10.10(b) gives an example of the precision in neurite positioning above a SiNW by showing a neurite under development, ended by an active growth cone with multiple filopodia at its front, that crosses perpendicularly the Si channel.

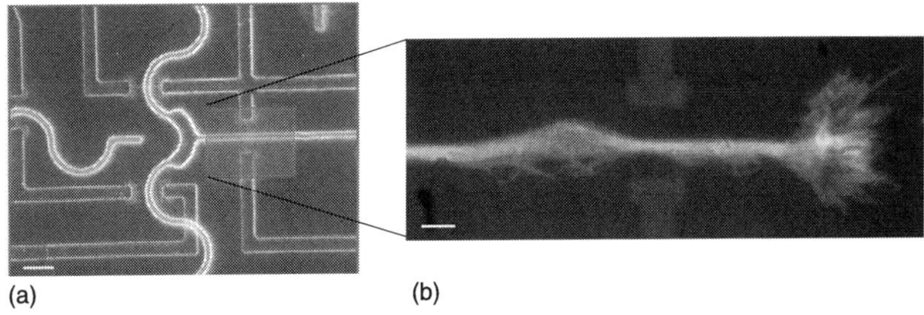

10.10 Neuron alignment on nanowires of the neuro field effect transistor (FET) chip. (a) Detail of the chip after photoresist spincoating and development, just before the poly-L-lysine deposition step. Scale bar: 10 μm. (b) Optical micrograph of a straight neurite above a silicon nanowire (SiNW)-FET after immunostaining. Scale bar: 5 μm. Courtesy of G. Bugmiroust.

10.4 Future trends

10.4.1 Requirements for neuronal devices: filling all adhesive sites

We have seen in the previous section that once a soma is positioned at a correct location, i.e. on its dedicated adhesive spot, its development into polarized neurons can be at 100% controlled by adequate physical constraints provided by micropatterns, leading to organized networks. However, there is a remaining difficulty to overcome to be in possession of a reliable neuronal engineering process: to reproducibly fill each adhesion site with a single neuron during the seeding step. This issue is of primary importance considering the fabrication of entire cellular networks (and not only the positioning of single isolated cells) as a technological target. For example, the probability of finding three well-positioned soma on a triangular pattern, like the one shown in Fig. 10.5(d), is extremely low (less than 1%) considering the usual seeding cell density. Increasing the seeding density of cells would, however, not be a viable solution as it would lead to the occupancy of the adhesive lines that should be mandatorily kept free for neuritic development. Manual loading by sucking neurons up in a glass micro-pipette has been used as an efficient method (Fromhertz, 2003; Erickson *et al.*, 2008), but 'blind' and fast automatic strategies would also be desirable. A selective attraction of cells from an initial suspension should be envisioned, and the use of the magnetic interaction is one possible strategy. This method is as follows: cell tagging is performed, for example through the endocytosis of superparamagnetic nanoparticles (Osman *et al.*, 2012), or the adhesion of micro-beads on the external cell membrane (Fig. 10.11), then hard or soft micro-magnets fabricated on top of

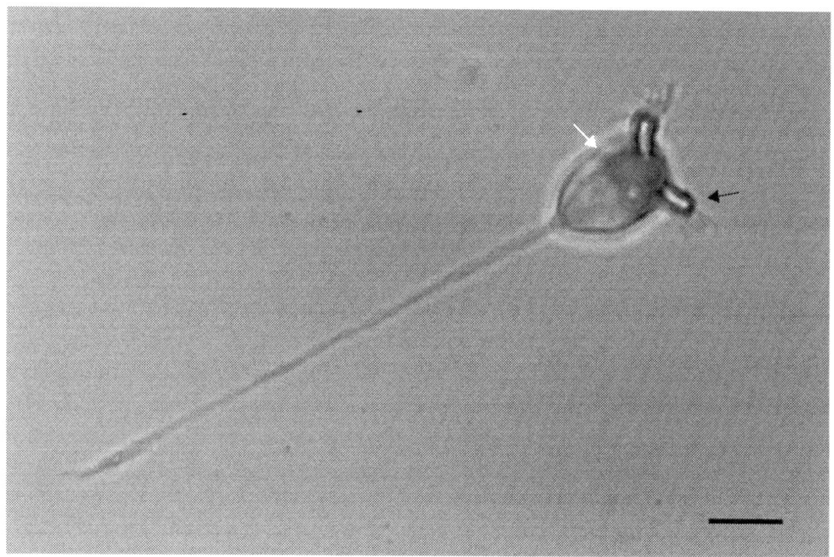

10.11 A magnetically tagged neuron attached on a NdFeB micro-magnet elaborated by lift-off and further annealed (black arrow). The white arrow points on three beads of 2.8 μm diameter. The cellular growth is constrained by a BmW micropattern (see Fig. 10.5) aligned with the micromagnet. Scale bar: 10 μm. Collaboration N. Dempsey and F. Dumas-Bouchiat, Institut Néel. Image: courtesy of G. Bugnicourt and C. Delacour, Institut Néel.

the culture substrates (Fig. 10.11) will attract both beads and cells. This step could benefit significantly from a microfluidic stage to force a cell circulation above micromagnets (Saliba *et al.*, 2010). Adhesive patterns have lastly to be aligned with (and on top of) the micromagnets, and eventually isolated from the magnets by a thin encapsulation layer like parylene, to ensure adhesion and cellular development.

This selective and precise cell positioning approach might ultimately be pushed toward the realization of heterogeneous networks made of different neuronal subtypes. Mixing excitatory and inhibitory neurons would, for example, lead to the fabrication of even more controlled and sophisticated living neuronal microcircuits that would in turn provide a unique experimental set of data useful for the development of neuromorphic architectures (Mitra *et al.*, 2009). On the other hand, fundamental biological problematics like the synaptic maturation in corticostriatal oriented networks, an issue relevant for the understanding of Huntington's and Parkinson's diseases (Penroda *et al.*, 2011), or the exploration of neuron–glia interaction through the fabrication of topographically controlled astrocyte–neuron assemblies (Kidambi *et al.*, 2008) should benefit from the single cell resolution in positioning and recording provided by the coupling of cellular networks and SiNW-FET arrays.

10.4.2 Three-dimensional engineering

The introduction during recent decades of integrated components for cellular detection has been accompanied by a shift from intracellular to extracellular recording methods. It is interesting to see this trend now reversed with the development of techniques based on the insertion of micro- and nanostructures through cell membranes. Vertical nanowires (nanopillars) can be used for their penetrative geometry rather than as active detection elements to perform intracellular recording. The purpose here is to access the small fluctuations of the membrane potential that carry the complexity and the wealth of neuronal computation, and intracellular recording shows by intrinsically higher signal to noise ratio and higher sensitivity to the subthreshold events than extracellular technologies.

This approach has been developed following the demonstration by Hai and Spira (2012) of the ability of giant *Aplysia californica* neurons to engulf micrometric mushroom-like electrodes, which operate, after a step of membrane permeabilization induced by a transient polarization (electroporation phenomena), as 3D sensors able to record intracellular potentials. Two papers were published in 2012 by groups from Harvard (Robinson *et al.*, 2012) (Fig. 10.12) and Stanford (Xie *et al.*, 2012) Universities using nanowires instead of mushroom-like electrodes to instrument mammalian neurons. The benefits of these technologies are, on one hand, a long-term intimate cell-detector relationship and, on the other hand, an impressively high signal to noise ratio of a few hundreds. Note that signal to noise

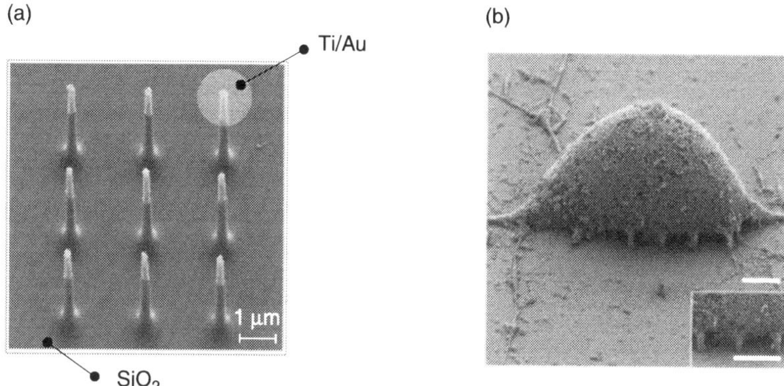

10.12 Neuronal interfacing with vertical silicon nanowires. (a) Scanning electron microscope (SEM) image of the nine silicon nanowires that constitute a recording/stimulation pad. Metal-coated tips and insulating silicon oxide are indicated. Scale bar: 1 µm. (b) SEM image of a rat cortical cell (3 days *in vitro* after seeding) on top of a 3×3 vertical nanowires array (scale bar: 2.5 µm), showing nanowires interfacing with the cellular membrane (inset: scale bar, 2.5 µm). Reprinted with permission from *Nature Nanotechnology* 7, 180 (2012). Copyright 2013 Nature Publishing Group.

ratios are one order of magnitude lower for extracellular recording using nanowires (Fig. 10.8), and even two orders of magnitude lower for conventional microelectrodes. In another work (Tian *et al.*, 2010), SiNW-FET functionalized with phospholipid bilayers to enable fusion with the cell membrane and positioned at the end of a kinked nanostructure provided an intracellular recording of the activity of cardiomyocyte HL-1 cells without an electroporation step.

In addition, there is a trend to build and instrument 3D cellular structures. To that aim, the integration of SiNW-FET into soft and porous scaffolds that can be folded to form 3D structures was achieved for the first time very recently (Tian *et al.*, 2012).

However, for all these approaches, the possibility of internalization of these nanowires in neuronal extensions (and not only the soma) or more generally the positioning of nanosensors along neuronal processes to access the propagation of the electrical signal at a sub-cellular scale, is an open, unresolved issue. There is again a need to align cells and sensors, and this challenge in 3D is even more demanding than in 2D.

10.5 Conclusion

Despite the amount of work and knowledge that have been accumulated during recent decades, and despite numerous achievements, the subject of interfacing neurons with electronics is only very partially unveiled. This chapter has described the recent advances in *in vitro* approaches that combine more and more sophisticated neuronal network engineering methods and the up-to-date silicon technologies to record and stimulate 2D, and soon 3D neuron networks with high signal to noise ratios. But beyond these *in vitro* methodologies, *in vivo* instrumentation is required for many applications in the vast field of brain–computer interfaces that tackles human health issues. Interfacing the brain is without a doubt a new natural outlet for the *in vitro* technologies that have been described in this chapter, and for all the knowledge of the neuron–electronic interface that has been jointly developed.

10.6 References

Barnes A.P. and Polleux F. (2009), 'Establishment of axon–dendrite polarity in developing neurons', *Annu. Rev. Neurosci.* 32, 347–81

Britland, S. and McCaig, C. (1996), 'Embryonic Xenopus neurites integrate and respond to simultaneous electrical and adhesive guidance cues', *Exp. Cell Res.* 226, 31–8.

Cui, Y. Lauhon, L.J., Gudiksen, M. S., Wang, J. and Lieber C.M. (2001), 'Diameter-controlled synthesis of single-crystal silicon nanowires', *App. Phys. Lett.* 78 (15), 2214–17. doi: 10.1063/1.1363692.

de Anda, F.C., Pollarolo, G., Da Silva, J.S., Camoletto, P.G., Feiguin, F., and Dotti, C.G. (2005), 'Centrosome localization determines neuronal polarity', *Nature* 436, 704–8.

Delacour, C., Marchand, R., Crozes, T., Ernst, T., Buckley, J., *et al.* (2012), 'Silicon nanowire field effect transistors to probe organized neuron networks'. In: Stett A (ed). *Proceedings MEA Meeting 2012.* Stuttgart: BIOPRO Baden-Wuerttemberg GmbH, 276–7.

Dertinger, S., Jiang, X., Murthy, Z.M., and Whitesides G.M. (2002), 'Gradients of substrate-bound laminin orient axonal specification of neurons' *Proc. Natl. Acad. Sci. USA* 99 (20), 12542–7.

Dotti, C.G. Sullivan, C.A., and Banker G.A. (1998), 'The establishment of polarity by hippocampal neurons in culture', *.J. Neurosci. 8*, 1454–68.

Erickson, J., Tooker, A., Tai, Y.C., and Pine, J. (2008), 'Caged neuron mea: a system for long- term investigation of cultured neural network connectivity', *J. Neurosci. Methods* 175 (1), 1–16.

Feinerman, O., Rotem, A., and Moses, E. (2008), 'Reliable neuronal logic devices from patterned hippocampal cultures', *Nat. Phys.* 4, 967–73. doi:10.1038/nphys1099.

Goslin, K. and Banker, G. (1991), 'Rat hippocampal neurons in low-density culture'. In *Culturing Nerve Cells.* Edited by Banker G, Goslin K. Cambridge, MA: MIT Press, 251–81.

Hai, A. and Spira, M.E. (2012), 'On-chip electroporation, membrane repair dynamics and transient in-cell recordings by arrays of gold mushroom-shaped microelectrodes', *Lab. Chip* 12 (16), 2865–73.

Hutzler, M., Lambacher, A., Eversmann, B., Jenkner, M., Thewes, R., and Fromherz, P. (2006), 'High-resolution multitransistor array recording of electrical field potentials in cultured brain slices', *J. Neurophysiol.* 96, 1638–45. doi: 10.1152/jn.00347.2006.

Katz, M.J. (1985), 'How straight do axons grow?', *J. Neurosci.* 5 (3), 589–95.

Kidambi, S., Lee, I., and Chan, C. (2008), 'Primary neuron/astrocyte co-culture on polyelectrolyte multilayer films: a template for studying astrocyte-mediated oxidative stress in neurons', *Adv. Funct. Mater.* 18, 294–301. DOI: 10.1002/adfm.200601237.

Kim, A., Ah, C.S., Yu, H.Y., Yang, J.H., Baek, I.B., *et al.* (2007), 'Ultrasensitive, label-free, and real-time immunodetection using silicon field-effect transistors', *Appl. Phys. Lett.* 91, 103901. doi:10.1063/1.2779965.

Kordeli, E., Lambert S. and Bennett V. (1995), AnkyrinG: a new ankyrin gene with neural-specific isoforms localized at the axonal initial segment and node of Ranvier, *J. Biol. Chem.* 270, 2352–9. doi: 10.1074/jbc.270.5.2352.

Maeda, E., Robinson, H.P., and Kawana, A. (1995), 'The mechanisms of generation and propagation of synchronized bursting in developing networks of cortical neurons', *J. Neurosci.* 15 (10), 6834–45.

Mitra, S., Fusi, S., and Indiveri, G. (2009), 'Real-time classification of complex patterns using spike-based learning in neuromorphic VLSI', *IEEE Transactions on Biomedical Circuits and Systems* 3, 32–42

Mattson M.P. (2004), 'Pathways towards and away from Alzheimer's disease', *Nature* 430 (7000), 631–9.

Meister, M., Wong, R., Baylor, D.A., and Shatz, C.J. (1991), 'Synchronous bursts of action potentials in ganglion cells of the developing mammalian retina', *Science* 252 (5008), 939–43.

Oliva, A.A., James, C.D.Jr, Kingman, C.E., Craighead, H.G., and Banker, G.A. (2003), 'Patterning axonal guidance molecules using a novel strategy for microcontact printing', *Neurochem.* 28 (11), 1639–48.

Nam, Y., Branch, D.W., and Wheeler, B.C. (2006), 'Epoxy-silane linking of biomolecules is simple and effective for patterning neuronal cultures'. *Biosens. Bioelectron.* 22, 589–97.

Osman, O., Zanini, L.F., Frénéa-Robin, M., Dumas-Bouchiat, F., Dempsey, N., *et al.* (2012), 'Monitoring the endocytosis of magnetic nanoparticles by cells using permanent micro-flux sources', *Biomedical Microdevices* 14, 947–54.

Patolsky, F., Timko, B., Yu, G., Fang, Y., Greytak, A.B., *et al.* (2006a), 'Fabrication of silicon nanowire devices for ultrasensitive, label-free, real-time detection of biological and chemical species', *Nature Protocole* 1 (4), 1711–24. doi:10.1038/nprot.2006.227.

Patolsky, F., Timko, B., Yu, G., Fang, Y., Greytak, A.B., *et al.* (2006b), 'Detection, stimulation, and inhibition of neuronal signals with high-density nanowire transistor arrays', *Science* 313 (5790), 1100–4.

Penroda, R.D., Kourrich, S., Kearney, E., Thomas, M., and Laniera, L.M. (2011), 'An embryonic culture system for the investigation of striatal medium spiny neuron dendritic spine development and plasticity', *J. Neurosci. Methods* 200 (1), 1–13. doi:10.1016/j.jneumeth.2011.05.029.

Peyrin, J.M., Saias, L., Vignes, M., Gougis, P., Magnifico, S., *et al.* (2011), 'Axon diodes for the reconstruction of oriented neuronal networks in microfluidic chambers', *Lab. Chip.* 11, 3663–73.

Prochiantz, A., Rousselet, A., and Chamak, B. (1990) 'Adhesion and the in vitro development of axons and dendrites', *Prog. Brain Res.* 86, 331–6.

Qinga, Q., Palb, S.K., Tiana, B., Duan, X., Timko, B., *et al.* (2010) 'Nanowire transistor arrays for mapping neural circuits in acute brain slices', *Proc. Natl. Acad. Sci. USA* 107 (5), 1882–7.

Robinson, H.P.C, Kawahara, M., Jimbo, Y., Torimitsu, K., Kuroda, Y., and Kawana, A. (1993) 'Periodic synchronized bursting and intracellular calcium transients elicited by low magnesium in cultured cortical neurons', *Neurophysiol.* 70 (4), 1606–16.

Robinson, J.T., Jorgolli, M., Shalek, A.K., Yoon, M.H., Gertner, R.S., and Park, H. (2012), 'Vertical nanowire electrode arrays as a scalable platform for intracellular interfacing to neuronal circuits', *Nature Nanotech.* 7 (3), 180–4.

Roth, S., Bugnicourt, G., Bisbal, M., Gory-Fauré, S., Brocard, J., and Villard, C. (2012a) 'Neuronal architectures with axo-dendritic polarity above silicon nanowires', *Small* 8 (5), 671–5. doi: 10.1002/smll.201102325

Roth, S., Bisbal, M., Brocard, J., Bugnicourt, G., Andrieux, A., *et al.* (2012b) 'How morphological constraints affect axonal polarity in mouse neurons', *PLoS One* 7 (3), e33623.

Shahaf, G. and Marom, S. (2001) 'Learning in networks of cortical neurons', *J. Neurosci* 21 (22), 8782–8.

Shinoe, T., Shi, J., Chen, Y., and Triller, A. (2010) 'Controlling neuronal networks, axo-dendritic polarities and synapse formation by microcontact printing and microchannel techniques', *Proc. Micro Total Analysis Systems* 14, 479–81.

Stern, E., Klemic, J.F., Routenberg, D.A., Wyrembak, P.N., Turner-Evans D.B., *et al.* (2006) 'Label-free immunodetection with CMOS-compatible semiconducting nanowires', *Nature* 445 (7127), 519–22.

Théry, M., Racine, V., Pépin, A., Piel, M., Chen, Y., *et al.* (2005) 'The extracellular matrix guides the orientation of the cell division axis', *Nature. Cell Biol.* 7, 947–53.

Tian, B., Cohen-Karni T., Qing, Q., Duan, X., Xie, P., and Lieber, C.M. (2010) 'Three-dimensional, flexible nanoscale field-effect transistors as localized bioprobes', *Science* 329, 830–4.

Tian, B., Liu, J., Dvir, T., Jin, L., Tsui, J.H., *et al.* (2012) 'Macroporous nanowire nanoelectronic scaffolds for synthetic tissues', *Nature Mater.* 11, 986–94. doi: 10.1038/nmat3404.

Wagenaar, D., Pine, J., and Potter, S. (2006), 'An extremely rich repertoire of bursting patterns during the development of cortical cultures', *BMC Neuroscience* 7 (11). doi: 10.1186/1471-2202-7-11.

Wyart, C., Ybert, C., Bourdieu, L., Herr, C., Prinz, C., and Chatenay, D. (2002) 'Constrained synaptic connectivity in functional mammalian neuronal networks grown on patterned surfaces', *J. Neurosci. Methods* 117 (2), 123–31.

Xie, C., Lin, Z., Hanson, L., Cui, Y., and Cui, B. (2012), 'Intracellular recording of action potentials by nanopillar electroporation', *Nature Nanotechnol.* 7, 185–90. doi:10.1038/nnano.2012.8.

Zeck, G. and Fromherz, P. (2001), 'Noninvasive neuroelectronic interfacing with synaptically connected snail neurons immobilized on a semiconductor chip', *Proc. Natl. Acad. Sci. USA* 98 (18), 10457–62.

10.7 Appendix: experimental section

This secion gives the basics of neuronal culture on adhesive patterns and of the different types of neuron fixations, including the protocol used for observations using a scanning electron microscope.

Basics of brain mammalian neuronal culture derived from Banker (Goslin and Banker, 1991): Tissues (hippocampus or cortex) are isolated from E18 embryos under a dissecting microscope using a fine forceps and scissors and then deposited in HBSS (Hank's Balanced Salt Solution). Cells undergo a chemical dissociation with trypsin (0.05% trypsin/EDTA 0.53 mM) at 37 °C for 15 minutes followed by mechanical dissociation by successive pipettings. After counting, the cell suspension is diluted and deposited on the adhesive substrates. A typical cell density of 15 cells per square millimetre, or 1500 cells per centimetre square, is usually used to grow organized neuronal networks. A few hours later, the initial MEM/10% foetal bovine (or horse) serum medium is replaced by a supplemented (usually with B-27 and Glutamax) serum-free Neurobasal medium. The culture is maintained in the incubator at 37 °C in an atmosphere saturated with humidity containing 5% of CO_2. The presence of glial cells must be limited in these preparations, as these cells secrete diffusive factors in the solution that attract neurons and provide them with an extracellular matrix that disrupts the initial adhesive pattern geometry. The anti-mitotic drug AraC (1 μg/mL) is therefore added before 3DIV. All products from Invitrogen (culture mediums) and Sigma (AraC).

Fixation and immunostaining: methods may differ depending on the antigens to label and on the method of observation (optical or electronic microscopies). Three types of fixation protocols are mainly used: 1 methanol: the coverslips are incubated for a few minutes in anhydrous methanol maintained at −20 °C. This organic solvent dissolves fats, dehydrates cells and precipitates the proteins in a way that respects the cellular architecture; 2 paraformaldehyde (PFA): the plates are incubats at 37 °C for about half an hour with a solution of PFA/sucrose (4% paraformaldehyde, 120 mM sucrose, PBS). After washing in PBS, cells are permeabilized for a few minutes in PBS/Triton X-100 0.2%. Triton is then eliminated by rinsing with PBS. PFA form intermolecular bridges and better preserve membrane structures than methanol; 3 in the particular case of imaging with a scanning electron microscope, the PFA protocol is used without the step of membrane permeabilization (i.e. without Triton). The sample is then dehydrated by successive dippings in 50:50 volume of water and acetone then pure acetone (for a few minutes) followed by immersion in 50:50 solution of HMDS (hexamethyldisilazane) and acetone before a last step in pure HMDS. The samples are then allowed to dry slowly under a hood. HMDS is a volatile solvent whose surface tension is particularly low, and therefore can evaporate without altering the cell's structure.

Two examples of axonal primary antibodies: mouse mAbs against Ankyrin G (Santa Cruz) to label the initial axonal segment (Kordeli *et al.*, 1995; see Fig. 10.5(d)), Tau (clone tau-1, Millipore) to label the axon outside the initial segment. Secondary antibodies can be purchased from Molecular Probes, USA.

11

Semiconducting silicon nanowires and nanowire composites for biosensing and therapy

E. SEGAL and Y. BUSSI, Technion,
Israel Institute of Technology, Israel

DOI: 10.1533/9780857097712.3.214

Abstract: Over the past decade, significant research effort has been directed towards the synthesis and characterization of silicon nanowires. Because of their many advantageous properties, these nanowires have emerged as promising materials for both sensing and biosensing applications. This chapter provides an up-to-date overview of the fabrication routes of silicon nanowires, two-dimensional silicon nanowire arrays and their application as biosensing platforms. In addition, we present recent progress on the design and fabrication of silicon nanowire–polymer composites and some new exciting biomedical-relevant applications, which we believe have the potential to be developed and adapted in future biosensing platforms.

Key words: semiconductor nanowires, silicon nanowire-based biosensors, silicon nanowire–polymer composites.

11.1 Introduction

In recent years, silicon nanowires (SiNW) have emerged as promising materials for a multitude of nanotechnology-related applications because of their unique structural, electrical, optical and thermoelectric properties, in addition to their compatibility with current silicon-based microelectronic devices (Peng and Lee, 2011; Kenry and Lim, 2013). Specifically, SiNW provide a promising platform that combines tailored sensitivity and selectivity for realizing biosensing applications (Ramgir *et al.*, 2010).

The following sections will address issues of SiNW fabrication, and their organization into aligned two-dimensional arrays and complex architectures. Thereafter, we present the basic concepts in design of SiNW-based biosensors focusing on SiNW field effect transistors, highlighting their real-time, label-free, multiplexing and femtomolar level accuracy features. Finally, we focus on the emerging topic of SiNW–polymer composites in terms of their fabrication, device and integration, and present some of their biomedical applications.

11.2 Fabrication of silicon nanowires (SiNWs) and two-dimensional SiNW architectures

Silicon nanowires, having different diameters and lengths, can be synthesized and arranged on a variety of substrates (Joshi and Schneider, 2012). There are two main approaches for the fabrication of SiNW: bottom-up and top-down (Bashouti *et al.*, 2013). The bottom-up approach is an assembly process in which Si atoms are joined to form SiNW and it includes the famous vapor–liquid–solid (VLS) mechanism of single-crystal growth (Bashouti *et al.*, 2013; Schmidt *et al.*, 2009). The VLS method introduced by Wagner and Ellis in 1964 (Wagner and Ellis, 1964) is still the most commonly practised procedure to synthesize SiNW (Schmidt *et al.*, 2009). In this process, a metal catalyst initiates and guides growth of the nanowire. Briefly, a metal catalyst, for example Au, forms liquid alloy droplets at a high temperature by adsorbing the precursor vapor components. The catalyst liquid droplet is intentionally introduced to promote the anisotropic crystal growth of the nanowires and, simultaneously, suppress their lateral growth (Wu and Yang, 2001; Kenry and Lim, 2013; Ramgir *et al.*, 2010). Once the one-dimensional (1D) crystal growth begins, it continues for as long as the vapor components are supplied. The diameter and position of the resulting nanowires can be controlled by the size and position of the catalyst, as the liquid phase is confined to the area of the precipitated solid phase (Choi, 2012). A large number of process variations have been introduced in the case of VLS, including epitaxial chemical vapor deposition (CVD) (Joshi and Schneider, 2012), laser ablation and carbothermal reduction (Choi, 2012; Ramgir *et al.*, 2010). It should be noted that the major limitations associated with these 'growth' methods are the need for high temperature or high vacuum, templates and complex equipment, and the use of hazardous silicon precursors (Peng *et al.*, 2002). These and related issues are described in additional detail in Chapter 2.

When considering the top–down approach, one should distinguish between the fabrication of horizontal nanowires, that is nanowires lying in the substrate plane, or the fabrication of vertical nanowires that are oriented perpendicular to the surface (Schmidt *et al.*, 2009). The latter nanowires are commonly fabricated by conventional silicon technology processes, such as reactive ion etching (RIE) to etch vertical SiNW out of a silicon wafer, whereas the diameter of the NW is defined by a preceding lithography step (Bashouti *et al.*, 2013; Schmidt *et al.*, 2009). Horizontal SiNW are mostly fabricated from either silicon-on-insulator wafers or bulk silicon wafers using a sequence of lithography and etching steps (Joshi and Schneider, 2012; Schmidt *et al.*, 2009).

To realize the full potential of SiNW, the nanowires must be integrated efficiently and economically into a variety of device configurations and architectures. Indeed, devices have been constructed around single, or several dispersed SiNW (Cui *et al.*, 2001) and methods have been developed to manipulate nanowires into geometries amenable to large-scale device fabrication (Whang *et al.*, 2003; Kelzenberg *et al.*, 2010; Mulvihill *et al.*, 2005). Several approaches to assemble nanowire arrays on various substrates have been reported in the literature. The main routes will be briefly presented in the following paragraphs. These methods have been recently reviewed by Joshi and Schneider (2012), Long *et al.* (2012) and Liu *et al.* (2012), and are also described in detail in Chapter 9 of this book.

Horizontally aligned SiNW arrays can be prepared by flow-assisted alignment in microchannels (Huang *et al.*, 2001), the Langmuir-Blodgett method and bubble-blown techniques (Liu *et al.*, 2012). Vertically aligned SiNW arrays have been fabricated with low cost and high throughput by several catalytic templated etching processes (Huang *et al.*, 2009; 2007) or metal-assisted chemical etching techniques (Huang *et al.*, 2012; Zhang *et al.*, 2008). Vertically grown SiNW have been selectively functionalized and transferred to different substrates to form 2D thin films of ordered nanowires. This can be achieved by transferring the SiNW from the original growth substrate by dry transfer processes (Shiu *et al.*, 2009; Joshi and Schneider, 2012). Weisse *et al.* (2011) demonstrated vertical transfer of SiNW with uniform length onto adhesive substrates by creating a horizontal crack throughout SiNW, which enables the breakage of SiNW with uniform lengths. Another method developed by Javey and co-workers (Fan *et al.*, 2007) is the contact printing process, which enables direct transfer and positioning of NW from a donor substrate to a chemically modified receiver support.

These transfer processes, allowing the integration of nanowires with flexible substrates, could enable exciting avenues in fundamental research and novel applications and, as such, are of vital relevance in chemical and biological sensing applications (Joshi and Schneider, 2012; Mcalpine *et al.*, 2007). Three-dimensional nanowire architectures also have been constructed by sequential assembly of NW into vertically stacked device layers (Javey *et al.*, 2007).

11.3 SiNWs for biosensing applications

The 1D configuration of SiNW together with their high aspect ratios have significant advantages for biosensing applications because the interaction between the target analytes and the SiNW surface are rapidly translated to electrical outputs (Joshi and Schneider, 2012; Lee *et al.*, 2012). Indeed, extensive research has been directed towards the design of simple, fast, label-free and low-cost SiNW-based sensors and biosensors with high sensitivity and selectivity (Joshi and Schneider, 2012; Kenry and Lim, 2013; Lee *et al.*, 2012; Ramgir *et al.*, 2010; Vu *et al.*, 2010).

The most common biosensor configuration to date is the SiNW field effect transistor (SiNW-FET) (Chen *et al.*, 2011; Ramgir *et al.*, 2010; Vu *et al.*, 2010). Highly sensitive and selective SiNW-FET biosensors have been reported, demonstrating the detection of DNA hybridization (Gao *et al.*, 2007), proteins (Cui *et al.*, 2001), cell signaling, viruses (Patolsky *et al.*, 2004) and several other target analytes of interest (Lee *et al.*, 2012; Chen *et al.*, 2011). The following section will present the different designs of SiNW-FET biosensors and discuss some of the important aspects in the future utilization of these devices.

The SiNW-FET biosensor is a three-electrode system which includes source, drain and gate electrodes. The SiNW, which serves as the semiconductor channel, is placed between the source and the drain electrodes with a gate electrode on the bottom (Chen *et al.*, 2011; Ramgir *et al.*, 2010). The first SiNW-FET biosensor was reported by the Lieber group (Cui *et al.*, 2001) and demonstrated sensitive protein binding. Figure 11.1(a) schematically illustrates the general concept of their biosensing scheme. SiNW are functionalized with biotin, which is used as a capture probe for streptavidin (SA). The conductance of the biotin-modified SiNW rapidly increases (Fig. 11.1(b)) on the introduction of SA solution. The conductance increase is attributed to the binding of negatively charged species to the *p*-type SiNW surface. Control experiments with unmodified SiNW (Fig. 11.1(c)) did not produce any change in conductance and confirmed that observed signals are caused by the specific interaction between SA and the surface-bound biotin.

Following this seminal work, many other SiNW-FET biosensors were reported (Patolsky *et al.*, 2006; Kenry and Lim, 2013; Lee *et al.*, 2012; Ramgir *et al.*, 2010; Roy and Gao, 2009), in which specific binding events occurring on the

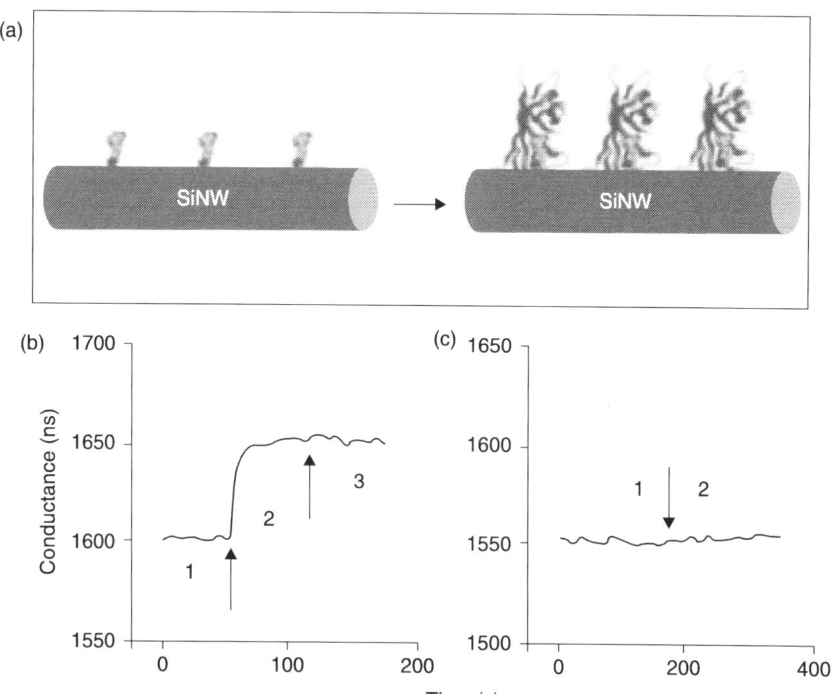

11.1 Detection of protein binding to biotin-modified silicon nanowires (SiNW). (a) Schematic illustration of streptavidin (SA) binding to biotin-modified SiNW surface. (b) Conductance versus time showing an increase in conductance upon exposure to 250 nM of SA at point 2. (c) Control experiment, conductance versus time of non-modified SiNW to the same concentration of SA. Adapted with permission from Cui *et al.*, 2001.

nanowire surface led to depletion or accumulation of carriers onto the SiNW, as schematically illustrated in Fig. 11.2, and consequently, varying the surface potential of the semiconductor channel and modulating the channel conductance.

In terms of SiNW-FET device fabrication, SiNW are prepared according to the methods described in the previous section, using both bottom-up and top-down approaches. The top-down approach is more complex when compared with bottom-up procedures because the process relies on high-resolution lithography, and thus requires expensive equipment and facilities (Chen *et al.*, 2011). Indeed, the advantages of the bottom-up approach are cost-effective fabrication of SiNW with high crystallinity, designated dopant density and easily controlled diameters. Nonetheless, the uniform and reproducible assembly of the synthesized SiNW onto the support substrate is still a challenge. Detailed fabrication techniques for preparing SiNW-FET biosensors were recently reviewed by Chen *et al.* (Chen *et al.*, 2011), with salient details of these processes covered in Chapter 9 of this book.

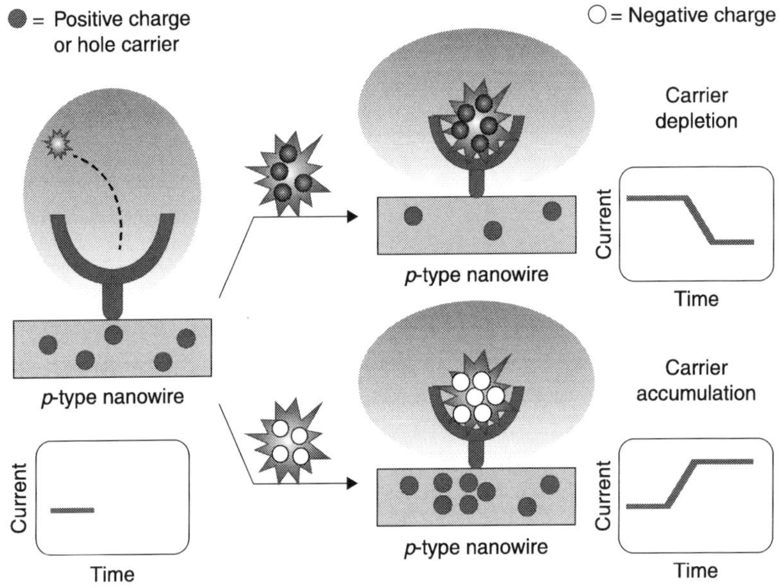

11.2 Schematic illustration of silicon nanowires (SiNW)-field effect transistor (FET) biosensor behavior on binding of charged target molecules to the receptor-modified SiNW. When positively charged target molecules bind the receptor modified on a *p*-type NW, positive carriers are depleted in the NW, resulting in a conductance decrease. Conversely, when negatively charged target molecules are captured, an accumulation of hole carriers occurs, resulting in a conductance increase. Reproduced with permission from Chen *et al.*, 2011.

The general concept in the design of SiNW-FET biosensors is to anchor an appropriate bioreceptor, for example antibody (Patolsky *et al.*, 2006), DNA sequence (Hahm and Lieber, 2003; Li *et al.*, 2004), enzyme (Chen *et al.*, 2011), onto the SiNW surface. The surface functionalization of SiNW with the different bioreceptors commonly employs the diverse chemistry of silicon oxide, which naturally forms on the surface of the SiNW (Ramgir *et al.*, 2010; Lee *et al.*, 2012; Wanekaya *et al.*, 2006). Typically, before surface modification, the surface of the SiNW is treated with water-vapor plasma, to clean the nanowire surface and generate a hydrophilic surface by hydroxyl-terminating the silicon oxide surface. Then this hydroxide layer is activated via silanol chemistry to achieve a specific surface functionality (Wanekaya *et al.*, 2006). Another possible surface functionalization route is to use hydrosilylation chemistry to form Si–C bonds with different alkyl chains (Bashouti *et al.*, 2013; Bunimovich *et al.*, 2006). Several studies have pointed out that SiNW without the native oxide exhibit improved solution-gated FET characteristics and significantly enhanced sensitivity in biologically relevant media (Bashouti *et al.*, 2013; Bunimovich *et al.*, 2006). These results are ascribed to the higher surface coverage and improved stability.

SiNW arrays have been extensively utilized for developing SiNW-FET biosensors. Arrays of SiNW have been used to detect protein cancer markers (Zheng *et al.*, 2005). Label-free, real-time, multiplexed detection of these markers with high selectivity and sensitivity is demonstrated using antibody-functionalized integrated SiNW arrays. Negatively charged proteins lead to a change in electrical conductance in both *n*- and *p*-type FET devices either by accumulation or by depletion of charge carriers. The device sensitivity is as low as femtomolar concentrations, significantly lower than other biosensing schemes such as surface plasmon resonance (SPR). Another example by the same group (Patolsky *et al.*, 2004) is the ability to detect a single virus by using SiNW arrays modified with antibodies for *influenza A*. In addition to protein–protein interactions, SiNW-FET arrays were also adapted for the detection of DNA and RNA (Chen *et al.*, 2011). Label-free detection of DNA at femtomolar levels has also been demonstrated using arrays of highly ordered *n*-type SiNW (Gao *et al.*, 2007). Again the sensitivity of this biosensing scheme is far better than other label-free techniques, such as SPR.

An interesting recent work by Yang's group (Jeong *et al.*, 2013) reports binding of a model Gram-negative bacteria species, *Shewanella oneidensis*, to a SiNW array platform. Nanoscale topographies on these surfaces were shown to play an important role during the early stage of biofilm formation. It should be noted that this work did not demonstrate biosensing of bacteria. However, it reveals the potential of SiNW arrays as future biosensing platforms for microorganisms.

Crucial challenges in the realization of these novel biosensing schemes in practical rapid laboratory-based or 'point-of-care' diagnostic or therapeutic applications are: 1 measurements in clinically relevant samples, for example whole blood or serum, caused by the screening effect of counter-ions (Debye screening). SiNW-FET sensitivity dependence on the buffer composition is an important limitation for future applications of nano-biosensors when fast detection is required (Bunimovich *et al.*, 2006; Chen *et al.*, 2011); 2 developing efficient methods for sample/target analyte delivery to the biosensor interface and proper device integration (Joshi and Schneider, 2012); 3 studying device stability and baseline drifts; and 4 ensuring reliability and reproducibility in cost-effective device fabrication. Some of these critical issues are addressed in greater detail in Chapter 12 of this book.

11.4 Fabrication of SiNW-polymer composite systems

The incorporation of organic materials with nanostructured inorganic scaffolds opens opportunities to engineer advanced materials with highly tunable properties, for example mechanical, chemical, optical and electrical properties (Bonanno and Segal, 2011). Specifically, integration of SiNW with polymers to form nanocomposites may result in new materials that combine the unique properties of SiNW with the flexibility and processability of polymers (Fan *et al.*, 2007; Plass *et al.*, 2009). These nanowire-based composites can be synthesized and fabricated in a variety of designs and architectures, as schematically illustrated in a Fig. 11.3.

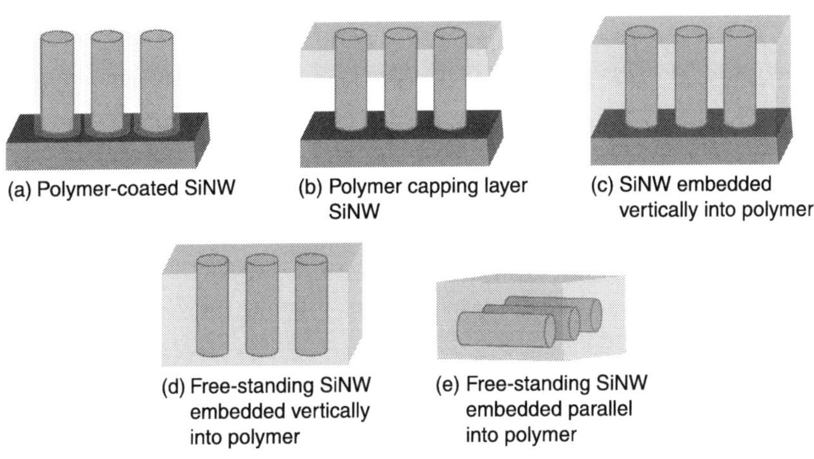

11.3 Strategies for the design of silicon nanowire (SiNW)–polymer composites. Please note that schematics are not drawn to scale. (a) Polymer-coated SiNW; (b) polymer capping layer SiNW; (c) SiNW embedded vertically into polymer; (d) free-standing SiNW embedded vertically into polymer; and (e) free-standing SiNW embedded parallel into polymer.

For many applications, for example solar cells, sensing, optical and electronics devices, it is beneficial to embed the highly fragile nanowires within a polymeric matrix, which provides structural stability to the array. For example, polymer-supported arrays of SiNW (as illustrated in Fig. 11.3(c)) were embedded within a transparent, mechanically and chemically robust, polydimethyl-si-loxane (PDMS) (Plass *et al.*, 2009). An important consideration is that the chosen organic material (either a polymer or a monomer) would be deposited in such a manner that it properly fills the space between the high aspect ratio nanowires while maintaining their shape and architecture. Another device requirement is to ensure that the polymer possesses low thermal conductivity, preferably one or more orders of magnitude lower than that of Si. This would maintain an overall low thermal conductivity of the nanowire composite, a criterion essential for superior thermoelectric device performance (Feser *et al.*, 2012). Moreover, as the device fabrication requires that contacts be deposited on at least the top surface, the polymeric component must be compatible with standard microfabrication processes (e.g. chemical resistance to standard lithography reagents), and particularly able to withstand a high processing temperature up to 300 °C (Winther-Jensen *et al.*, 2008; Somboonsub *et al.*, 2010). In order to achieve the above, several techniques that induce a high density of organic molecule attachment to the Si surface have been suggested. These techniques can be categorized into two main approaches: 1 subsequent chemical modification of the SiNW by grafting methodologies (Wang *et al.*, 2011), including vapor-phase polymerization (Li *et al.*, 2013; Abramson *et al.*, 2004); 2 incorporation of polymers as bridging components directly and specifically onto the SiNW walls/matrix array. The latter is carried out by casting (Lu *et al.*, 2011), electrodeposition, printing or spray coating (Morozova *et al.*, 2013) methods.

The first approach includes a promising method for controlling the surface properties of SiNW by multiple grafting steps via chlorination–alkylation processes, to create, for example, Si–O, Si–C, Si–N bonds (Bashouti *et al.*, 2013). Additionally, surface-initiated polymerization emerges as an important tool to functionalize the surface of SiNW by providing a route to covalently attach polymer chains in a well-controlled fashion to tailor the properties of the resulting composite material, such as stability, wetting and aggregation of the nanostructures. Atom transfer radical polymerization (ATRP) has been used to grow polymers from a variety of surfaces (Mulvihill *et al.*, 2005; Yuan *et al.*, 2008). An additional route is to form conformal coating of protecting layers on SiNW arrays by vapor-phase polymerization. Li *et al.* (2013) have recently fabricated SiNW/PEDOT (poly(3,4-ethylenedioxythiophene)) composites in which the polymer coats the nanowires, as illustrated in Fig. 11.3(a). First, the oxidant iron(III) *p*-toluenesulfonate hexahydrate is spin-coated onto the surface of wet chemically etched SiNW arrays. Then, the monomer 3,4-ethylenedioxythiophene in the vapor phase is polymerized on the surface, forming a conformal layer of PEDOT on the SiNW array. Subsequently, an APTES (3-aminopropyltriethoxysilane) monolayer is used to modify the SiNW surface before PEDOT polymerization to improve polymer adhesion.

The second category includes a wide variety of physical/mechanical techniques aimed at depositing the polymer component to either embedded (Plass *et al.*, 2009) or coated (Moiz *et al.*, 2012) nanowire arrays to form different architectures. As these methodologies are more common in solar cell technologies, they will not be discussed in this chapter.

11.5 Biomedical applications of SiNW-polymer composites

The successful assembly of Si-based nanowires into hierarchical structures is a key step necessary towards commercially attractive materials and devices displaying various electrical, optical, optoelectronic and energy applications. Ongoing research is mostly aimed towards improvements in the synthesis and fabrication of high performance solar cells (Shen *et al.*, 2011; Zhang *et al.*, 2011; Syu *et al.*, 2012; Lu *et al.*, 2011; Zhang *et al.*, 2012; Wright and Uddin, 2012). To the best of our knowledge, SiNW-polymer systems have been scarcely studied as potential materials for biosensing applications. The main challenge is the ability to assemble these nanomaterials into a functional biosensor and to individually address each nanostructured sensing element with the desired bioreceptor, a requirement necessary for the successful fabrication of nanosensor arrays (Roy and Gao, 2009). Thus, in the following sections we will present some exciting new biomedical applications for SiNW–polymer systems, which may in the future be integrated or developed into biosensing schemes.

11.5.1 Protein patterning and microarray-based immunoassays

Han *et al.* (2013) have fabricated multiscale substrates based on micropatterned nanostructures by overlaying SiNW with poly(ethylene glycol) (PEG) hydrogels. Figure 11.4 schematically outlines the fabrication of these composite arrays. A vertically aligned SiNW array is micropatterned with PEG hydrogel to define micro-wells. Thus, the bottom of the well consists of APTES-modified SiNW (Fig.11.4(a)), while the PEG hydrogel forms the walls, which surround each modified-SiNW well (Fig. 11.4(b)). The non-adhesive character of PEG hydrogels towards proteins allows for proteins to be selectively immobilized on the surface-modified SiNW regions to create protein micropatterns. These composite arrays are used for performing simple immunobinding assays (between IgG and anti-IgG or IgM and anti-IgM). The SiNW micropatterns are found to emit higher fluorescence intensity and show higher sensitivity in comparison with corresponding planar silicon substrates.

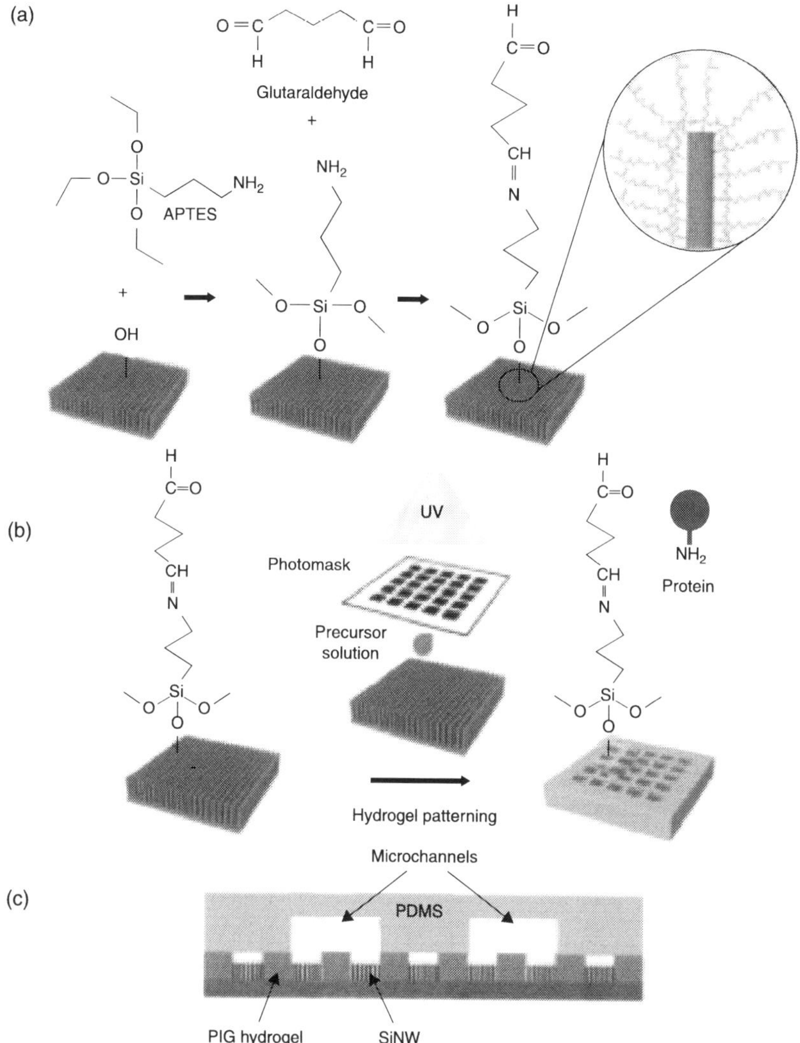

11.4 Schematic fabrication of PEG hydrogel micropatterns and protein immobilization. Reproduced with permission from Han *et al.*, 2013. (a) Proteins are immobilized onto the surface of the SiNWs. (b) Hydrogel micropatterns are fabricated on surface-modified SiNWs. (c) PDMS-based microchannels are integrated with micropatterned SiNWs.

11.5.2 Applications of SiNW field effect transistor (FET) arrays fabricated on polymeric supports

An emerging research area is the development of electronic artificial skin systems (commonly termed as e-skin) (Takei *et al.*, 2010). Flexibility is a crucial character for enabling the fabrication of this kind of curved transducer array surface, especially for mimicking the tactile sensing properties of human skin for prosthetic devices (Park *et al.*, 2012). The ability to effectively align, assemble and transfer SiNW on flexible and stretchable substrates is thus of great importance. Takei *et al.* (2010) have fabricated a new flexible pressure-sensor array which functions as an electro-mechanical system for e-skin. Although organic base materials systems can show potential for mechanical flexible electronics, unfortunately their poor semiconductor properties often require large operating voltages (Park *et al.*, 2012; Takei *et al.*, 2010). Thus, the combination of a large-scale uniform assembly SiNW array laminated on a flexible polyimide substrate, assembled using a contact-printing method, is the key to the e-skin's operation. The process scheme for the e-skin device with an integrated nanowire active-matrix backplane is shown on Plate VIII(a) (see colour section between pp. 94 and 95), and an optical image of a fully fabricated e-skin is depicted on Plate VIII(b). The structure can easily be bent or rolled to a small radius of curvature, demonstrating the superior mechanical flexibility of the substrate and its integrated electronic components. Future attractive applications for these e-skin systems include mobile biomonitoring in long-term medical diagnostics and to restore the sense of touch to patients with prosthetic limbs.

Another exciting application of flexible SiNW-FET arrays was demonstrated by Timko *et al.* (2009). In this work, SiNW-FET arrays fabricated on both planar and flexible polymeric substrates are interfaced with spontaneously beating embryonic chicken hearts in both planar and bent conformations. Multiplexed arrays recording from SiNW-FET arrays can be mapped and yielded signal propagation times across the myocardium with high spatial resolution. The signals are synchronized with the beating heart, and the signal amplitude is directly related to the device sensitivity. Another unique feature of these flexible and transparent SiNW-FET arrays is they enable simultaneous optical imaging and electronic recording in configurations that are not readily accessible with traditional planar device chips, yet advantageous for producing diverse, functional tissue–device interfaces (Joshi and Schneider, 2012; Timko *et al.*, 2009).

11.6 Conclusions and future trends

Over the past decade we have witnessed significant progress in the development of SiNW-based biosensors. Much work has been focused towards the fabrication of SiNW, SiNW arrays and their integration in the device level to demonstrate biosensing functions towards various relevant targets. These biosensing platforms exhibit several significant advantages in comparison to current methodologies, including ultrahigh sensitivity, label-free and real-time electrical detection. Clearly, the technological and scientific potential of these nanomaterials is immense. We have only started to see how multi-phase (nanowires and organic) composites can enhance the functionality of these nanomaterials. Thus, we anticipate that these new composites will have a profound impact on new applications.

In the near future, we expect to see further research aimed at overcoming sensitivity and specificity challenges posed by diagnostic testing in complex biological fluids (including serum, blood and urine). The latter may pave the way to realizing the vision of personalized medicine.

11.7 References

Abramson, A. R., Woo Chul, K., Huxtable, S. T., Haoquan, Y., Yiying, W., *et al.* (2004) Fabrication and characterization of a nanowire/polymer-based nanocomposite for a prototype thermoelectric device, *Microelectromechanical Systems*, 13, 505–13.

Bashouti, M. Y., Sardashti, K., Schmitt, S. W., Pietsch, M., Ristein, J., *et al.* (2013) Oxide-free hybrid silicon nanowires: from fundamentals to applied nanotechnology, *Progress in Surface Science*, 88, 39–60.

Bonanno, L. M. and Segal, E. (2011) Nanostructured porous silicon-polymer-based hybrids: from biosensing to drug delivery, *Nanomedicine*, 6, 1755–70.

Bunimovich, Y. L., Shin, Y. S., Yeo, W.-S., Amori, M., Kwong, G. and Heath, J. R. (2006) Quantitative real-time measurements of DNA hybridization with alkylated nonoxidized silicon nanowires in electrolyte solution, *Journal of the American Chemical Society*, 128, 16323–31.

Chen, K.-I., Li, B.-R. and Chen, Y.-T. (2011) Silicon nanowire field-effect transistor-based biosensors for biomedical diagnosis and cellular recording investigation, *Nano Today*, 6, 131–54.

Choi, H.-J. (2012) In 'Vapor–Liquid–Solid Growth of Semiconductor Nanowires'. *Semiconductor nanostructures for optoelectronic devices*. Springer: Berlin, Heidelberg.

Cui, Y., Wei, Q., Park, H. and Lieber, C. M. (2001) Nanowire nanosensors for highly sensitive and selective detection of biological and chemical species, *Science*, 293, 1289–92.

Fan, Z., Ho, J. C., Jacobson, Z. A., Yerushalmi, R., Alley, R. L., *et al.* (2007) Wafer-scale assembly of highly ordered semiconductor nanowire arrays by contact printing, *Nano Letters*, 8, 20–5.

Feser, J. P., Sadhu, J. S., Azeredo, B. P., Hsu, K. H., Ma, J., *et al.* (2012) Thermal conductivity of silicon nanowire arrays with controlled roughness, *Journal of Applied Physics*, 112, 114306–7.

Gao, Z., Agarwal, A., Trigg, A. D., Singh, N., Fang, C., *et al.* (2007) Silicon nanowire arrays for label-free detection of DNA, *Analytical Chemistry*, 79, 3291–7.

Hahm, J.-I. and Lieber, C. M. (2003) Direct ultrasensitive electrical detection of DNA and DNA sequence variations using nanowire nanosensors, *Nano Letters*, 4, 51–4.

Han, S. W., Lee, S., Hong, J., Jang, E., Lee, T. and Koh, W.-G. (2013) Multiscale substrates based on hydrogel-incorporated silicon nanowires for protein patterning and microarray-based immunoassays, *Biosensors and Bioelectronics*, 45, 129–35.

Huang, Y., Duan, X., Wei, Q. and Lieber, C. M. (2001) Directed assembly of one-dimensional nanostructures into functional networks, *Science*, 291, 630–3.

Huang, Z., Fang, H. and Zhu, J. (2007) Fabrication of silicon nanowire arrays with controlled diameter, length, and density, *Advanced Materials*, 19, 744–8.

Huang, Z., Shimizu, T., Senz, S., Zhang, Z., Zhang, X., *et al.* (2009) Ordered arrays of vertically aligned [110] silicon nanowires by suppressing the crystallographically preferred <100> etching directions, *Nano Letters*, 9, 2519–25.

Huang, Z., Wang, R., Jia, D., Maoying, L., Humphrey, M. G. and Zhang, C. (2012) Low-cost, large-scale, and facile production of Si nanowires exhibiting enhanced third-order optical nonlinearity, *ACS Applied Materials and Interfaces*, 4, 1553–9.

Javey, A., Nam, S., Friedman, R. S., Yan, H. and Lieber, C. M. (2007) Layer-by-layer assembly of nanowires for three-dimensional, multifunctional electronics, *Nano Letters*, 7, 773–7.

Jeong, H. E., Kim, I., Karam, P., Choi, H.-J. and Yang, P. (2013) Bacterial recognition of silicon nanowire arrays, *Nano Letters*, 13, 2864–9.

Joshi, R. K. and Schneider, J. J. (2012) Assembly of one dimensional inorganic nanostructures into functional 2D and 3D architectures. Synthesis, arrangement and functionality, *Chemical Society Reviews*, 41, 5285–312.

Kelzenberg, M. D., Boettcher, S. W., Petykiewicz, J. A., Turner-Evans, D. B., Putnam, M. C., *et al.* (2010) Enhanced absorption and carrier collection in Si wire arrays for photovoltaic applications, *Nature Materials* 9, 239–44.

Kenry and Lim, C. T. (2013) Synthesis, optical properties, and chemical–biological sensing applications of one-dimensional inorganic semiconductor nanowires, *Progress in Materials Science*, 58, 705–48.

Lee, S., Sung, J. and Park, T. (2012) Nanomaterial-based biosensor as an emerging tool for biomedical applications, *Ann Biomed Eng*, 40, 1384–97.

Li, X., Lu, W., Dong, W., Chen, Q., Wu, D., *et al.* (2013) Si/PEDOT Hybrid core/shell nanowire array as photoelectrode for photoelectrochemical water-splitting, *Nanoscale*, 5, 5257–61.

Li, Z., Chen, Y., Li, X., Kamins, T. I., Nauka, K. and Williams, R. S. (2004) Sequence-specific label-free DNA sensors based on silicon nanowires, *Nano Letters*, 4, 245–7.

Liu, X., Long, Y.-Z., Liao, L., Duan, X. and Fan, Z. (2012) Large-scale integration of semiconductor nanowires for high-performance flexible electronics, *ACS Nano*, 6, 1888–900.

Long, Y.-Z., Yu, M., Sun, B., Gu, C.-Z. and Fan, Z. (2012) Recent advances in large-scale assembly of semiconducting inorganic nanowires and nanofibers for electronics, sensors and photovoltaics, *Chemical Society Reviews*, 41, 4560–80.

Lu, W., Wang, C., Yue, W. and Chen, L. (2011) Si/PEDOT:PSS core/shell nanowire arrays for efficient hybrid solar cells, *Nanoscale*, 3, 3631–4.

McAlpine, M. C., Ahmad, H., Wang, D. and Heath, J. R. (2007) Highly ordered nanowire arrays on plastic substrates for ultrasensitive flexible chemical sensors, *Nature Materials*, 6, 379–84.

Moiz, S. A., Nahhas, A. M., Um, H.-D., Jee, S.-W., Cho, H. K., *et al.* (2012) A stamped PEDOT:PSS–silicon nanowire hybrid solar cell, *Nanotechnology*, 23, 145401.

Morozova, M., Kluson, P., Dzik, P., Vesely, M., Baudys, M., *et al.* (2013) The influence of various deposition techniques on the photoelectrochemical properties of the titanium dioxide thin film, *J Sol-Gel Sci Technol*, 65, 452–8.

Mulvihill, M. J., Rupert, B. L., He, R., Hochbaum, A., Arnold, J. and Yang, P. (2005) Synthesis of bifunctional polymer nanotubes from silicon nanowire templates via atom transfer radical polymerization, *Journal of the American Chemical Society*, 127, 16040–1.

Park, S., Wang, G., Cho, B., Kim, Y., Song, S., *et al.* (2012) Flexible molecular-scale electronic devices, *Nature Nanotechnology*, 7, 438–42.

Patolsky, F., Zheng, G., Hayden, O., Lakadamyali, M., Zhuang, X. and Lieber, C. M. (2004) Electrical detection of single viruses, *Proceedings of the National Academy of Sciences of the United States of America*, 101, 14017–22.

Patolsky, F., Zheng, G. and Lieber, C. M. (2006) Fabrication of silicon nanowire devices for ultrasensitive, label-free, real-time detection of biological and chemical species, *Nature Protocols*, 1, 1711–24.

Peng, K.-Q. and Lee, S.-T. (2011) Silicon nanowires for photovoltaic solar energy conversion, *Advanced Materials*, 23, 198–215.

Peng, K. Q., Yan, Y. J., Gao, S. P. and Zhu, J. (2002) Synthesis of large-area silicon nanowire arrays via self-assembling nanoelectrochemistry, *Advanced Materials*, 14, 1164–7.

Plass, K. E., Filler, M. A., Spurgeon, J. M., Kayes, B. M., Maldonado, S., *et al.* (2009) Flexible polymer embedded Si wire arrays, *Advanced Materials*, 21, 325–8.

Ramgir, N. S., Yang, Y. and Zacharias, M. (2010) Nanowire-based sensors, *Small*, 6, 1705–22.

Roy, S. and Gao, Z. (2009) Nanostructure-based electrical biosensors, *Nano Today*, 4, 318–34.

Schmidt, V., Wittemann, J. V., Senz, S. and Gösele, U. (2009) Silicon nanowires: a review on aspects of their growth and their electrical properties, *Advanced Materials*, 21, 2681–702.

Shen, X., Sun, B., Liu, D. and Lee, S.-T. (2011) Hybrid heterojunction solar cell based on organic–inorganic silicon nanowire array architecture, *Journal of the American Chemical Society*, 133, 19408–15.

Shiu, S.-C., Hung, S.-C., Chao, J.-J. and Lin, C.-F. (2009) Massive transfer of vertically aligned Si nanowire array onto alien substrates and their characteristics, *Applied Surface Science*, 255, 8566–70.

Somboonsub, B., Invernale, M. A., Thongyai, S., Praserthdam, P., Scola, D. A. and Sotzing, G. A. (2010) Preparation of the thermally stable conducting polymer PEDOT – sulfonated poly(imide), *Polymer*, 51, 1231–6.

Syu, H.-J., Shiu, S.-C. and Lin, C.-F. (2012) Silicon nanowire/organic hybrid solar cell with efficiency of 8.40%, *Solar Energy Materials and Solar Cells*, 98, 267–72.

Takei, K., Takahashi, T., Ho, J. C., Ko, H., Gillies, A. G., *et al.* (2010) Nanowire active-matrix circuitry for low-voltage macroscale artificial skin, *Nature Materials*, 9, 821–6.

Timko, B. P., Cohen-Karni, T., Yu, G., Qing, Q., Tian, B. and Lieber, C. M. (2009) Electrical recording from hearts with flexible nanowire device arrays, *Nano Letters*, 9, 914–18.

Vu, X. T., Ghoshmoulick, R., Eschermann, J. F., Stockmann, R., Offenhäusser, A. and Ingebrandt, S. (2010) Fabrication and application of silicon nanowire transistor arrays for biomolecular detection, *Sensors and Actuators B: Chemical*, 144, 354–60.

Wagner, R. S. and Ellis, W. C. (1964) Vapor–liquid–solid mechanism of single crystal growth, *Applied Physics Letters*, 4, 89–90.

Wanekaya, A. K., Chen, W., Myung, N. V. and Mulchandani, A. (2006) Nanowire-based electrochemical biosensors, *Electroanalysis*, 18, 533–50.

Wang, H., Wang, L., Zhang, P., Yuan, L., Yu, Q. and Chen, H. (2011) High antibacterial efficiency of pDMAEMA modified silicon nanowire arrays, *Colloids and Surfaces B: Biointerfaces*, 83, 355–9.

Weisse, J. M., Kim, D. R., Lee, C. H. and Zheng, X. (2011) Vertical transfer of uniform silicon nanowire arrays via crack formation, *Nano Letters*, 11, 1300–05.

Whang, D., Jin, S., Wu, Y. and Lieber, C. M. (2003) Large-scale hierarchical organization of nanowire arrays for integrated nanosystems, *Nano Letters*, 3, 1255–9.

Winther-Jensen, B., Winther-Jensen, O., Forsyth, M. and Macfarlane, D. R. (2008) High rates of oxygen reduction over a vapor phase–polymerized PEDOT electrode, *Science*, 321, 671–4.

Wright, M. and Uddin, A. (2012) Organic–inorganic hybrid solar cells: a comparative review, *Solar Energy Materials and Solar Cells*, 107, 87–111.

Wu, Y. and Yang, P. (2001) Direct observation of vapor–liquid–solid nanowire growth, *Journal of the American Chemical Society*, 123, 3165–6.

Yuan, J., Xu, Y., Walther, A., Bolisetty, S., Schumacher, M., *et al.* (2008) Water-soluble organo-silica hybrid nanowires, *Nature Materials*, 7, 718–22.

Zhang, F., Song, T. and Sun, B. (2012) Conjugated polymer–silicon nanowire array hybrid Schottky diode for solar cell application, *Nanotechnology*, 23, 1–9.

Zhang, F., Sun, B., Song, T., Zhu, X. and Lee, S. (2011) Air stable, efficient hybrid photovoltaic devices based on poly(3-hexylthiophene) and silicon nanostructures, *Chemistry of Materials*, 23, 2084–90.

Zhang, M.-L., Peng, K.-Q., Fan, X., Jie, J.-S., Zhang, R.-Q., *et al.* (2008) Preparation of large-area uniform silicon nanowires arrays through metal-assisted chemical etching, *Journal of Physical Chemistry C*, 112, 4444–50.

Zheng, G., Patolsky, F., Cui, Y., Wang, W. U. and Lieber, C. M. (2005) Multiplexed electrical detection of cancer markers with nanowire sensor arrays, *Nat Biotech*, 23, 1294–301.

12

Probe-free semiconducting silicon nanowire platforms for biosensing

A. DE and S. CHEN, University of Twente, The Netherlands and
E. T. CARLEN, University of Tsukuba, Japan and
University of Twente, The Netherlands

DOI: 10.1533/9780857097712.3.229

Abstract: This chapter discusses implementation of silicon nanowires (SiNW) for probe-free biosensing and their integration with microfluidic chambers for small volume sample confinement and delivery to an automated biosensor platform. The SiNW biosensor platform consists of three key elements: SiNW sensors, the probe layer for specific binding to a particular target molecule and gate voltage generation, and controlled delivery of small sample volumes that enables quantitative hybridization kinetics assessment directly from the real-time measurements. Advantages and disadvantages of using the nanoscale SiNW biosensors, and approaches to reduce the limitations are discussed. The chapter concludes with future trends and overall outlook.

Key words: probe-free silicon nanowire biosensor, integrated microfluidic biosensor platform, real-time DNA hybridization, peptide nucleic acid (PNA) probe, Debye screening, DNA condensation, streaming potential.

12.1 Introduction

Probe-free, also called label-free, silicon nanowire (SiNW) biosensors have been reported extensively over the past decade for the detection of biomolecules binding to the sensor surface, such as protein-protein and antibody-antigen (Cui *et al.*, 2001) binding, DNA hybridization (Bunimovich *et al.*, 2006), and for the stimulation and recording of cellular bioelectricity (Patolsky *et al.*, 2006a). All of the SiNW biosensing platforms contain four essential elements that are integrated together to form a functional analysis system, which includes SiNW sensor arrays, probe layers that are chemically attached to the sensor surface that bind specifically to a particular target molecule in the sample solution, an integrated sample delivery system, and the electrical bias and signal measurement instrumentation.

SiNW-field effect transistor (FET) biosensors measure surface charge density changes on the gate sensor surface, from potential determining surface ions and/or the intrinsic electronic charge of hybridized biomolecule complexes attached to the sensor surface. The surface charge density changes electrostatically induce a gate voltage change, which is transduced into a measurable conductance change in the body of the SiNW sensors via a field effect. The small cross-sectional area and three-dimensional geometry of SiNW-FET sensors result in a large surface-to-volume ratio and multi-gate interface, which can give an increased sensitivity to surface charges compared with a microscale planar single gate configuration. As SiNW-FET sensors have a smaller sensor surface area compared with microscale sensors, a reduced number of biomolecule complexes are required to generate a comparable gate voltage, which can further increase their sensitivity.

However, the nanoscale dimension has a disadvantage in terms of transport of the target molecule from the sample solution to the small-area sensor surface, which requires a balance of convection, diffusion and sensor surface area to avoid impractically long hybridization times (Sheehan and Whitman, 2005). Hence, optimization of target molecule collection on the SiNW surface is required in order to take advantage of the nanoscale sensitivity increase effect (Squires *et al.*, 2008; Zheng *et al.*, 2008; Gong, 2010). Therefore, careful design and control of each of the essential elements of the SiNW biosensing platform is necessary (Nair and Alam, 2007; De *et al.*, 2013a), which will be described in more detail throughout the remainder of the chapter.

12.2 Silicon nanowire (SiNW) biosensors

Over the past decade there has been continued interest in electronic biosensors, with many reports of functional nanoscale (~10–100 nm) electrical field effect devices such as carbon nanotubes (Tans *et al.*, 1997; Avouris *et al.*, 2003) and semiconductor nanowires (Lieber, 1989; Beckman *et al.*, 2005). SiNW sensors are particularly interesting for the label-free electrical detection of biomolecules (Cui *et al.*, 2001; Kong *et al.*, 2000; Li *et al.*, 2004; Beckman *et al.*, 2004; Bunimovich *et al.*, 2006; Kim *et al.*, 2007a, b; Stern *et al.*, 2007a, b; Gao *et al.*, 2007), making them suitable for large-scale multiplexed electrical recording (Beckman *et al.*, 2005), and high-density integration with electronic instrumentation computers and handheld devices. From a technological point of view, SiNW sensors are important because they exploit mature planar microfabrication technologies and well-established techniques for surface passivation and modification.

A biosensor typically consists of a sensor element that transduces binding of a target biomolecule to specific receptor molecules, or probe molecules, directly attached to the sensor surface into a measurable signal that can be recorded using conventional instrumentation. The probe recognition molecules can be designed to bind specifically to either chemical or biological moieties. SiNW biosensors measure surface charge density changes on the sensor surface, from potential determining surface ions and/or the intrinsic electronic charge of hybridized biomolecule complexes attached to the sensor surface, in the form of conductance changes in the body of the sensor caused by a field effect transduction effect.

12.2.1 Conventional field effect sensors

Silicon field effect sensors were first introduced more than 40 years ago, with reports of the ion-sensitive silicon field effect transistor (ISFET) sensor to detect ions in solution adsorbed to the gate of a conventional metal oxide field effect transistor device with the metal gate layer removed (Bergveld, 1970). When the gate is used as a sensing electrode in aqueous solution, a charge density change at the gate electrode surface will ideally result in an equal change of charge density with opposite polarity in the semiconductor layer, as was described in the pioneering work on field effect devices (Bardeen, 1947; Shockley and Pearson, 1948).

For example, consider the electrolyte–insulator–semiconductor (EIS) structure shown in Fig. 12.1(a), and assume an ideal interface capacitance (i.e. the electrical double layer capacitance C_{dl} is constant over the measurement time) and a low ionic strength electrolyte (i.e. $C_{dl}=C_{st}$, where C_{st} is the Stern layer capacitance). According to charge neutrality across the electrolyte–gate interface, the surface potential change $\Delta\psi_o$ induced at the gate surface caused by a net charge density change $\Delta\sigma_o$ at the gate electrode surface is related to the gate dielectric capacitance C_o and C_{dl} as $\Delta\psi_o \approx \Delta\sigma_o(C_{dl}+C_o)^{-1}$. For a 5 nm thick SiO_2 gate layer, $C_o \approx 0.18\,\mu F\ cm^{-2}$, and assuming $C_{dl} \approx 20\ \mu F\ cm^{-2}$ and $\Delta\sigma_o \approx 10^{14}$ electrons cm^{-2} results in $\Delta\psi_o \approx 400\,mV$, which can be readily measured with a conventional field effect sensor.

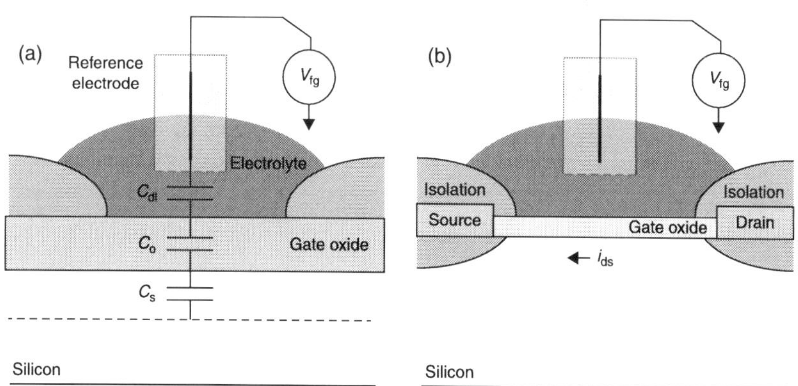

12.1 Conventional field effect sensors. (a) Electrolyte–insulator–semiconductor (EIS) sensor. (b) Ion-sensitive silicon field effect transistor (ISFET) sensor configuration with front gate (fg) using a reference electrode, and drain-source current i_{ds}.

Unfortunately, the detection of intrinsic charge on a conventional gate electrode surface using a field effect sensor is far from ideal and problems arise because of counter-ion screening when measurements are conducted at, or near, physiological conditions (i.e. 100–200 mM ionic concentration) where the biomolecule complex formation is preferable, which will be described in more detail later in the chapter.

As silicon field effect devices measure charge density changes on the sensor surface, ideally they can directly measure the intrinsic electronic charge of adsorbed biomolecules, and specifically bound biomolecule complexes on the sensor surface. The prospect of the real-time electrical detection of charged biomolecules using semiconductor field effect sensors is compelling in that it offers a direct electrical readout in real time in a format that can be mass produced at low cost using conventional integrated circuit technology. It was proposed more than 30 years ago that field effect semiconductor sensors could be used to detect surface polarization effects caused by the formation of an antibody–antigen complex on the sensor surface (Schenck, 1978).

Conventional ISFET sensors measure the surface potential changes $\Delta\psi_0$ at a single gate interface in the form of change in the device conductance ΔG, where $G \equiv \partial i_{ds}/\partial v_{ds}$ and i_{ds} is the measurand, shown in Fig. 12.1(b) (Bergveld, 1970). The charge density change $\Delta\sigma_0$ on the gate surface, an effective dipole because of the polar monolayer and/or dipole (Shalev *et al.*, 2008), provides the $\Delta\psi_0$, which is effectively the applied gate potential and the associated measurand i_{ds}. Similarly, biosensing is achieved when molecular binding at the sensing surface induces $\Delta\psi_0$, which results in a measurable conductance change ΔG via the field effect across the gate dielectric layer. As the target biomolecules binding to the probe molecules on the sensing surface result in a change in the net surface charge density $\Delta\sigma_o$, then the Grahame equation can be used as a first order to estimate the surface potential (Israelachvili, 1992).

$$\psi_0 = \gamma^{-1}\sinh^{-1}(\sigma_0(8\varepsilon_w\varepsilon_0 kT n_0)^{-1/2}) \qquad [12.1]$$

for small values of ψ_0 in a monovalent electrolyte, where ε_w is the dielectric constant of water, ε_0 is the permittivity of free space, k is the Boltzmann constant, T is absolute temperature and n_0 is the buffer ionic concentration.

12.2.2 SiNW field effect sensors

In principle, SiNW-FET sensors perform the same measurement as conventional ISFET sensors as both device configurations measure surface potential changes in the form of a conductance change. However, SiNW devices are very different from their planar predecessors because of their small size and three-dimensional shape, and multi-gate configuration. Most reported SiNW devices operate in a majority carrier configuration, which is advantageous compared with conventional ISFET devices as this simplifies the fabrication technology (Carlen and van den

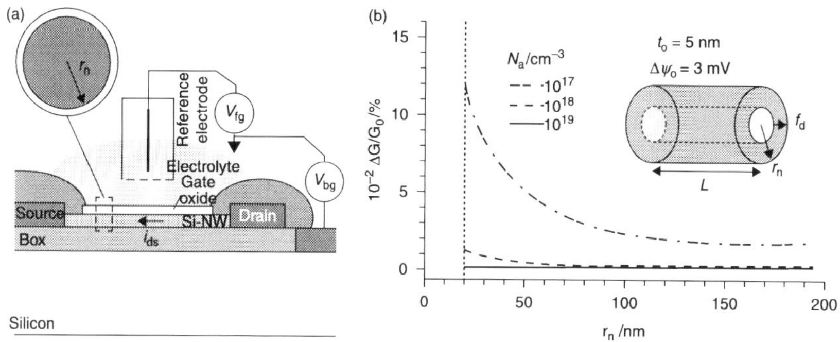

12.2 (a) Silicon nanowire (SiNW) sensor configuration with front gate (fg) connected to a reference electrode, backgate (bg) electrode connected to the silicon substrate, measurand i_{ds}. (b) Calculated relative conductance $\Delta G/G_o$ as a function of the cross-section radius of a circular nanowire for a fixed gate surface potential change $\Delta\psi_o$=3 mV, gate dielectric SiO_2 thickness t_o=5 nm, length L=10 μm, doping dependent mobility μ_b for three different impurity doping concentrations N_a.

Berg, 2007). Most SiNW-FET biosensors consist of either a *p*-doped or *n*-doped SiNW connected to source and drain contacts, as shown in Fig. 12.2(a). The thin SiNW body is electrically isolated from the silicon substrate by a buried oxide layer (BOX), and the front-gate (fg) and back-gate (bg) contacts are used to control the conductance of the SiNW electrostatically across the gate dielectric, which is assumed to be SiO_2 and/or BOX layers, respectively. Two common device configurations are normally on depletion-mode where the device body is doped with an impurity (Cui *et al.*, 2001; Li *et al.*, 2004; Bunimovich *et al.*, 2006; Chen *et al.*, 2011), and normally off enhancement-mode with a non-doped device body (Stern *et al.*, 2007b).

12.2.3 Nanoscale advantages

The small cross-sectional area and three-dimensional geometry of the SiNW field effect sensors result in a large surface-to-volume ratio and multi-gate interface to the electrolyte solution, which can produce an increased sensitivity to surface charges compared with a planar single gate configuration (Fig. 12.2a). For example, consider the relative conductance change $\Delta G/G_o$ of three depletion-mode SiNW sensors with circular cross-sections and different impurity doping concentrations N_a, as a function of the device radius r_n, that all undergo the same surface potential change of $\Delta\psi_o$=3 mV, where the conductance is calculated with

$$G=\pi q \mu_b N_a L^{-1}(r_n-f_d)^2, \qquad [12.2]$$

where the majority carrier hole mobility is defined as

$$\mu_b = \mu_{max} + \mu_{max} - \mu_{min}/1 + (N_a/N_{ref})^{\alpha} \text{ with } \mu_{max} = 44.9\,\text{cm}^2\,\text{V}^{-1}\,\text{s}^{-1}, \quad \mu_{max} = 470.5\,\text{cm}^2\,\text{V}^{-1}\,\text{s}^{-1} \qquad [12.3]$$

$N_{ref} = 2.23 \times 10^{17}$ cm^{-3} and $\alpha = 0.719$ (Sze, 1981).

The sensor length is L and the depletion distance function is defined as (Masood *et al.*, 2010)

$$f_d \approx ((t_{ox}\,\varepsilon_{Si}/\varepsilon_{ox})^2 + 2\varepsilon_{Si}\varepsilon_{ox}\psi_o / qN_a)^{1/2} - t_{ox}\,\varepsilon_{Si}/\varepsilon_{ox} \qquad [12.4]$$

with t_{ox} the gate dielectric thickness and permittivities of silicon $\varepsilon_{Si} = 11.9$ and SiO_2 $\varepsilon_{ox} = 3.9$, respectively. The quiescent bias conductance G_o is the sensor conductance prior to a sensing event. From Fig. 12.2(b), the $\Delta G/G_o$ increases as the device radius is decreased for a particular impurity doping concentration N_a, which clearly demonstrates that scaling depletion-mode SiNW sensors to smaller dimensions can increase the relative conductance change. It is interesting to note that large radius nanowires $r_n = 200$ nm with moderate doping $N_a \approx 10^{17}$ cm^{-3} have similar or greater relative conductance change compared with heavily doped devices $N_a \geq 10^{18}$ cm^{-3} with small radius $r_n = 20$ nm. It should be noted that the electrolyte–gate SiO_2 and gate SiO_2–silicon interfaces have been treated as ideal.

Another important advantage of nanoscale potentiometric sensors is that a smaller number of molecules can generate the same surface potential change compared with macroscale sensors. For example, consider that a particular surface attachment protocol results in surface coverage density of $\kappa = 1$ probe molecule per 2 nm^2, with a surface potential change $\Delta\psi_o$. A circular nanowire with radius $r_n = 25$ nm and length $L = 10\,\mu$m has a surface area of $A_n = 1.6 \times 10^6$ nm^2, which results in $\kappa A_n \approx 8 \times 10^5$ total number of probe molecules on the nanowire surface. A macroscale sensor with surface area $A_m = 50\,\mu\text{m}^2$ will have $\kappa A_n \approx 2.5 \times 10^7$ probe molecules on the surface. The nanoscale sensor requires ~32× fewer probe molecules to produce the same $\Delta\psi_o$, which was shown to produce an increased relative conductance response $\Delta G/G_o$ as the SiNW cross-sectional area is decreased. Therefore, if we assume the same hybridization efficiency, then the nanoscale sensors can offer significantly improved sensitivity. In this simple example, we have assumed that probe molecules cover the entire SiNW surface area, which is clearly not possible when the nanowire sensor surface is partially in contact with the silicon substrate, and, therefore, reduces the previously described enhancement factor.

12.2.4 Nanoscale disadvantages

Although SiNW biosensors can provide advantages over their planar predecessors, because of their small size and multi-gate structure, there are also disadvantages for chemical and biochemical sensing applications that must be addressed in order to make the most of the advantages. An obvious disadvantage of nanoscale biosensors is that the sensing surface is significantly reduced compared with microscale sensors, which reduces the number of biomolecules that can be captured from the sample volume. For ultra-low biomolecule concentrations, for example picomolar or smaller, the time that is required to transport the target molecules to the nanoscale sensors can be impractically long and on the scale of hours to days (Sheehan and Whitman, 2005; Squires *et al.*, 2008). Therefore, special attention is required to collect the biomolecules on the SiNW surface, which will be discussed later in the chapter.

The charge sensitivity of SiNW sensors increases as the cross-sectional area is decreased (Fig. 12.2(b)), however, they are also increasingly sensitive to trapped charges in the gate dielectric layer and electrical imperfections at the gate–silicon interface consisting of donor and acceptor interface states, which can lead to device instability, reduced sensitivity and failure. Although high temperature annealing can reduce these imperfections, precautions taken during device fabrication produce more stable and reliable devices (Deal, 1974). Radiation–induced traps created during certain fabrication steps, such as contact metal deposition, and fast surface states can be reduced by moderately high temperature hydrogen annealing (Deal, 1974). All the charges and surface states contribute to the sensor transconductance, and therefore, can strongly influence the response of small SiNW devices (Schmidt *et al.*, 2007).

12.2.5 Fabrication technology

The SiNW fabrication techniques can be classified into two categories: bottom-up and top-down. The bottom-up approach assembles molecules and small solid structures from atoms that are combined into a large variety of shapes and functions. There have been many reports of synthesized SiNW structures over the past decade (Morales and Lieber, 1998; Westwater *et al.*, 1998; Ho *et al.*, 2008); however, suitable methods for accurate nanowire alignment are lacking, and electrical contact formation is problematic, making it difficult to construct functional device arrays (Tong *et al.*, 2009).

Therefore, the majority of SiNW biosensing platforms are implemented with top-down fabrication technology, which remains the standard technique for nanoscale semiconductor manufacturing, and is based on standard microfabrication methods consisting of deposition, patterning and etching and standard nanopatterning techniques, such as deep-UV photolithography (Gao *et al.*, 2007). Direct-writing nanopatterning techniques, such as electron-beam lithography and focused-ion-beam lithography can realize sub-10 nm feature sizes; however, they are not suitable for wafer-scale fabrication and are typically used in a research setting. Other top-down nanopatterning techniques available for realizing SiNW sensors and sensor arrays, such as nanoimprint lithography (Chou *et al.*, 1996; Heath, 2008), spacer patterning (Flanders and Efremow, 1983; Choi *et al.*, 2002; Choi *et al.*, 2003), lithographically patterned electrodeposition (Menke *et al.*, 2006) and shadow edge lithography (Bai *et al.*, 2009). These strategies, applied to biosensor fabrication, are described in greater detail in Chapter 9 of this book.

Once the nanopatterned features are written, the patterns are transferred to the device silicon layer by selectively removing the silicon using either dry or wet anisotropic etching. The most common technique for dry anisotropic etching is reactive ion etching, which is the standard etching method that has high reproducibility, uniformity and aspect ratios (Fischer and Chou, 1993). However, reactive ion etching can induce substrate damage and degrade device performance (Choi *et al.*, 2002; Melosh *et al.*, 2003). Wet anisotropic etching employs crystal plane-dependent processes on silicon surfaces where hydroxide ions in alkaline solutions, such as potassium hydroxide or tetra-methyl-ammonium hydroxide, etch the (111) planes at a very slow etch rate compared with the (100) and (110) crystal planes and results in a cost-effective method to form SiNW sensors and sensor arrays (Chen *et al.*, 2009).

12.3 Probe layers

The transformation of a nanowire into a biosensor requires the formation of a biomolecule interface, such that a biologically active probe, which binds specifically to a target biomolecule of interest, is directly tethered to the sensor surface (van der Voort *et al.*, 2005; Bunimovich *et al.*, 2004). The biomolecule interface is a key component to ensure that a biosensor attains both high detection specificity and sensitivity. The detection specificity depends on how well the target binds to the probe molecule, as well as the probe surface density, such that the steric hindrance does not interfere with the hybridization process. A conventional biomolecule interface probe layer consists of a covalently attached probe moiety tethered at one end to a solid surface via a linker molecule, as shown in the illustration of different probe–target complexes in Fig. 12.3.

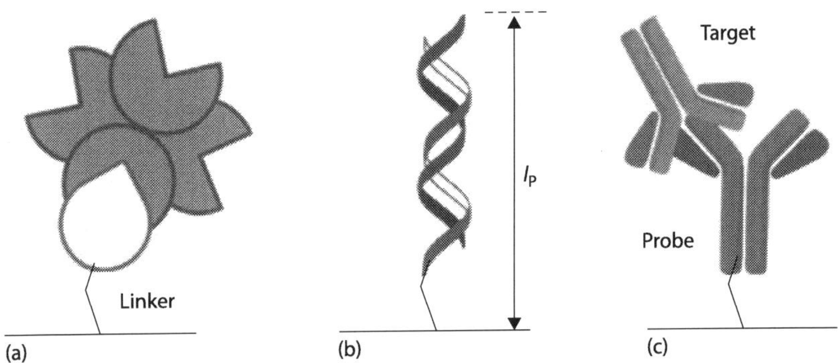

12.3 Different probe–target complexes tethered to a surface with a linker. (a) Biotin–streptavidin protein complex. (b) DNA–DNA duplex. (c) Antigen (target)–antibody (probe) protein complex.

The probe molecule can be any moiety ranging from DNA, RNA, peptide nucleic acid (PNA), antibodies and proteins that hybridize to their respective complementary target. Unlike DNA, which has a stable structure and fixed charge distribution, proteins, especially soft proteins, are intrinsically disordered, and are conformable upon adsorption to solid surfaces (Norde, 1986; Wahlgren and Arnebrant, 1991). The irregular charge distribution of many soft proteins presents a challenge to the surface charge estimation and needs to be considered when designing the SiNW biosensing assay. The linker layer with functional chemical groups at both ends, such as amine (NH_2), aldehyde (COH) and carboxyl (COOH) groups, covalently attaches the probe to the solid surface in one or more chemical processing steps. These important functionalization strategies are also discussed in detail in Chapter 3 of this book

The majority of SiNW biosensors have a SiO_2 surface layer, and therefore studies in this area benefit greatly from the vast knowledge of surface preparation techniques of glass surfaces developed for microarrays, which is a well-established and mature field for genomics applications (Southern, 1975; Southern *et al.*, 1999). For covalent attachment to glass surfaces, functional amine groups are attached to the glass surface using a covalent bond between the primary amine and chemical moieties on the glass surface, as oligonucleotides cannot be coupled directly to the silanol groups (Si–OH) to the silicate glass surface. The formation of probe oligonucleotides on glass microarray surfaces involves tethering one end of the probe to the surface, which affects the duplex formation with a target in the sample solution as the bases nearest the surface are less accessible than those farthest away. The incorporation of a linker layer, which separates the probe from the glass surface, has been used to overcome the limitation of base accessibility, and hybridization yields have been reported to increase by up to two orders of magnitude (Southern *et al.*, 1999; Duggan *et al.*, 1999; Shchepinov *et al.*, 1997). However, if the linker length is increased, the hybridization yield decreases because of strand coiling (Southern *et al.*, 1999). For DNA hybridization and other biological complexes, the binding specificity also depends on the ionic strength of the hybridization buffer, which is preferred to be similar to physiological conditions for high hybridization efficiency.

12.3.1 Probe layers for SiNW biosensors

A suitable probe layer is characterized by a surface coverage that balances probe surface coverage to generate the maximum surface potential while avoiding steric hindrances that can reduce the probe–target binding efficiency, and the linker and probe length must be suitable to match the requirements of the biosensor transduction method, which are all important to optimize the biosensor sensitivity. Although the lessons learned from conventional microarray assays are useful for the basic probe tethering techniques, SiNW biosensors have additional functional constraints that affect their biosensing performance.

12.3.2 Probe surface density

As SiNW-FET biosensors detect the intrinsic surface charge density of probe–target complexes on the sensor surface, the probe attachment density is extremely important. From the use of the Grahame equation described in Eq. 12.1, the magnitude of the surface potential applied to the SiNW biosensor $\Delta\psi_o$ increases with increasing surface charge density $\Delta\sigma_o$. Therefore, the surface attachment density should be as large as possible in order to maximize the surface potential changes. However, the probe–target binding efficiency drastically decreases for high probe surface densities because of steric hindrance, which counteracts the benefit from the high probe surface density. It has been shown that probe surface densities of $\sim 10^{12}$ probes cm^{-2} result in optimal binding efficiencies of many probe–target complexes (Bajaj, 2005). Strategies of inserting linkers have improved hybridization yields by up to two orders of magnitude by introducing spacers between the surface and the oligonucleotides. The length of the spacer has a marked effect and there appears to be an optimum linker length beyond which hybridization yield declines (Southern *et al.*, 1999).

The estimation of the surface charge density of a biomolecule complex on the sensor electrode surface requires consideration of how the molecular complexes are arranged on the surface, length of any linker molecules, pH of the bulk solution as the gate dielectric is intrinsically charged and intrinsic electronic charge of the probe molecule. For example, it has been estimated that a surface potential change $\Delta\psi_o = 3$ mV resulted from the hybridization of 3×10^{12} oligonucleotide (12-mer) molecules cm^{-2} (corresponding to a target concentration of approximately 20 nM) from complementary DNA probe molecules adsorbed to the gate electrode that is covered with a positively charged poly-L-lysine layer (Fritz *et al.*, 2002).

12.3.3 Counter-ion screening

The detection of intrinsic charge on a conventional SiO_2 gate electrode surface using a field effect sensor is far from ideal and problems arise because of counter-ion screening when measurements are conducted at or near physiological conditions where the probe–target complex formation is often optimal. The Debye length is a measure of the distance that the counter-ions extend away from the sensor surface, which is defined as (Israelachvili, 1992)

$$\lambda_D = (\varepsilon_w \varepsilon_o kT\,(2z^2q^2I)^{-1})^{1/2} \qquad [12.5]$$

for a symmetric monovalent electrolyte; z is the valence ionic charge and I is the ionic strength. The Debye length of a buffer solution with ionic strength similar to a physiological solution in contact with the gate electrode is $\lambda_D \sim 1$ nm, and electronic charge added to the sensing surface at distances beyond λ_D, with respect to the gate interface, will be effectively screened and not sensed by the silicon layer. Therefore, measurements of biomolecule complexes in high ionic strength solutions require the biomolecule complex to be located very near the sensor surface. Another electrolyte-related problem is counter-ion screening where small inorganic counter-ions present in the electrolyte solution will screen the intrinsic biomolecule charge, thus reducing its net effective charge. The counter-ion screening of the intrinsic charge of an antibody–antigen complex for the measurement of binding using a field effect sensor was reported more than 30 years ago (Janata, 1986).

Counter-ion screening, or counter-ion condensation, has also been reported for DNA duplexes where the negative charges on the phosphate backbone are screened by cations in the electrolyte, thus reducing the intrinsic DNA charge by up to 80% for high ionic strength solutions (Manning, 1978). Recently, the significance of the Debye length for the detection of DNA hybridization of SiNW surfaces has been reported (Zhang *et al.*, 2008). In general, the charge screening effect can be reduced when the Debye length, $\lambda_D > l_P$, where l_P is the total length of the probe–target complex, as shown in Fig. 12.3 (Stern *et al.*, 2007a; Nair and Alam, 2008; Zhang *et al.*, 2008). For biomolecules, l_P falls between ~2 nm (e.g. short DNA oligomers) to 10 nm and beyond (e.g. proteins). The net effective charge density of target–probe complexes located at a distance beyond λ_D will be effectively screened. In order to electrically detect the target molecules sensitively, the SiNW biosensor system needs to be carefully designed with the use of short probe and linker molecules, for example by using a short DNA oligomer, a small antibody probe (Elnathan *et al.*, 2012), or in low ionic concentration solution, for example 10 mM NaCl, for biotin–streptavidin detection (Cui *et al.*, 2001) and 10 μM KCl for virus detection (Patolsky *et al.*, 2004). However, the biomolecules favour binding in physiological buffers where λ_D~1 nm. For example, it has been reported that DNA condenses in the presence of multivalent cations and about 90% of its intrinsic charge is neutralized (Bloomfield, 1996). Higher probe coverage can be used in high ionic strength solution and complementary DNA strands can get closer to each other to hybridize; however, the hybridization efficiency decreases for high probe surface densities (Peterson *et al.*, 2001). Detection of the intrinsic molecular charge by the SiNW-FET biosensors performs optimally in the absence of the screening effect, that is in a low ionic strength buffer (Fritz *et al.*, 2002; Li *et al.*, 2004; Bunimovich *et al.*, 2006). Synthetic probe biomolecules, such as PNA, have been developed and reported to have good hybridization properties to complementary DNA sequences in low ionic strength buffers (Nielsen and Haaima, 1997; Wang *et al.*, 1996). In this case, the ion concentration can be significantly reduced, which in turn increases the Debye length λ_D~10 nm and relaxes the requirements for short linker and target lengths.

Generally, it can be concluded that the practical realization of field effect devices for detection of intrinsic biomolecule charge in electrolyte solutions with ionic strength near physiological conditions is problematic and remains a primary challenge to sense biomolecule complexes tethered directly to the gate electrode surface. Despite these challenges, there have been numerous reports of conventional field effect sensors for measuring antibody–antigen binding (Schasfoort *et al.*, 1990; Lud *et al.*, 2006) and DNA hybridization (Souteyrand *et al.*, 1997; Berney *et al.*, 2000; Fritz *et al.*, 2002; Wei *et al.*, 2003; Kim *et al.*, 2004; Uslu *et al.*, 2004). In any case, a suitable probe interface layer requires a high density of probes on the gate sensing electrode, a short linker between sensor surface and probe to reduce the counter-ion screening effects, while at the same time ensuring a suitable hybridization efficiency (Poghossian *et al.*, 2005).

12.3.4 Covalent attachment to silicon dioxide (SiO_2) SiNW surfaces

The probe molecules can be tethered to the SiNW SiO_2 gate surfaces using covalent linkages (Li *et al.*, 2004; Hahm and Lieber, 2004; Gao *et al.*, 2007; Cattani-Scholz *et al.*, 2008; Gao *et al.*, 2011; Elnathan *et al.*, 2012) and electrostatic adsorption (Bunimovich *et al.*, 2006; Dorvel *et al.*, 2012). Covalent attachment to the SiO_2 surface is preferred and has been adapted directly from methods developed for microarray assays, as previously described. A common approach is to use direct attachment of DNA probes to the SiNW SiO_2 surface by tethering it covalently to an end-functionalized siloxane layer that is grafted onto the SiO_2 surface. Typically, 3-amino-propyl-triethoxy silane (APTES) is used for the SiO_2 surface, which is terminated with an amine group that can be further attached to many different biological moieties. For example, the thiol group from the cysteine moiety from a conventional PNA probe can be attached to the amine of the APTES functionalized SiO_2 surface using a hetero-bifunctional cross-linker sulfo-succinimidyl-cyclohexane-carboxylate (Stern *et al.*, 2007a). The thiolated-amine attachment is very robust and results in a surface probe density of 10^{13} molecules cm^{-2} (Chrisey *et al.*, 1996). APTES grafted onto the SiO_2 surface and coupled to aminated-DNA is reported to have a surface density coverage of 4×10^{11} molecules cm^{-2} (Dugas *et al.*, 2004). An APTES attachment scheme based on carboxylated PNA has also been reported for SiNW sensor probe surface preparation (Gao *et al.*, 2011). APTES on the SiO_2 surface with N-ethyl-N′-(3-(dimethylamino) propyl)-carbodiimide (EDC) and N-hydroxy-succinimide (NHS) results in vertical orientation of immobilized antibodies at their C-end (Vacic *et al.*, 2011). Additionally, APTES followed by glutaraldehyde can couple antibodies via the N-end, resulting in horizontal orientation of the immobilized antibodies (Vacic *et al.* 2011). The sulfo-NHS-SS-biotin attachment to APTES–treated SiO_2 surfaces is a one-step surface biotinylation method for streptavidin binding (Stern *et al.*, 2007b). The biotinylation of SiNW surfaces has been reported using biotinyl *p*-nitrophenyl ester, with 4-(dimethylamino) pyridine (DMAP) in pyridine (99.9%), which was subsequently followed by the addition of avidin, and subsequent attachment of a biotinylated PNA probe layer (Cai *et al.*, 2004). Biotinylation has also been performed using biotin-amidocaproyl labelled bovine serum albumin on SiNW (Cui *et al.*, 2001). Using trimethoxy-silane-aldehyde, direct aminated PNA have been attached to the SiO_2 surface (Gao *et al.*, 2007; Patolsky *et al.*, 2006b). Silane aldehyde, followed by the reduction of the amine for biomolecule attachment via secondary amine bonds on the SiO_2 surface of SiNW, has been reported for the detection of a variety of biological target molecules such as tyrosine kinase (Wang *et al.*, 2005), prostate specific antigen and telomerase (Zheng *et al.*, 2005). However, the basic siloxane coupling to SiO_2 surfaces has associated problems of reduced hydrolytic stability, which depend on the availability of hydroxyl binding sites on the SiO_2, and the inherent risk of multilayer formation (Cattani-Scholz *et al.*, 2008). Self-assembled monolayers of organophosphonates based on physisorption have been reported to be more stable in that regard (Hanson *et al.*, 2003).

12.3.5 Covalent attachment to silicon (Si) SiNW

The probe moieties can be directly attached to the silicon surface without the intervening SiO_2 layer with the motivation to keep the probe layer closer to the gate electrode surface and to create a heterogeneous sensor surface for SiNW sensing applications (Masood *et al.*, 2010). As previously described, conventional SiNW biosensors have the probe attached directly to the SiO_2 sensor surface using silane-based attachment chemistry. However, the attachment to the SiO_2 layer reduces the sensor selectivity because the entire substrate surface is most often a homogeneous oxide surface. Selective functionalization of the SiNW surfaces with SiO_2 gate has been reported (Park *et al.*, 2007a); however, the preferred method is the formation of a hydrogen-terminated silicon surface that can be used to directly tether the linker molecules, which results in a heterogeneously functionalized surface without polymer residue remaining from selective sensor masking (Bunimovich *et al.*, 2006; Masood *et al.*, 2010; Zhang *et al.*, 2008).

The direct molecular attachment to the silicon surface is obtained by hydrogen termination of silicon following removal of the native SiO_x layer and subsequent hydrosilylation. The Si(111) surface is preferred for the covalent alkylation of organic monolayers because of the surface atomic arrangement, resulting in densely packed layers with a low density of non-terminated bonds. Once a stable H–Si surface is formed, the C–Si monolayer is prepared with a suitable precursor involving free radical initiation, such as ultraviolet irradiation (Masood *et al.*, 2010). Hydrogen-terminated SiNW, with a mixture of surface planes, have been prepared using gas phase hydrosilylation (Bunimovich *et al.*, 2006). Further, hydrosilylation has been used for the attachment of PNA probes using a glutaraldehyde homo-bifunctional cross-linker that binds to the aminated-PNA (Zheng *et al.*, 2008).

12.3.6 Non-covalent attachment to SiNW surfaces

Non-covalent adsorption of the probe layer on the gate electrodes based on the electrostatic attraction of poly-L-lysine polymers onto negatively charged SiO_2 at neutral pH has been reported. A negatively charged DNA probe is then adsorbed onto the poly-L-lysine–covered SiO_2 surface. This method has been used to measure complementary DNA hybridization to silicon and SiNW surfaces in different ionic strength buffers (Bunimovich *et al.*, 2006; Fritz *et al.*, 2002).

12.4 Integrated sample delivery

Analytical assays for the detection of biomolecule interactions at very low sample concentrations and volumes are becoming increasingly important for applications requiring high-throughput biomolecule analyses, such as genomics screening, protein assays and disease diagnostics. In recent years there has been considerable effort in developing probe-free measurement techniques, which can directly measure biomolecule complex binding events in real time using small sample volumes in microfluidic channels that are integrated into automated lab-on-chip systems.

Currently, surface plasmon resonance sensors (Homola *et al.*, 1999) and surface plasmon resonance imaging sensors (Krishnamoorthy *et al.*, 2010) are well-established label-free techniques, but they require sophisticated optical instrumentation for signal readout and have limited sensitivity, especially for small molecular weight molecules. All electrical readout SiNW biosensor platforms are emerging as the next generation of probe-free biosensor platforms because of their reported high label-free detection sensitivities to biomolecules and their suitability for large-scale and high-density integration that can readily be interfaced with conventional electronic systems.

We have recently developed an integrated microfluidic label-free analysis platform that uses a disposable analysis cartridge consisting of all-electrical SiNW sensor arrays, for (bio)chemical detection, integrated with a small volume microfluidic flow-cell (De *et al.*, 2013a). The integration of the all-electrical SiNW sensor arrays with a small volume microfluidic reaction flow-cell requires special consideration of many aspects related to the type of bioassay and the type of target–probe biomolecule complexes that are to be measured.

The integrated biological analysis platform uses an analysis cartridge that consists of a SiNW biosensor array chip, which is pre-functionalized with specific probe molecules designed for a particular experiment, directly integrated with a microfluidic flow-cell using a simple and non-destructive bonding method. The analysis cartridge is interconnected to the automated multi-sample injection system that consists of a precision sample injection switch (Cheminert Nanovolume valve, Valco Instruments Inc.) for precise sample dosing, a regulated pressure source (MFCS-8C, Fluigent) for simultaneous sample transport, and electrical instrumentation for sensor readout, as shown in Fig. 12.4.

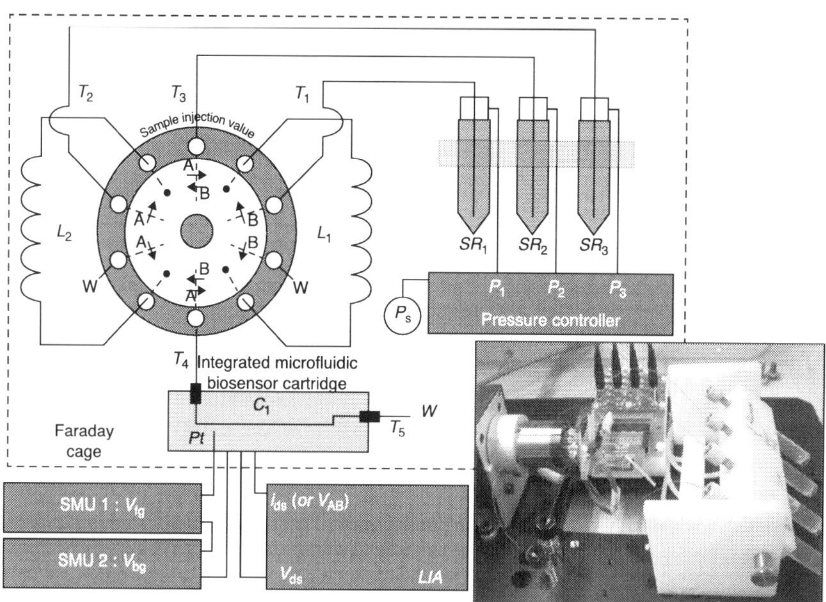

12.4 Schematic diagram of the integrated silicon nanowire (SiNW) biosensor system with automated multi-sample delivery to the microfluidic chip. Microflow tubing is used for all hydraulic connections T_1–T_5 and sample loops L_1 and L_2. The integrated microfluidic chip is labeled as C_1. The pressure sources P_1–P_3 control the flow rate in tubes T_1–T_3, respectively. The dotted enclosure represents the Faraday cage. The electrical biasing and measurement equipment are connected directly to the SiNW chip. Inset: Image of analysis system. The sample injection valve has two switch positions: A and B (reproduced from De *et al.*, 2013 with permission).

The electrical biasing and measurement system, which will be described in more detail later in the chapter, are connected to the SiNW biosensors, denoted by SMU 1 for the application of the front-gate voltage V_{fg} and the measurement of the front-gate leakage current i_{fg}, SMU 2 for the application of the back-gate voltage V_{bg} and the measurement of the back-gate leakage current i_{bg}, and a lock-in-amplifier (LIA) for the application of the drain-source voltage and to measure the drain-source current i_{ds}. The microfluidic analysis cartridge and samples are housed in a Faraday cage to eliminate external electrical and ambient light noise sources. The sample injection valve is used to selectively deliver different sample solutions from sample reservoirs SR_1, SR_2 and SR_3, to the microfluidic flow-cell from pre-loaded sample loops, L_1 or L_2. We now describe the components of the integrated measurement system and its use for meaningful data acquisition for real-time biosensing applications.

12.4.1 Flow rate control

Two types of biosensing measurements are performed with the integrated microfluidic SiNW biosensor platform: diffusion-based measurements that use a very small sample flow rate and the target biomolecules diffuse through the sample volume to the sensor surface with the probe layer; and flow-based measurements, which use volumetric flow rates that are large enough to perform real-time hybridization kinetics estimations of the target–probe complex formation. As a range of the flow rate conditions are used, a well-controlled sample delivery system is required for quantitative measurements that prevent erroneous sensor responses and provide meaningful data acquisition for real-time biosensing applications with SiNW biosensors.

Typically, diffusion-based measurements use small sample reservoirs that are interfaced directly with polymer tubing for sample introduction, or often the reservoirs are filled with sample solutions using conventional pipetting (Elfstrom *et al.*, 2008; Chen and Zhang, 2011; Dorvel *et al.*, 2012) A two-step antigen capture from blood and subsequent detection in sensing buffer has been reported (Stern *et al.*, 2012). In some cases, long hybridization times (approximately 60 minutes) have been monitored over the SiNW sensors (Gao *et al.*, 2007), which requires careful calibration of the significant SiNW sensor drift.

Flow-based measurements are most often used for quantitative real-time hybridization kinetics estimations (i.e. equilibrium dissociation constant K_D) of the target–probe complex formation where it is necessary to control the flow velocity such that the transport of the target to probes on the surface operates in the reaction-limited transport regime. Flow-based measurements are typically done using pressure-driven flow of the sample using a syringe pump that is connected directly to a microchannel (Hahm and Lieber, 2004; Bunimovich *et al.*, 2006; Stern *et al.*, 2007a). Different types of chemical and biological sensing experiments have been reported in different ionic strength buffer solutions (Cui *et al.*, 2001; Patolsky *et al.*, 2006a; Stern *et al.*, 2007a; De *et al.*, 2013a). The transport of the target to the sensor surface has been enhanced using different mixing strategies (Stern *et al.*, 2007a, b; Vacic *et al.*, 2011; Bunimovich *et al.*, 2006; Duan *et al.*, 2012). The detection efficiency in unmixed fluids with varying sensor sizes from the micrometer to the nanometer scale has been estimated to be in the order of picomolar sensitivity for practical time scales (Sheehan and Whitman, 2005).

Three physical processes govern the transport of the target in pressure-driven sample flow to the sensor surface. The first is diffusion across the stagnant layer at the walls of the microchannel where the fluid velocity is ideally zero because of the parabolic flow profile in the microchannel. The second is the convection flux of molecules into this layer because of flow of targets in the microchannel, and third the balance of the reaction rate or molecular affinity with the flux of transport with the combined transport of the target to the sensor (Squires *et al.*, 2008; Dong and Xiaolin, 2008). A flow velocity higher than about 0.5 mm s^{-1} (Zimmermann *et al.*, 2005) is required to operate in the reaction-limited transport regime for the target concentration to arrive at the sensor surface.

12.4.2 Sample switching

Sample delivery in the integrated analytical instrument is performed with an automated multi-sample injection system that eliminates erroneous sensor responses from sample switching as a result of flow rate fluctuations and provides precise sample volumes. The SiNW sensors are sensitive to many different sources of surface charge fluctuations, such as pH and ion concentration (Chen *et al.*, 2011), reference electrode fluctuations (Minot *et al.*, 2007) and flow rate-induced fluctuations (Kim *et al.*, 2009). The flow rate-induced fluctuations are primarily caused by the sample switching process where sample solutions from different containers, for example SR_1–SR_3 (Fig. 12.4), are transferred to the integrated microchannel, which can cause large erroneous sensor responses if not properly controlled (Bunimovich *et al.*, 2006; De *et al.*, 2013a).

Many SiNW biosensing measurements are conducted in dilute ionic strength buffers (<10 mM) as the Debye length is larger in high ionic strength buffer, as previously described. However, the surface potential of the SiNW sensors ψ_o is strongly affected by flow rate variations in low ionic strength buffers caused by the streaming potential of ions (Schoch *et al.*, 2008), and, in fact, SiNW sensors have been reported to be good flow rate sensors (Kim *et al.*, 2009). The streaming potential of low ionic strength buffers has been reported in microchannels (Kim *et al.*, 2007b), as well as bulk nanotubes (Ghosh *et al.*, 2003). The effect of the streaming potential is greater in low ionic strength buffers and becomes constant at an ionic strength of about 10 mM. Measurements with buffers with ionic strength larger than about 10 mM are relatively free from effects of baseline shifts as a result of sample switching. Flow rate fluctuations have been reported to strongly affect the SiNW sensor response while using a peristaltic pump to transport the sample in the microchannel (Lehoucq *et al.*, 2012). Syringe pumps have been reported to have a reduced effect on the SiNW sensor response compared with peristaltic pumping (Stern *et al.*, 2007a; Lehoucq *et al.*, 2012; Park *et al.*, 2007a). A sequential experimental protocol has been reported where the hybridization is done with a high ionic strength buffer, followed by detection at lower ionic strength (Lehoucq *et al.*, 2012).

12.4.3 Microfluidic channel integration

The integration of the microfluidic reaction or flow vessel is extremely critical such that the channel is well aligned to the SiNW sensor or sensor arrays, the microchannel is leak-free, and bonding process of attaching the SiNW sensor and microfluidic chips does not damage or contaminate the pre-functionalized sensor surface. With the critical dimensions of the microfluidic chip in the microscale and the SiNW sensor chips in the nanoscale, the bonding process often causes problems such as part-to-part misalignment, damage to the sensors and leakage from the channels. Alternative approaches have been investigated to obtain strongly bonded substrates, such as thermoplastics (Sunkara *et al*., 2011) and epoxy polymers (Li *et al*., 2012; Zhang *et al*., 2011). Photo-definable polymers provide alignment capabilities with a microscope during the exposure process (Li *et al*., 2012).

The microfluidic packaging on the SiNW sensors is crucial for simultaneous fluid delivery and electrical measurements (Choi *et al*., 2012). Typically, the microfluidic channel is fabricated by bonding a plasma-treated polydimethylsiloxane (PDMS) microfluidic chip to the SiNW sensor substrates. This method is quite effective in forming sealed microfluidic channels on silicon and glass substrates. However, the assembly of PDMS chips requires direct bonding of sensors and electronic components. A fabrication technique and assembly process have been reported that consist of a two-layer PDMS chip for solution injection in one layer and SiNW sensing in a different layer with separate microchannels, automated injection and changing of target solutions to get reduced erroneous signals because of streaming potential variations (Bunimovich *et al*., 2006). More recently, an autosampler system has been reported that uses a conventional single layer microfluidic channel and precision sample switching to eliminate flow rate variations (De *et al*., 2013a). The isolation of the measurement and elimination of noise caused by light, pumping, air-bubbles and buffer switching are crucial for real-time data acquisition for kinetic estimation of the target–probe complex formation (Bunimovich *et al*., 2006; Fritz *et al*., 2002; De *et al*., 2013a).

12.5 Electrical biasing and signal measurement

The SiNW biosensor measurement system consists of many components that include: the SiNW biosensor, sample solution, reference electrode, voltage sources for applying the front-gate (V_{fg}) and back-gate (V_{bg}) bias voltages, drain-source voltage (v_{ds}) and an instrument to measure the conductance changes Δi_{ds}, which is shown in the schematic diagram of Fig. 12.5. The basic sensor operation and basic components will be described in detail in the following sections.

12.5.1 Dual gate biasing

SiNW-field effect devices operate in three distinct regimes where the accumulation, depletion and inversion space charge regions are created in the semiconductor layer by controlling the potential of the silicon surface ψ_s. In the depletion regime, bound ionized impurity charge dominates, while in the accumulation and inversion regimes free charge at the surface dominates. The device current–voltage and sensitivity characteristics are very different in each of the different operation regimes. As previously described, there are two basic SiNW device configurations:

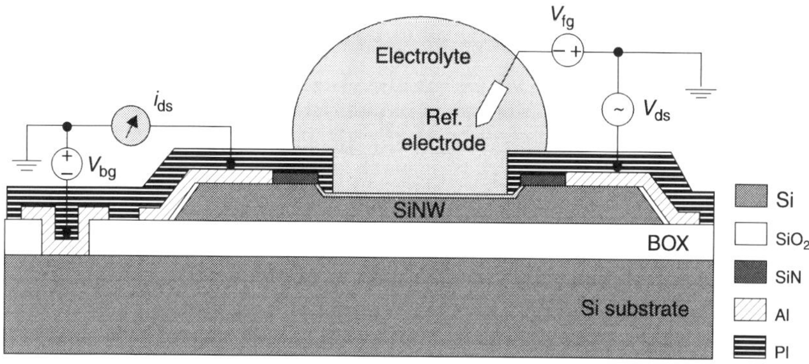

12.5 Silicon nanowire (SiNW) sensor configuration in solution with reference electrode with dual-gate bias control with applied voltages at the front-gate (fg) via the reference electrode and the back-gate (bg). The drain-source is biased with voltage v_{ds} and the current change Δi_{ds} is measured as a function of surface potential change $\Delta\psi_o$.

depletion-mode and enhancement-mode, and we have developed our measurement system with our depletion-mode SiNW biosensors (Chen *et al.*, 2009; 2011; De *et al.*, 2013a).

Depletion-mode SiNW devices are similar to recently reported homogeneously doped transistors (Lee *et al.*, 2009), and ideally operated as normally- on devices for an applied v_{ds}. In the depletion region, the device conductance depends on all dimensions, that is width W, length L and height h, as well as the doping concentration N_a, dopant concentration-dependent bulk hole mobility μ_b, as well as the electronic surface states of the interfaces (Sze, 1981; Schmidt *et al.*, 2007; Chen *et al.*, 2009). Gating the depletion-mode devices is similar to a junction field effect transistor (Sze, 1981), with the exception that the depletion layer width is modulated with dual gate configuration using the field effect at the fg and bg. The conductance of the SiNW device can be changed with the fg voltage source V_{fg} and/or with the surface potential ψ_o of the fg dielectric, which is a function of the surface charge density. As V_{fg} and/or ψ_o (for a given applied V_{bg} and v_{ds}) are varied positively and negatively, the SiNW conductance spans the depletion and accumulation modes, respectively. With a multi-gate structure, the bg ideally controls the depletion behaviour of the bottom side of the channel, and the fg controls the upper sides in contact with the solution using a calibrated reference electrode.

The SiNW-field effect biosensors in solution form an EIS structure (Fig. 12.1). The reference electrode electrically biases the solution, which results in a modified flatband voltage that represents the mechanisms responsible for energy band bending at the silicon/front-gate interface

$$V_{fb} = E_{ref} - \varphi_{Si} q^{-1} - \psi_o - Q_f/C_{ox} - Q_{it}/C_{ox} + \chi^{sol} \qquad [12.6]$$

where E_{ref} is the reference electrode potential, $\varphi_{Si}q^{-1}$ is the silicon work function, Q_f and Q_{it} are the fixed charge and interface state density near the silicon/front-oxide interface, respectively, $C_{ox} = \varepsilon_{Si}\varepsilon_o t_{ox}^{-1}$ is the gate-oxide capacitance and χ^{sol} is the dipole potential of the solution (Bousse and Bergveld, 1983; Chen *et al.*, 2011). The reference electrode is critical for controlling the surface potential of SiNW by tuning the solution potential, and the sensitivity of the NW sensor can be effectively tuned (Chen *et al.*, 2009). The front-gate layers of the multi-gate structure have approximately the same thickness and are considered to be the same. As the BOX layer is typically much thicker (>10×) than the fg layer, bg bias voltages are typically much larger in order to produce a comparable field effect.

12.5.2 Reference electrode

The Ag/AgCl reference electrode is a commonly used reference electrode in electrochemistry; however, because of the size mismatch problem, pseudo-reference electrodes are typically used in microfluidic systems to provide liquid gating and stable solution potential (Chen *et al.*, 2009; 2011). For example, a Pt electrode is inserted into the flow channel to ground the solution (Bunimovich *et al.*, 2006) and as a liquid gate for device biasing (De *et al.*, 2013a), or a patterned Au electrode near the SiNW sensors for the liquid gate control (Gao *et al.*, 2010).

12.5.3 Conductance measurements

There are different methods for measuring the conductance change, such as the constant current DC feedback method commonly employed with conventional ISFET sensors or by using a synchronous AC detection measurement technique using a conventional quadrature phase-locked loop lock-in amplifier method. In either case, a small voltage v_{ds} (either AC or DC) is normally applied to the drain and source, and the drain-source current is measured and recorded. The advantage of using an AC measurement is that a small bandwidth measurement can be performed using a lock-in amplifier (Cui *et al.*, 2001; Bunimovich *et al.*, 2006). However, the frequency of the driving signal should be low, for example less than about 1000 Hz (Chen *et al.*, 2011), in order to have minimum influence of the low-pass filter formed at the input of the transconductance amplifier readout circuit caused by electrical capacitance between the SiNW and the substrate separated by the BOX layer.

12.6 Examples/applications of SiNW biosensor platforms

We have developed a SiNW fabrication technology that requires only conventional microfabrication processes including microlithography, oxidation and wet anisotropic plane-dependent etching (Chen *et al.*, 2009). Figure 12.6 shows

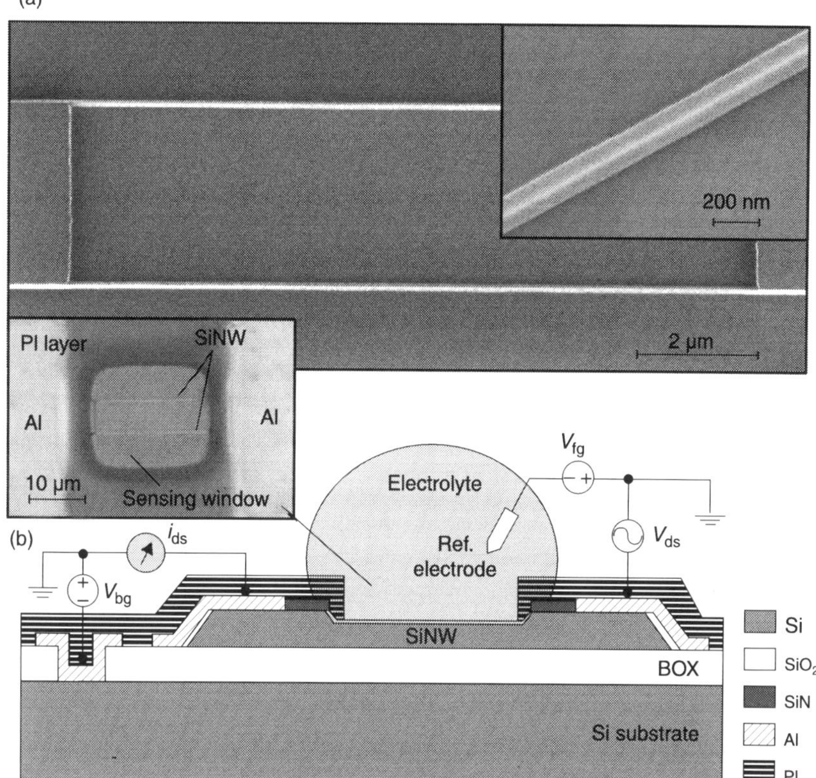

12.6 (a) High-resolution scanning electron microscopy images of fabricated silicon nanowires (SiNW). (b) Schematic of electrical measurement. Inset: microscope image of polyimide-encapsulated device prior to testing (SiN, low-stress silicon nitride; Al, aluminum; PI, polyimide). (Reprinted with permission from Chen *et al.*, 2011. Copyright 2011, American Chemical Society.)

representative microscopy images of SiNW fabricated with this technology. This scalable SiNW fabrication technology does not require expensive nanolithography to form sub-30 nm feature sizes.

The advantage of this technology is that moderately dense arrays of SiNW sensors, with precisely controlled dimensions and near atomically smooth surfaces, are simultaneously fabricated with thick microscale electrical contact regions from a continuous layer of single crystal silicon using a simple size reduction method. SiNW device arrays with lateral dimensions down to about 10 nm and lengths up to 100 μm can be consistently fabricated with high wafer-level yields.

12.6.1 SiNW electrochemical operation

SiNW-FET sensors measure surface potential changes $\Delta\psi_o$ on the sensor surface that originate from changes in the surface charge density $\Delta\sigma_o$ caused by changes in surface ions and/or the intrinsic charge of attached biomolecule complexes, as previously described. The surface potential change $\Delta\psi_o$ is determined from conductance measurements from the depletion-mode SiNW. We recently reported that the conductance of depletion-mode SiNW can be operated in either the depletion or accumulation regions, and can be represented with $G \approx q\mu_b N_a L^{-1}\xi + 2\eta\mu_a W_a L^{-1} Q_a$, where q is the electronic charge, μ_b is the dopant concentration-dependent bulk hole mobility, N_a is the boron impurity dopant concentration in the NW body, L is the device length, ξ is a gating function that is dependent on ψ_o, μ_a is the field-dependent hole mobility in the accumulated layer, W_a is the total surface area of the accumulated surface and Q_a is the electronic charge in the accumulation layer (Chen *et al.*, 2011). The SiNW operated in the depletion region have $Q_a = 0$ as $|V_{fg}| < |V_{fb}|$. The metal/semiconductor contacts of these depletion-mode SiNW devices have negligible contribution to the total device conductance (Chen *et al.*, 2009).

The conductance change for SiNW biased in the depletion region can be approximated with

$$\Delta G \approx q\mu_b N_a L^{-1} \Delta\xi \quad [12.7]$$

where

$$\Delta\xi \approx (W - (2)^{1/2}\Delta f_d)(a - (2/3)^{1/2}\Delta f_d) \quad [12.8]$$

is the area of the cross-section of the triangular SiNW with width W and height a, which is modulated by a depletion length function

$$\Delta f_d \approx (\gamma^2 t_{ox}^{\ 2} + 2\varepsilon_{Si}\varepsilon_o(V_{fg} - \Delta V_{fb})/qN_a)^{1/2} - \gamma^2 t_{ox}^{\ 2} \quad [12.9]$$

with $\gamma = \varepsilon_{Si}\varepsilon_{ox}^{\ -1}$ (Masood *et al.*, 2010; Chen *et al.*, 2011). The change flatband voltage scales with the surface potential change as

$$\Delta V_{fb} = \zeta - \Delta\psi_o \text{ with } \zeta = E_{ref} - q^{-1}\phi_{Si} - Q_f C_o^{\ -1} - Q_{ss} C_o^{\ -1} + \chi^{sol} \quad [12.10]$$

as previously described. The flatband voltage provides the physical link between $\Delta\psi_o$ and ΔG. It has been assumed that the buried-oxide/silicon interface is not depleted and the back-gate is not electrostatically coupled to the front-gate, which is accomplished in practice by choosing an appropriate V_{bg}. From this description we can summarize the important aspects of device biasing and measurements: the dual gate biasing configuration with a reference electrode in the sample buffer ensures reproducible device behaviour; the sensor surfaces should be well cleaned and surface charge effects from Q_f and Q_{it} minimized through thermal annealing; and the contact resistance of the drain and source contacts should be minimized such that it represents a small fraction of the total quiescent sensor resistance.

12.6.2 SiNW response characteristics

The SiNW were characterized prior to experiments to assess the front-gate $V_{fg}-|i_{ds}|$ and $g_m-|i_{ds}|$ response curves, where $g_m=\partial|i_{ds}|/\partial V_{fg}$ (at fixed v_{ds}) is the sensor transconductance. Front-gate biasing affects the depletion region differently from back-gate biasing, as previously described, and we expect higher surface charge sensitivity for biosensor applications with this configuration (Tong *et al.*, 2009). The sensors can be biased in the linear $V_{fg}-|i_{ds}|$ range while maintaining a small front-gate leakage current $i_{fg} < 1$ nA. The back-gate bias V_{bg} is determined and fixed during the initial calibration. Figure 12.7 shows a typical example of a $V_{fg}-|i_{ds}|$ response curve of a SiNW where the front-gate voltage is scanned with a platinum reference electrode in a low ionic strength buffer. The operation transitions from depletion to accumulation as V_{fg} is scanned to larger negative voltages. The decrease in transconductance for V_{fg} beyond the maximum is caused by a decrease in field-dependent effective mobility μ_a and charge screening.

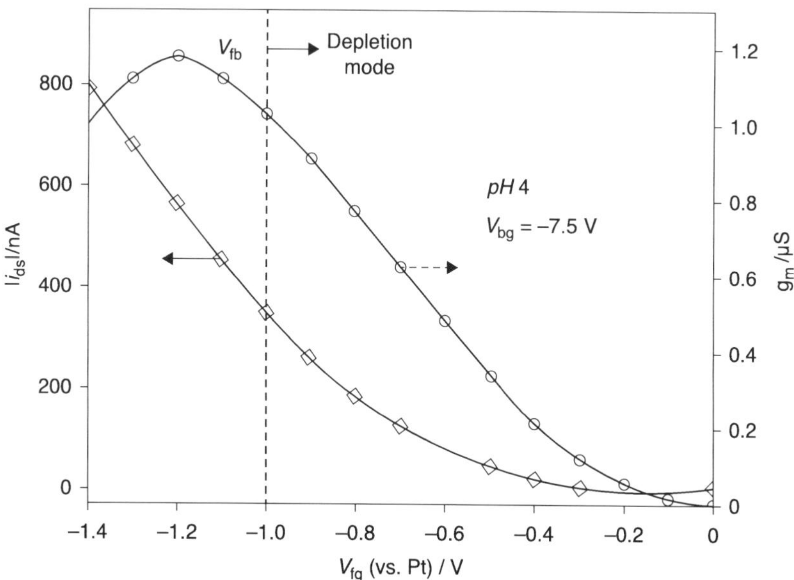

12.7 Integrated biosensor analysis platform measurements. Silicon nanowire (SiNW) front-gate biasing characteristics in pH 4 buffer solution with platinum reference electrode (reproduced from De *et al.*, 2013 with permission).

12.6.3 SiNW DNA hybridization measurements

As previously described, the ionic strength of the buffer is important for DNA hybridization measurements on SiNW sensors. Binding of negatively charged DNA biomolecules occurs near the SiNW gate surface provided that the charge is not screened because of the electrical double layer, or counter-ion condensation. As a low ionic strength hybridization buffer is preferred for the high sensitivity detection, we use synthetic PNA probes for the DNA–PNA hybridization measurements as they are less dependent on the ionic strength of the hybridization buffer compared with DNA–DNA duplex formation. The SiNW sensor surface consists of a thin SiO_2 layer, and the previously described techniques using an APTES silane attachment scheme with further conjugation to the PNA probe.

Real-time 15-mer DNA–PNA duplex hybridization measurements of complementary DNA target molecules (cDNA) to PNA probe molecules covalently attached to the gate surface are shown in Fig. 12.8. The measurements have been performed with sample flow speeds of $v_s \approx 0.6\,\mathrm{mm\,s^{-1}}$ in the reaction-limited transport regime. A first-order Langmuir binding model is used to model the equilibrium dissociation constant K_D of the DNA–PNA duplex hybridization process, which is defined as $K_D = k_d/k_a$, where k_a is the association rate constant that represents the speed of the second-order probe–target interaction and k_d is the dissociation rate constant that represents the speed of the first-order breakdown of the probe–target complex. The time-dependent surface density formation of a single type of probe–target complex on the surface is monitored in real time according to the response function

$$R(t) = R_o(1 - \exp[-(k_a[\mathrm{C}] + k_d)t]) \qquad [12.11]$$

where

$$R_o = (k_a R_{max}[\mathrm{C}])/(k_a[\mathrm{C}] + k_d) \qquad [12.12]$$

R_{max} is the maximum sensor signal response for a given probe density N_p and $[\mathrm{C}]_i$ is the concentration of the target sample. The drain-source current is the measured response from the SiNW sensors, thus $R(t) \equiv |i_{ds}|(t)$ for the particular bias arrangement of V_{fg}, V_{bg} and v_{ds}, and target concentration $[\mathrm{C}]_i$ for each measurement, as previously described.

The DNA–PNA hybridization measurements, shown in Fig. 12.8, were conducted in a 1 mM phosphate buffer using sample concentrations of cDNA of $[C]_{c1} = 500\,\mathrm{nM}$ and $[C]_{c2} = 100\,\mathrm{nM}$, and as a control experiment a non-complementary DNA sample with $[C]_{nc} = 100\,\mathrm{nM}$ was measured and is used as the baseline noise of the measurement. The increase in conductance with an increase of negatively charged molecules on the gate sensor surface of the *p*-type depletion mode sensor is expected from the sensor characterization (Fig. 12.7). We have previously characterized the density of DNA–PNA duplexes formed on

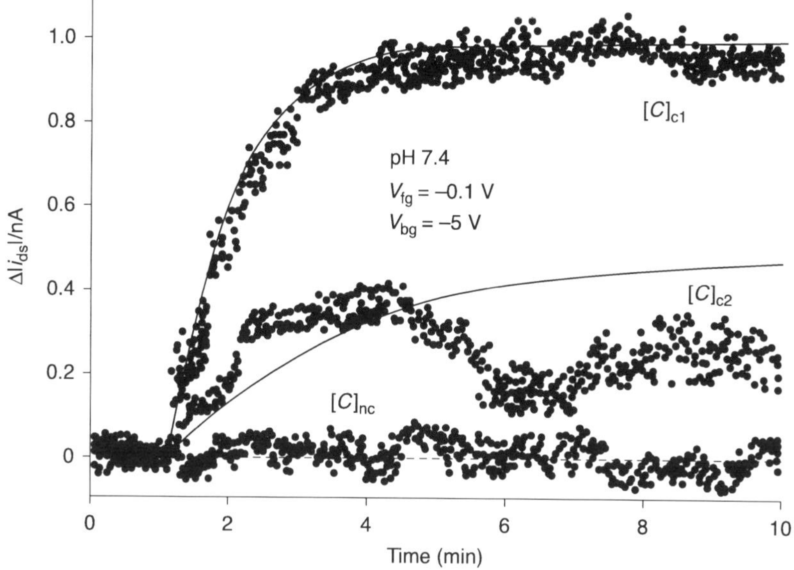

12.8 DNA–PNA duplex hybridization detection with $[C]_{c1}=500\,nM$, $[C]_{c2}=100\,nM$ and $[C]_{nc}=100\,nM$ target concentrations and flow speed $v_s \approx 0.6\,mm\,s^{-1}$ (reproduced from De *et al.*, 2013 with permission).

SiO_2 surfaces in equilibrium to about $N_d \sim 10^{12}$ duplexes cm^{-2} (De *et al.*, 2013a). Therefore the surface potential change can be estimated with

$$\Delta\psi_o \approx (2kTq^{-1})(sinh^{-1}\,((\sigma_{DNA}-\sigma_o)/\beta)-sinh^{-1}\,(\sigma_o/\beta)) \qquad [12.13]$$

where $\beta=(kT\varepsilon_w\varepsilon_o c_o)^{1/2}$, k is the Boltzmann constant, T is temperature, ε_o is the permittivity of free space, $\varepsilon_w=80$, $c_o=1\,mM$ and $\sigma_o=0.2\,C\,m^{-2}$ for a buffer with pH 7 (Chen *et al.*, 2011), and $\sigma_{DNA} \approx N_d \times 15 \times q \times 10^4\,Cm^{-2}$ for a saturated 15-mer DNA–PNA surface, which corresponds to a lower limit $\Delta\psi_o \sim 10\,mV$ and much larger than what is measured.

The association and dissociation rate constants, k_a and k_d, are extracted by fitting the measured current responses for all of the experimental concentrations. The solid lines in Fig. 12.8 represent the model fit to the measured data using $R_{max} \approx 1\,nA$, $k_a \approx 3\times10^4\,M^{-1}s^{-1}$ and $k_d \approx 4\times10^{-3}s^{-1}$, which gives $K_D=140\,nM$. The R_{max} value is consistent with $[C]_{c2} > K_D$, where all PNA binding sites are assumed to be occupied. The extracted K_D is much larger than the dissociation constants extracted from previously reported measurements of DNA–PNA duplex hybridization (Liu *et al.*, 2006; Park *et al.*, 2007b; Yu *et al.*, 2004). The discrepancy between the measurements can be caused by differences in the PNA attachment scheme (Park *et al.*, 2007b). Extracted K_D values of DNA–DNA duplex hybridization measured using SiNW sensors in a 165 mM ionic strength buffer are similar to our measurements (Bunimovich *et al.*, 2006).

12.6.4 SiNW flow rate dependence

As the SiNW sensors are sensitive to surface potential changes, flow rate changes during sample injections can lead to significant and erroneous sensor responses caused by changes in the streaming potential at the gate surface induced by flowing electrolyte solutions (Kim *et al*., 2009). Low ionic strength samples, for example less than 10 mM ionic strength, are most problematic where the electrical response of the SiNW sensors can change significantly as the flow rates change, as shown in Fig. 12.9. The SiNW sensor responds to changes in the sample flow rate in the integrated microfluidic flow-cell under different conditions. Figure 12.9(a) shows that the measured current can change by about 80% when the driving pressure (P_1 from Fig. 12.4) is switched from 1 bar to 0 bar using a deionized water sample. Figure 12.9(b) shows a ~14% change in the current response as the driving pressure is switched from 1 bar to P_1=0.1 bar in a 1 mM NaCl buffer solution; this observation is consistent with previous reports that decreases in the flow rate result in a decrease in conductance of the depletion-mode SiNW devices and increased ionic strength reduces the effect of conductance changes from sample flow rate changes (Kim *et al*., 2009). Figure 12.9(c) shows that flow rate changes caused by sample switching can be eliminated with the automated multi-sample injection system and pressure driven flow. The precision sample injection valve (Fig. 12.4) and pressure–driven sample transport provides stable and rapid sample switching when combined with a flow network that has balanced hydraulic resistances and provides stable baseline measurements (De *et al*., 2013a).

12.6.5 Differential measurements

Sensor drift is the systematic increase or decrease of the sensor response as a function of measurement time, and it is well known that the output response of

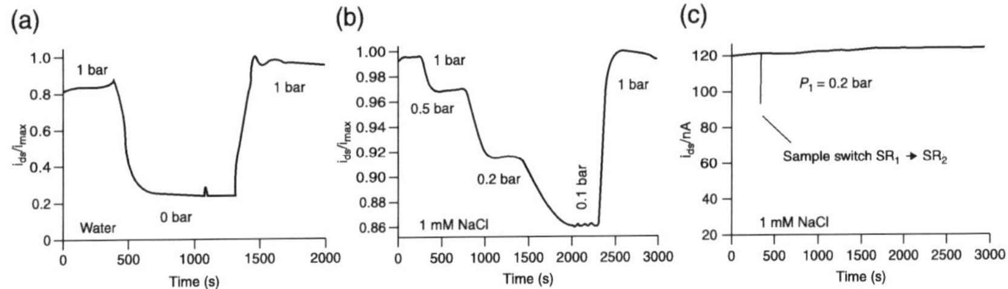

12.9 Measured Si-NW sensor current as a function of flow rate changes and buffer ionic strength in a microfluidic flow-cell. (a) Deionized water sample. (b) 1 mM NaCl buffer. (c) Constant pressure and flow rate of 1 mM NaCl buffer as sample is switched from sample reservoir SR_1 to SR_2 (reproduced from De *et al*., 2013 with permission).

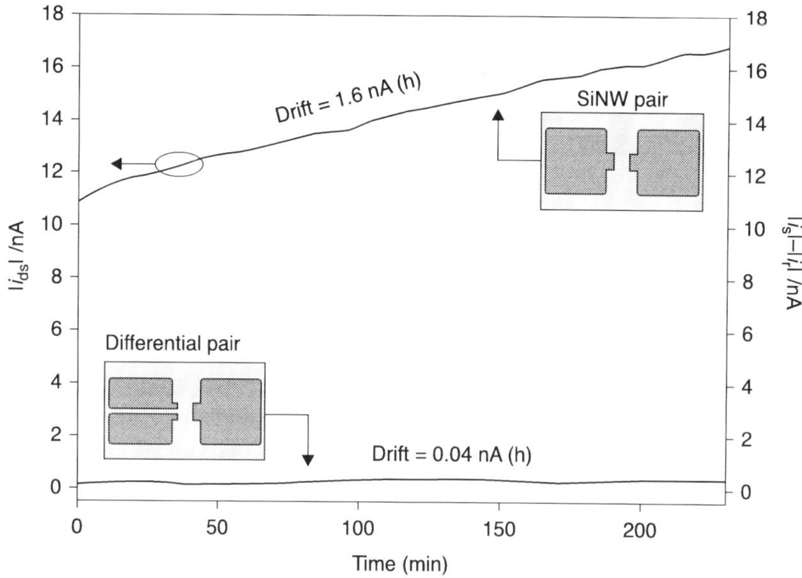

12.10 Silicon nanowire (SiNW) sensor drift measurements in 10 mM ionic strength buffer solution. Upper trace: Response from conventional sensor configuration. Lower trace: Response from differential sensor configuration (reproduced from De *et al.*, 2013 with permission).

ISFET sensors, the microscale predecessors to SiNW FET sensors, suffer from drift caused by ion migration at the gate surface (Bergveld, 2003). A typical response of a SiNW sensor with a SiO_2 gate surface in a 1 mM NaCl buffer solution over a period of about 4 hours is shown in Fig. 12.10 (upper trace); this drift rate of 1.6 nA h^{-1} is significant as the quiescent current of the sensor is $|i_{ds}|_o \approx 10$ nA and changes from biomolecule hybridization are typically in the order of $0.1|i_{ds}|_0$.

A differential measurement configuration can significantly improve the output response of the SiNW sensors. Figure 12.11 shows an example of the differential measurement configuration where a pair of identical SiNW sensors are driven by the same drain-source voltage v_{ds}. However, the source contacts are separated and the current is divided into reference and sensor currents, i_s and i_r respectively. The currents are converted to voltages v_s and v_r with external transconductance amplifiers (TA) and the lock-in amplifier instrument produces the difference output, or null output value $v_o = v_s - v_r$.

A typical differential response is shown in the lower trace of Fig. 12.10, where the drift rate has been significantly reduced by 30× to a drift rate of approximately 0.04 nA h^{-1}. The output signal is ideally nulled to zero (in this case the offset current is 0.1 nA because of device mismatch), and therefore, changes in the

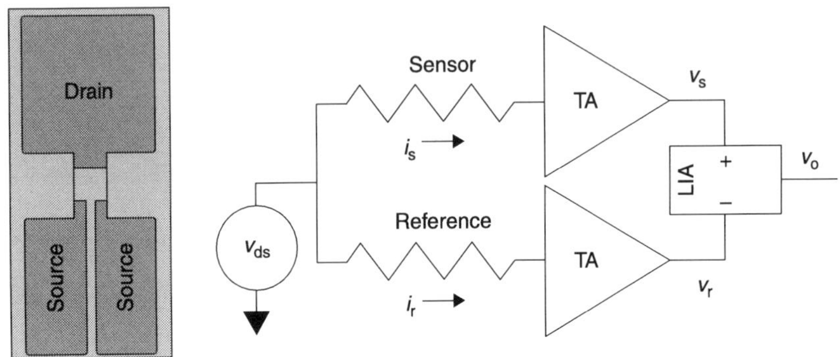

12.11 Differential silicon nanowire (SiNW) sensor configuration with common drain contacts and split source contacts where $v_o = v_s - v_r = A(i_s - i_r)$ and A is the gain of the TA. LIA represents 'look-in amplifier.' The left inset image shows a fabricated differential mode device with split source contacts (reproduced from De *et al.*, 2013 with permission).

sensor signal Δi_s can be detected with a higher sensitivity as the detection range of the instrument can be reduced.

12.7 Conclusions

We have presented a system level description of the probe-free SiNW biosensor platforms in an attempt to provide insight into the main components of the complete system, which includes the SiNW biosensors, probe layers required for specific binding to targets in solution, the sample delivery system that is integrated with a microfluidic flow reaction vessel bonded to the SiNW sensors and, finally, the SiNW electrical biasing and measurement configurations commonly used. The critical components of each part of the SiNW biosensor platform have been identified and described.

Finally, we describe an integrated SiNW biosensor platform that uses a SiNW sensor cartridge which is integrated with a microfluidic flow-cell into a simple disposable cartridge format mounted into the all-electrical readout instrument. An automated multi-sample injection system was presented that provides fast and reliable sample switching without spurious flow rate fluctuations that can produce erroneous sensor signals. The integrated microfluidic probe-free analytical platform that can be used for small volume chemical analysis, such as pH or ion sensing, as well as diagnostics for the specific measurement of biomolecules and for the kinetic estimation of biomolecule hybridization.

Real-time DNA–PNA duplex hybridization measurements have been presented with different sample concentrations in a low ionic strength buffer, and the equilibrium dissociation constant $K_D \approx 140\,\text{nM}$ has been extracted from the experimental data using the first order Langmuir binding model. We also presented a differential sensor configuration and a 30× reduction in sensor drift has been shown. The integrated microfluidic label-free analysis platform with measurement cartridge based on an electronic SiNW biosensor array, integrated with a simple microfluidic flow-cell, has been demonstrated by measuring different sample types.

12.8 Future trends

Probe-free SiNW biosensor platforms have great potential for the all-electrical detection of biomolecules because of their suitability for large-scale multiplexed electrical recording, and suitability to exploit conventional microfabrication and nanofabrication technologies for the their large-scale and high-density integration with existing electronic instrumentation of microcomputers. However, there are some fundamental problems that must be overcome in order for the technique to move from the device and sensor research laboratories into the hands of the analytical chemistry and clinical medicine communities.

We have previously described the problem of counter-ion screening at the gate sensor surface and counter-ion condensation, and this remains a primary challenge to detect the intrinsic biomolecule charge in electrolyte solutions with ionic strength near physiological conditions of biomolecule complexes tethered directly to the gate electrode surface using the conventional field effect mechanism. Despite the huge investment in terms of research effort over the past decade to apply SiNW-FET biosensors to the measurement of charged biomolecules at the sensor surface, there has been little progress relative to that of ISFET first described nearly three decades ago.

We have recently explored the feasibility of horizontally attached PNA probes that could better accommodate short Debye lengths, that is large ionic strength buffers, for DNA hybridization detection (De *et al.*, 2013b). The horizontal γ-PNA–DNA duplex will be less affected by counter-ion screening than the conventional vertical configuration. Both vertical PNA–DNA duplexes and horizontal γ-PNA–DNA duplexes hybridize more specifically with higher ionic strength buffers. In addition, γ-PNA could prevent non-specific binding of non-complementary DNA in a low ionic strength buffer. More progress is needed to overcome the charge screening limitations of electronic field effect detection devices.

The remaining disadvantage of the SiNW biosensors is that their sensing surface is extremely small compared with microscale sensors, which significantly reduces the number of biomolecules that can be captured from the sample volume for sub-nM sample concentration such that the time required to transport the target molecules to the nanoscale sensors is not prohibitively long, as we previously described. Therefore, special attention is required to collect the biomolecules on the SiNW surface, which essentially translates into the situation where we want an extremely large area sensor surface, that is macroscale, while maintaining the high sensitivity of the nanoscale devices. One method that may provide this capability is the differential sensor configuration (Fig. 12.11), where the single sensor pair is replaced by a pair of large area SiNW sensor arrays where each individual SiNW sensor in both arrays is connected in parallel (thus N-parallel connected SiNW sensor arrays). Therefore, the sensor and reference devices are each replaced by N-parallel connected SiNW devices and will ideally result in an amplified sensor response of $N\times$. For example, when the single sensor and reference devices are replaced by a parallel array of N-SiNW, the ideally nulled output signal is $i_s - i_r = \Delta i_s$, where Δi is current change from the NW sensor induced by a surface potential change $\Delta\psi_o$. In order for this method to work, the nulled output response is required for the LIA (lock-in amplifier) measurement

$$\text{updated: } v_o = v_s - v_r \approx -Nv_{ds}(R_f/R_{NW})(\Delta R/R_{NW}) \qquad [12.14]$$

where $A = R_f$ is the gain of the TA, R_{NW} is the quiescent resistance of the NW and ΔR is the change in the NW resistance induced by $\Delta\psi_o$. The differential measurement configuration produces an output signal that is proportional to ΔR and can be configured to reduce the measurement range of the LIA and increase the detection sensitivity compared to a single output configuration. This kind of the differential measurement with banks of reference and sensing SiNW can improve the limit of detection for biosensing applications.

In conclusion, the field of SiNW biosensor development has grown tremendously over the past decade with many different integrated platforms reported for measuring small quantities of the DNA and proteins in small volume microfluidic flow cells and reaction chambers. In the coming years, more development effort should be directed towards optimizing the overall SiNW biosensor system performance and systematic characterization for a select group of the target analytes.

12.9 References

Avouris, P., Appenzeller, J., Martel, R. and Wind, S. J. (2003) 'Carbon nanotube electronics', *Pr Inst Electr Elect*, 91, 1772–84. Doi: 10.1109/Jproc.2003.818338.

Bai, J. G., Yeo, W. H. and Chung, J. H. (2009) 'Nanostructured biosensing platform–shadow edge lithography for high-throughput nanofabrication', *Lab Chip*, 9, 449–55. Doi: 10.1039/b811400e.

Bajaj, M. G. (2005) 'DNA hybridization: fundamental studies and applications in directed assembly', MIT Thesis, Chemical Engineering.

Bardeen, J. (1947) 'Surface states and rectification of a metal semi-conductor contact', *Phys Rev*, 71, 717–27. Doi: 10.1103/PhysRev.71.717.

Beckman, R. A., Johnston-Halperin, E., Melosh, N. A., Luo, Y., Green, J. E. and Heath, J. R. (2004) 'Fabrication of conducting Si nanowire arrays', *J Appl Phys*, 96, 5921–3. Doi: 10.1063/1.1801155.

Beckman, R. A., Johnston-Halperin, E., Luo, Y., Green, J. E. and Heath, J. R. (2005) 'Bridging dimensions: demultiplexing ultrahigh-density nanowire circuits', *Science*, 310, 465–8. Doi: 10.1126/science.1114757.

Bergveld, P. (1970) 'Developments of ion-sensitive solid-state device for neurophysiological measurements', *IEEE Trans Biomed Eng*, 17, 70–1. Doi: 10.1109/TBME.1970.4502688.

Bergveld, P. (2003) 'Thirty years of ISFETOLOGY: what happened in the past 30 years and what may happen in the next 30 years', *Sensor Actuat B-Chem*, 8, 1–20. Doi: 10.1016/S0925–4005(02)00301–5.

Berney, H., West, J., Haefele, E., Alderman, J., Lane, W. and Collins, J. K. (2000) 'A DNA diagnostic biosensor: development, characterisation and performance', *Sensor Actuat B-Chem*, 68, 100–8. Doi: 10.1016/S0925–4005(00)00468–8.

Bloomfield, V.A. (1996) 'DNA condensation', *Curr Opin Struc Biol*, 6, 334–41. Doi: 10.1016/S0959–440X(96)80052–2.

Bousse, L. and Bergveld, P. (1983) 'Operation of chemically sensitive field-effect sensors as a function of the insulator–electrolyte interface', *IEEE T Electron Dev*, ED-30, 1263–70. Doi: 10.1109/T-ED.1983.21284.

Bunimovich, Y. L., Ge, G., Beverly, K. C., Ries, R. S., Hood, L. and Heath, J. R. (2004) 'Electrochemically programmed, spatially selective biofunctionalization of silicon nanowires', *Langmuir*, 20, 10630–8. Doi: 10.1021/la047913h.

Bunimovich, Y. L., Shin, Y. S., Yeo, W.-S., Amori, M., Kwong, G. and Heath, J. R (2006) 'Quantitative real-time measurements of DNA hybridization with alkylated nonoxidized silicon nanowires in electrolyte solution', *J Am Chem Soc*, 128, 16323–31. Doi: 10.1021/ja065923u.

Cai, W., Peck, J. R., van der Weide, D. W. and Hamers, R. J. (2004) 'Direct electrical detection of hybridization at DNA-modified silicon surfaces', *Biosens Bioelectron*, 19, 1013–19. Doi: 10.1016/j.bios.2003.09.009.

Carlen, E.T. and van den Berg, A. (2007) 'Nanowire electrochemical sensors: can we live without labels?' *Lab Chip*, **7**, 19. Doi: 10.1039/B616805C.

Cattani-Scholz, A., Pedone, D., Dubey, M., Neppl, S., Nickel, B., *et al.* (2008) 'Organophosphonate-based PNA-functionalization of silicon nanowires for label-free DNA detection', *ACS Nano*, 2, 1653–60. Doi: 10.1021/nn800136e.

Chen, S., Bomer, J. G., van der Wiel, W.G., Carlen, E. T. and van den Berg, A. (2009) 'Top-down fabrication of sub-30 nm monocrystalline silicon nanowires using conventional microfabrication', *ACS Nano*, 3, 3485–92. Doi: 10.1021/nn901220g.

Chen, S. and Zhang, S. L. (2011) 'Contacting versus insulated gate electrode for Si nanoribbon field-effect sensors operating in electrolyte', *Anal Chem*, 83, 9546–51. Doi: 10.1021/ac2023316.

Chen, S., Bomer, J. G., Carlen, E. T. and van den Berg, A. (2011) 'Al_2O_3/Silicon nanoISFET with near ideal Nernstian response', *Nano Lett*, 11, 2334–41. Doi: 10.1021/nl200623n.

Choi, S., Park, I., Hao, Z., Holman, H.-Y. N. and Pisano, A. P. (2012) 'Quantitative studies of long-term stable, top-down fabricated silicon nanowire pH sensors', *Applied Physics A: Materials Science and Processing*, 107, 421–8. Doi: 10.1557/opl.2011.84.

Choi, Y. K., King, T. J. and Hu, C. M. (2002) 'Spacer FinFET: nanoscale double-gate CMOS technology for the terabit era', *Solid State Electron*, 46, 1595–601. Doi: 10.1016/S0038–1101(02)00111–9.

Choi, Y. K., Zhu, J., Jeff Grunes, J., Bokor, J. and Somorjai, G. A. (2003) 'Fabrication of sub-10-nm silicon nanowire arrays by size reduction lithography', *J Phys Chem B*, 107, 3340–3. Doi: 10.1021/jp0222649.

Chou, S. Y., Krauss, P. K. and Renstrom, P. J. (1996) 'Nanoimprint lithography', *J Vac Sci Tech B*, 14, 4129–33. Doi: 10.1116/1.588605.

Chrisey, L. A., Lee, G.U. and O'Ferrall, C. E. (1996) 'Covalent attachment of synthetic DNA to self-assembled monolayer films', *Nucleic Acids Res*, 24, 3031–9. Doi: 10.1093/nar/24.15.3031.

Cui, Y., Wei, Q., Park, H. and Lieber, C. M. (2001) 'Nanowire nanosensors for highly sensitive and selective detection of biological and chemical species', *Science*, 293, 1289–92. Doi: 10.1126/science.1062711.

De, A., van Nieuwkasteele, J., Carlen, E.T. and van den Berg A. (2013a) 'Integrated label-free silicon nanowire sensor arrays for (bio)chemical analysis', *Analyst*, 138, 3221–9. Doi: 10.1039/c3an36586g.

De, A., Souchelnytskyi, S., van den Berg, A., and Carlen, E.T. (2013b) 'Peptide nucleic acid (PNA)–DNA duplexes: Comparison of hybridization affinity between vertically and horizontally tethered PNA probes', *ACS Appl Mater Interfaces*, 5, 4607–12. Doi: 10.1021/am4011429.

Deal, B.E. (1974) 'The current understanding of charges in the thermally oxidized silicon structure', *J Electrochem Soc*, 121, 198C–205C. Doi: 10.1149/1.2402380.

Dong, R. K. and Xiaolin, Z. (2008) 'Numerical characterization and optimization of the microfluidics for nanowire biosensors', *Nano Lett*, 2008, 8, 3233–7. Doi: 10.1021/nl801559m.

Dorvel, B. R., Reddy Jr., B., Go, J., Guevara, C. D., Salm, E., *et al.* (2012) 'Silicon nanowires with high-k hafnium oxide dielectrics for sensitive detection of small nucleic acid oligomers', *ACS Nano*, 6, 6150–64. Doi: 10.1021/nn301495k.

Duan, X., Li, Y., Rajan, N. K., Routenberg, D. A., Modis, Y. and Reed, M. A. (2012) 'Quantification of the affinities and kinetics of protein interactions using silicon nanowire biosensors', *Nat Nanotechnol*, 7, 401–7. Doi: 10.1038/nnano.2012.82.

Dugas, V., Depret, G., Chevalier, Y., Nesme, X. and Souteyrand, E. (2004) 'Immobilization of single-stranded DNA fragments to solid surfaces and their repeatable specific hybridization: Covalent binding or adsorption?', *Sensor Actuat B-Chem*, 101, 112–21. Doi: 10.1016/j.snb.2004.02.041.

Duggan, D. J., Bittner, M., Chen, Y., Meltzer, P. and Trent, J. M. (1999) 'Expression profiling using cDNA microarrays', *Nature Genetics*, 21, 10–14. Doi: 10.1038/4434.

Elfstrom, N., Karlstrom, A. E. and Linnros, J. (2008) 'Silicon nanoribbons for electrical detection of biomolecules', *Nano Lett*, 8, 945–9. Doi: 10.1021/nl080094r.

Elnathan, R., Kwiat, M., Pevzner, A., Engel, Y., Burstein, L., *et al.* (2012) 'Biorecognition layer engineering: overcoming screening limitations of nanowire-based FET devices', *Nano Lett*, 12, 5245–54. Doi: 10.1021/nl302434w.

Fischer, P. B. and Chou, S. Y. (1993) 'Sub-50 nm high aspect-ratio silicon pillars, ridges, and trenches fabricated using ultrahigh resolution electron-beam lithography and reactive ion etching', *Appl Phys Lett*, 62, 1414–16. Doi: 10.1063/1.108696.

Flanders, D. C. and Efremow, N. N. (1983) 'Generation of <50 nm period gratings using edge defined techniques', *J Vac Sci Technol B*, 1, 1105–8. Doi: 10.1116/1.582643.

Fritz, J., Cooper, E. B., Gaudet, S., Sorger, P. K. and Manalis, S. R. (2002) 'Electronic detection of DNA by its intrinsic molecular charge', *P Natl Acad Sci USA*, 99, 14142–6. Doi: 10.1073/pnas.232276699.

Gao, A., Lu, N., Dai, P., Li, T., Pei, H., *et al.* (2011) 'Silicon-nanowire-based CMOS-compatible field-effect transistor nanosensors for ultrasensitive electrical detection of nucleic acids', *Nano Lett*, 11, 3974–8. Doi: 10.1021/nl202303y.

Gao, X.P.A., Zheng, G.F. and Lieber, C.M. (2010) 'Subthreshold regime has the optimal sensitivity for nanowire FET biosensors', *Nano Lett*, 10, 547–552. Doi: 10.1021/nl9034219.

Gao, Z., Agarwal, A., Trigg, A. D., Singh, N., Fang, C., *et al.* (2007) 'Silicon nanowire arrays for label-free detection of DNA', *Anal Chem*, 79, 3291–7. Doi: 10.1021/ac061808q.

Ghosh, S., Sood, A. K. and Kumar, N. (2003) 'Carbon nanotube flow sensors', *Science*, 299, 1042–4. Doi: 10.1126/science.1079080.

Gong, J. R. (2010) 'Label-free attomolar detection of proteins using integrated nanoelectronic and electrokinetic devices', *Small*, 6, 967–73. Doi: 10.1002/smll.200902132.

Hahm, J. and Lieber, C. M. (2004) 'Direct ultrasensitive electrical detection of DNA and DNA sequence variations using nanowire nanosensors', *Nano Lett*, 4, 51–4. Doi: 10.1021/nl034853b.

Hanson, E. L., Schwartz, J., Nickel, B., Koch, N. and Danisman, M. F. (2003) 'Bonding self-assembled, compact organophosphonate monolayers to the native oxide surface of silicon', *J Am Chem Soc*, 125, 16074–80. Doi: 10.1021/ja035956z.

Heath, J. R. (2008) 'Superlattice nanowire pattern transfer (SNAP)', *Accounts Chem Res*, 41, 1609–17. Doi: 10.1021/ar800015y.

Ho, T. T., Wang, Y., Eichfeld, S., Lew, K.-K., Liu, B., *et al.* (2008) 'In situ axially doped n-channel silicon nanowire field-effect transistors', *Nano Lett*, 8, 4359–64. Doi: 10.1021/nl8022059.

Homola, J., Yee, S. S. and Gauglitz, G. (1999) 'Surface plasmon resonance sensors: review', *Sens Act B: Chem*, 54, 3–15. Doi: 10.1016/S0925-4005(98)00321-9

Israelachvili, J. (1992) *Intermolecular and Surface Forces*, 2nd ed., Academic Press: London.

Janata, J. (1986) 'Chemical selectivity of field effect transistors'. In *Proceedings of the 2nd International Meeting on Chemical Sensors*, ed. J. L. Aucouturier *et al.*, Bordeaux, France, 25–31.

Kim, D.-S., Jeong, Y.-T., Lyu, H.-K., Park, H.-J., Shin, J.-K., *et al.* (2004) 'An FET-type charge sensor for highly sensitive detection of DNA sequence', *Biosens Bioelectron*, 20, 69–74. Doi: 10.1016/j.bios.2004.01.025.

Kim, A., Seong Ah, C., Young Yu, H., Yang, J.-H., Baek, I.-B., *et al.* (2007a) 'Ultrasensitive, label-free, and real-time immunodetection using silicon field-effect transistors', *Appl Phys Lett*, 91, 103901. Doi: 10.1063/1.2779965.

Kim, D.K., Majumdar, A. and Kim, S. J. (2007b) 'Electrokinetic flow meter', *Sensor Actuat A-Phys*, 136, 80–9. Doi: 10.1016/j.sna.2006.10.022.

Kim, D. R., Lee, C. H. and Zheng, X. (2009) 'Probing flow velocity with silicon nanowire sensors', *Nano Lett*, 9, 1984–8. Doi: 10.1021/nl900238a.

Kong, J., Franklin, N. R., Zhou, C. W., Chapline, M. G., Peng, S., *et al.* (2000) 'Nanotube molecular wires and chemical sensors', *Science*, 287, 622–5. Doi: 10.1126/science.287.5453.622.

Krishnamoorthy, G., Carlen, E. T., deBoer, H. L., van den Berg, A. and Schasfoort, R.B.M. (2010) 'Electrokinetic lab-on-a-biochip for multi-ligand/multi-analyte biosensing', *Anal Chem* 82, 4145–50. Doi: 10.1021/ac1003163.

Lee, C.W., Afzalian, A., Akhavan, N.D., Yan, R., Ferain, I. and Colinge, J.P. (2009) 'Junctionless multigate field-effect transistor'. *Appl Phys Lett*, 94, 053511. Doi: 10.1063/1.3079411.

Lehoucq, G., Bondavalli, P., Xavier, S., Legagneux, P., Abbyad, P., *et al.* (2012) 'Highly sensitive pH measurements using a transistor composed of a large array of parallel silicon nanowires', *Sensor Actuat B-Chem*, 171–72, 127–34. Doi: 10.1016/j.snb.2012.01.054.

Li, P., Lei, N., Sheadel, D. A., Xu, J. and Xue, W. (2012) 'Integration of nanosensors into a sealed microchannel in a hybrid lab-on-a-chip device', *Sensor Actuat B-Chem*, 166–7, 870–7. Doi: 10.1016/j.snb.2012.02.047.

Li, Z., Chen, Y., Li, X., Kamins, T. I., Nauka, K. and Williams, R. S. (2004) 'Sequence-specific label-free DNA sensors based on silicon nanowires', *Nano Lett*, 4, 245–7. Doi: 10.1021/nl034958e.

Lieber, C. M. (1989) 'One dimensional nanostructures: chemistry, physics & applications', *Solid State Commun*, 107, 607–16. Doi: 10.1016/S0038–1098(98)00209–9.

Liu, J., Tiefenauer, L., Tian, S., Nielsen, P.E. and Knoll, W. (2006) 'PNA–DNA hybridization study using labeled streptavidin by voltammetry and surface plasmon fluorescence spectroscopy' *Anal Chem*, **78**, 470. Doi: 10.1021/ac051299c.

Lud, S. Q., Nikolaides, M. G., Haase, I., Fischer, M. and Bausch, A. R. (2006) 'Field effect of screened charges: electrical detection of peptides and proteins by a thin-film resistor', *ChemPhysChem*, 2006, 7, 379–84. Doi: 10.1002/cphc.200500484.

Manning, G. S. (1978) 'The molecular theory of polyelectrolyte solution with applications to the properties of polynucleotides', *Quart Rev Biophys*, 11, 179–246. Doi: 10.1017/S0033583500002031.

Masood, M., Chen, S., Carlen, E. T. and van den Berg, A. (2010) 'All-(111) surface silicon nanowires: selective functionalization for biosensing applications', *ACS Appl Mater Interfaces*, 2, 3422–8. Doi: 10.1021/am100922e.

Melosh, N. A., Boukai, A., Diana, F., Gerardot, B., Badolato, A., *et al.* (2003) 'Ultrahigh-density nanowire lattices and circuits', *Science*, 300, 112–15. Doi: 10.1126/science.1081940.

Menke, E. J., Thompson, M. A., Xiang, C., Yang, L. C. and Penner, R. M. (2006) 'Lithographically patterned nanowire electrodeposition', *Nat Mater*, 914–19. Doi: 10.1038/nmat1759.

Minot, E. D., Janssens, A. M., Heller, I., Heerin, H. A., Dekker, C. and Lemay, S. G. (2007) 'Carbon nanotube biosensors: the critical role of the reference electrode', *Appl Phys Lett*, 91, 093507. Doi: 10.1063/1.2775090.

Morales, A. M. and Lieber, C.M. (1998) 'A laser ablation method for the synthesis of crystalline semiconductor nanowires', *Science*, 279, 208–11. Doi: 10.1126/science.279.5348.208.

Nair, P.R. and Alam, M. A. (2007) 'Design considerations of silicon nanowire biosensors', *IEEE T Electron Dev*, 54, 3400–8. Doi: 10.1109/TED.2007.909059.

Nair, P. R. and Alam, M. A. (2008) 'Screening-limited response of nanobiosensors', *Nano Lett*, 8, 1281–5. Doi: 10.1021/nl072593i.

Nielsen, P. E. and Haaima, G. (1997) 'Peptide nucleic acid (PNA). A DNA mimic with a pseudopeptide backbone', *Chem Soc Rev*, 26, 73–8. Doi: 10.1039/CS9972600073.

Norde, W. (1986) 'Adsorption of proteins from solution at the solid–liquid interface', *Adv Colloid Interfac*, 25, 267–340. Doi: 10.1016/0001–8686(86)80012–4.

Poghossian, A., Cherstvy, A., Ingebrandt, S., Offenhausser, A. and Schoning, M. J. (2005) 'Possibilities and limitations of label-free detection of DNA hybridization with field-effect-based devices', *Sensor Actuat B-Chem*, 111–12, 470–80. Doi: 10.1016/j.snb.2005.03.083.

Park, I., Li, Z., Pisano, A. P. and Williams, R. S. (2007a) 'Selective surface functionalization silicon nanowires via nanoscale joule heating', *Nano Lett*, 7, 3106–11. Doi: 10.1021/nl071637k.

Park, H., Germini, A., Sforza, S., Corradini, R., Marchelli, R. and Knoll, W. (2007b) 'Effect of ionic strength on PNA–DNA hybridization on surfaces and in solution', *Biointerphases*, 2, 80–8. Doi: 10.1116/1.2746871.

Patolsky, F., Zheng, G., Hayden, O., Lakadamyali, M., Zhuang, X. and Lieber, C. M. (2004) 'Electrical detection of single viruses', *P Natl Acad Sci USA*, 101, 14017–22. Doi: 10.1073/pnas.0406159101.

Patolsky, F., Timko, B. P., Yu, G., Fang, Y., Greytak, A. B., *et al.* (2006a) 'Detection, stimulation, and inhibition of neuronal signals with high-density nanowire transistor arrays', *Science*, 313, 1100–4. Doi: 10.1126/science.1128640.

Patolsky, F., Zheng, G. and Lieber, C.M. (2006b) 'Fabrication of silicon nanowire devices for ultrasensitive, label-free, real-time detection of biological and chemical species', *Nat Protoc*, 1, 1711–24. Doi: 10.1038/nprot.2006.227.

Peterson, A. W., Heaton, R. J. and Georgiadis, R. M. (2001) 'The effect of surface probe density on DNA hybridization', *Nucleic Acids Res*, 29, 5163–8. Doi: 10.1093/nar/29.24.5163.

Schasfoort, R. B. M., Kooyman, R. P . H., Bergveld, P. and Greve, J. (1990) 'A new approach to immunoFET operation', *Biosens Bioelectron*, 5, 103–24. Doi: 10.1016/0956–5663(90)80002-U.

Schmidt, V., Senz, S. and Gösele, U. (2007) 'Influence of the Si/SiO_2 interface on the charge carrier density of silicon nanowires', *Appl Phys A-Mater*, 86, 187–91. Doi: 10.1007/s00339–006–3746–2.

Schenck, J. F. (1978) 'Technical difficulties remaining to the application of ISFET devices', In *Theory, Design and Biomedical Application of Solid State Chemical Sensors*, Ed. P. W. Cheung. CRC Press Inc., 1978, 165–73.

Schoch, R. B., Han, J. and Renaud, P. (2008) 'Transport phenomena in nanofluidics', *Rev Mod Phys*, 80, 839–83. Doi: 10.1103/RevModPhys.80.839.

Shalev, G., Doron, A., Virobnik, U., Cohen, A., Sanhedrai, Y. and Levy, I. (2008) 'Gain optimization in ion sensitive field-effect transistor based sensor with fully depleted silicon on insulator', *Appl Phys Lett*, 93, 083902. Doi: 10.1063/1.2977476.

Sheehan, P.E. and Whitman, L. J. (2005) 'Detection limits for nanoscale biosensors', *Nano Lett*, 5, 803–7. Doi: 10.1021/nl050298x.

Shchepinov, M. S., Case-Green, S. C. and Southern, E. M. (1997) 'Steric factors influencing hybridisation of nucleic acids to oligonucleotide arrays', *Nucleic Acids Res*, 25, 1155–61. Doi: 10.1093/nar/25.6.1155.

Shockley, W. and Pearson, G.L. (1948) 'Modulation of conductance of thin films of semi-conductors by surface charges', *Phys Rev*, 74, 232–3. Doi: 10.1103/PhysRev.74.232.

Souteyrand, E., Cloarec, J. P., Martin, J. R., Wilson, C., Lawrence, I., *et al.* (1997) 'Direct detection of the hybridization of synthetic homo-oligomer DNA sequences by field effect', *J Phys Chem B*, 101, 2980–5. Doi: 10.1021/jp963056h.

Southern, E.M. (1975) 'Detection of specific sequences among DNA fragments separated by gel electrophoresis', *J Mol Biol*, 98, 503–17. Doi: 10.1016/S0022–2836(75)80083–0.

Southern, E., Mir, K. and Shchepinov, M. (1999) 'Molecular interactions on microarrays', *Nature Genetics*, 21 (1 SUPPL.), 5–9. Doi: 10.1038/4429.

Squires, T. M., Messinger, R. J. and Manalis, S.R. (2008) 'Making it stick: convection, reaction and diffusion in surface-based biosensors', *Nat Biotechnol*, 26, 417–26. Doi: 10.1038/nbt1388.

Stern, E., Wagner, R., Sigworth, F. J., Breaker, R., Fahmy, T. M. and Reed, M. A. (2007a) 'Importance of the Debye screening length on nanowire field effect transistor sensors', *Nano Lett*, 7, 3405–9. Doi: 10.1021/nl071792z.

Stern, E., Klemic, J. F., Routenberg, D. A., Wyrembak, P. N., Turner-Evans, D. B., *et al.* (2007b) 'Label-free immunodetection with CMOS-compatible semiconducting nanowires', *Nature*, 445, 519–22. Doi: 10.1038/nature05498.

Stern, E., Vacic, A., Rajan, N. K., Criscione, J. M., Park, J., *et al.* (2012) 'Label-free biomarker detection from whole blood', *Nat Nanotechnol*, 5, 138–42. Doi: 10.1038/nnano.2009.353.

Sunkara, V., Park, D. K., Hwang, H., Chantiwas, R., Soper, S. A. and Cho, Y. K. (2011) 'Simple room temperature bonding of thermoplastics and poly(dimethylsiloxane)', *Lab Chip*, 11, 962–5. Doi: 10.1039/c0lc00272k.

Sze, S. M. (1981) *Physics of Semiconductor Devices*, 2nd ed., John Wiley and Sons, Inc.: New York.

Tans, S. J., Devoret, M. H., Dai, H. J., Thess, A., Smalley, R. E., *et al.* (1997) 'Individual single-wall carbon nanotubes as quantum wires', *Nature*, 386, 474–7. Doi: 10.1038/386474a0.

Tong, H. D., Chen, S. G. van der Wiel, W., Carlen, E. T. and van den Berg, A. (2009) 'Novel top-down wafer-scale fabrication of single crystal silicon nanowires', *Nano Lett*, 2009, 1015–22. Doi: 10.1021/nl803181x.

Uslu, F., Ingebrandt, S., Mayer, D., Böcker-Meffert, S., Odenthal, M. and Offenhausser, A. (2004) 'Label-free fully electronic nucleic acid detection system based on a field-effect transistor device', *Biosens Bioelectron*, 19, 1723–31. Doi: 10.1016/j.bios.2004.01.019.

Vacic, A., Criscione, J. M., Rajan, N. K., Stern, E., Fahmy, T. M. and Reed, M. A. (2011) 'Determination of molecular configuration by Debye length modulation', *J Am Chem Soc*, 133, 13886–9. Doi: 10.1021/ja205684a.

van der Voort, D., McNeil, C. A., Renneberg, R., Korf, J., Hermens, W. T. and Glatz, J. F. C. (2005) 'Biosensors: basic features and application for fatty acid-binding protein, an early plasma marker of myocardial injury', *Sensor Actuat B-Chem*, 105, 50–9. Doi: 10.1016/j.snb.2004.02.035.

Wahlgren, M. and Arnebrant, T. (1991) 'Protein adsorption to solid-surfaces', *Trends Biotechnol*, 9, 201–8. Doi: 10.1016/0167–7799(91)90064-O.

Wang, J., Palecek, E., Nielsen, P.E., Rivas, G., Cai, X., *et al.* (1996) 'Peptide nucleic acid probes for sequence-specific DNA biosensors', *J Am Chem Soc*, 118, 7667–70. Doi: 10.1021/ja9608050.

Wang, W. U., Chen, C., Lin, K.-H., Fang, Y. and Lieber, C. M. (2005) 'Label-free detection of small-molecule–protein interactions by using nanowire nanosensors', *P Natl Acad Sci USA*, 102, 3208–12. Doi: 10.1073/pnas.0406368102.

Wei, F., Sun, B., Guo, Y. and Zhao, X. S. (2003) 'Monitoring DNA hybridization on alkyl modified silicon surface through capacitance measurement', *Biosens Bioelectron*, 18, 1157–63. Doi: 10.1016/S0956–5663(03)00002–2.

Westwater, J., Gosain, D. P. and Usui, S. (1998) 'Si nanowires grown via the vapour–liquid–solid reaction', *Phys Status Solidi A*, 165, 37–42. Doi: 10.1002/(SICI)1521–396X(199801)165:1<37::AID-PSSA37>3.0.CO;2-Z.

Yu, F., Yao, D. and Knoll, W. (2004) 'Oligonucleotide hybridization studied by a surface plasmon diffraction sensor (SPDS)', *Nucleic Acids Res*, 32, e75. Doi:10.1093/nar/gnh067.

Zhang, G.-J., Zhang, G., Chua, J. H., Chee, R.-E., Wong, E. H., *et al.* (2008) 'DNA sensing by silicon nanowire: charge layer distance dependence', *Nano Lett*, 8, 1066–70. Doi: 10.1021/nl072991l.

Zhang, Z., Zhao, P., Xiao, G., Watts, B. R. and Xu, C. (2011) 'Sealing SU-8 microfluidic channels using PDMS', *Biomicrofluidics*, 5, 046503. Doi: 10.1063/1.3659016.

Zheng, G. F., Qin, L.D. and Mirkin, C. A. (2008) 'Spectroscopically enhancing electrical nanotraps', *Angew Chem Int Edit* 47, 1938–41. Doi: 10.1002/anie.200705312.

Zheng, G., Patolsky, F., Cui, Y., Wang, W. U. and Lieber, C. M. (2005) 'Multiplexed electrical detection of cancer markers with nanowire sensor arrays', *Nat Biotechnol*, 23, 1294–301. Doi: 10.1038/nbt1138.

Zimmermann, M., Delamarche, E., Wolf, M. and Hunziker, P. (2005) 'Modeling and optimization of high-sensitivity, low-volume microfluidic-based surface immunoassays', *Biomed Microdevices*, 7, 99–110. Doi: 10.1007/s10544–005–1587-y.

Index

Zeitfracht Medien GmbH
Ferdinand-Jühlke-Straße 7
99095 Erfurt, Deutschland
produktsicherheit@kolibri360.de